Springer Collected Works in Mathematics

For further volumes:
http://www.springer.com/series/11104

Leigh Photographs, Princeton, 1984

Atle Selberg

Atle Selberg

Collected Papers II

Reprint of the 1991 Edition

Springer

Atle Selberg (1917 – 2007)
Institute for Advanced Study
Princeton, NJ
USA

ISSN 2194-9875
ISBN 978-3-642-41022-2 (Softcover)
978-3-540-50626-3 (Hardcover)
DOI 10.1007/978-3-642-15089-0
Springer Heidelberg New York Dordrecht London

Library of Congress Control Number: 2012954381

Mathematical Subject Classification: AMS Classification (1980): 00A10

Printed on acid-free paper

Springer is part of Springer Science+Business Media (www.springer.com)

Foreword

The early work of Atle Selberg lies in the fields of analysis and number theory. It concerns the Riemann zeta-function, Dirichlet's L-functions, the Fourier coefficients of modular forms, the distribution of prime numbers, and the general sieve method. It is brilliant, and unsurpassed, and in the finest classical tradition. His later work cuts across many fields: function theory, operator theory, spectral theory, group theory, topology, differential geometry, and number theory. It has enlarged and transfigured the whole concept and structure of arithmetic. It exemplifies the modern tradition at its sprightly best, and makes him one of the master mathematicians of our time.

Thanks are due to Springer-Verlag, particularly to Dr. Heinz Götze, for bringing out this publication, which will enable the reader to perceive the depth and originality of Atle Selberg's ideas and results, and sense the scale and intensity of their influence on contemporary mathematical thought.

E.T.H. Zürich
1 July 1988

K. Chandrasekharan

Table of contents
Volume II

Table of contents
Volume I

42.
Linear operators and automorphic forms

Presented at the Ramanujan Colloquium,
Tata Institut of Fundamental Research, Bombay, January 1988

1. We consider a bounded symmetric complex domain B in the sense of Elie Cartan[1] and denote the group of analytic mappings of B onto itself by G, points in B by z and the elements of G by $g : z \to gz$. By a multiplier (or automorphy factor) $\varrho_g(z)$ we understand a function defined on $G \times B$ which is analytic holomorphic in z and differentiable in g and such that

$$\varrho_{g_1 g_2}(z) = \varrho_{g_1}(g_2 z)\varrho_{g_2}(z) \; . \tag{1.1}$$

Any such multiplier defines a kernel function $k_\varrho(z, \bar{z})$ which transforms in the way

$$k_\varrho(gz, \overline{gz}) = \varrho_g(z)\overline{\varrho_g(z)}k_\varrho(z, \bar{z}) \; . \tag{1.2}$$

We need only to write for some fixed z_0 in B,

$$k_\varrho(z, \bar{z}) = |\varrho_g(z_0)|^2 \; ,$$

where g is a solution of $z = gz_0$, and it is clear that this does not depend on the particular g chosen, but only on the point z.

From (1.2) we get that

$$ds^2 = \sum_{i,j} \frac{\partial^2 \log k_\varrho(z, \bar{z})}{\partial z_i \partial \bar{z}_j} dz_i d\bar{z}_j$$

is an invariant metric on B under the actions of the group G. Thus, if B is irreducible, we get that this metric can differ only by a constant factor from the Bergmann metric. If B is reducible it must be a linear combination of the Bergmann metrics of the irreducible factors of B. For irreducible B one easily derives that up to a factor of the form $cf(z)\overline{f(z)}$ where $f(z)$ is analytic, $k_\varrho(z, \bar{z})$ coincides with a real power of the Bergmann kernel function and $\varrho_g(z)$ is apart from a "trivial" multiplier of the form $\frac{f(gz)}{f(z)}$, equal to a power of the jacobian $j_g(z)$ of the mapping g. Similarly, if B is reducible, $\varrho_g(z)$ is apart from a trivial factor $\frac{f(gz)}{f(z)}$ equal to a product

[1] See for instance Siegel [1], Chapter XI.

of powers of the jacobians of the mapping g with respect to the various irreducible factors of B.

We may mention that essentially the same conclusion could be drawn from the weaker premise that instead of (1.1) $\varrho_g(z)$ satisfies the relation

$$|\varrho_{g_1 g_2}(z)| = |\varrho_{g_1}(g_2 z)|\,|\varrho_{g_2}(z)| \,, \tag{1.3}$$

then, apart from a factor of the form $\varepsilon_g \frac{f(gz)}{f(z)}$ where f is analytic and $|\varepsilon_g| = 1$, $\varrho_g(z)$ is equal to a product of powers of the jacobians of the mapping $z \to gz$ with respect to the irreducible factors of B.

We shall study linear operators on functions defined on B, which have the property of transforming with a multiplier on each side under the mappings of the group G. Call the operator $L = L_z$ and define L_{gz} through the relation

$$L_{gz}F(z) = \left[L_z F(g^{-1}z)\right]_{z \to gz} \,,$$

then L should transform according to the rule

$$L_{gz} = \varrho_g(z) L_z \sigma_g^{-1}(z) \,, \tag{1.4}$$

where ϱ and σ are two multipliers.[2]

We ask the question: for which B and which choices of multipliers ϱ and σ do such operators exist; and when they exist: how to determine their form as explicitly as possible.

It is easily seen that we may restrict oursevles to the case that B is irreducible and then derive the results for the general case from those obtained for the irreducible factors of B in case B is reducible.

As is known[3] there are six types of irreducible bounded symmetric domains. If we denote a matrix with m rows and n columns by $Z^{(m,n)}$ and the (n,n) unit matrix by E or $E^{(n)}$, there are the four main types:

(I) $Z = Z^{(m,n)}$, $E - \bar{Z}'Z > 0$,
(II) $Z = Z^{n,n)}$, $Z' = -Z$, $E - \bar{Z}'Z > 0$,
(III) $Z = Z^{(n,n)}$, $Z' = Z$, $E - \bar{Z}'Z > 0$

and

(IV) $Z = Z^{(n,1)}$, $\bar{Z}'Z < \frac{1}{2}(1 + |Z'Z|^2) < 1$.

Here Z' denotes the transposed matrix. In addition there are the types V and VI, two exceptional bounded symmetric domains of complex dimension 16 and 27, respectively, we shall not give a definition here.

---

[2] From now on we disregard trivial multipliers and consider only products of powers of the jacobians for the irreducible factors of B. This is only an apparent restriction.

[3] See Siegel [1], Chapter XI for instance.

It should also be noted that for $n = 2$ the domain IV is reducible, and that there also is some overlapping between the four types for low dimension, the unit circle $|z| < 1$ in one complex variable is for instance a special case of all four types.

2. The question of linear operators that transform according to the rule (1.4) can be split in two:
(a) Operators that conserve the multiplier, that is, when (1.4) holds, but with $\varrho_g(z) \equiv \sigma_g(z)$.

We shall refer to such operators as invariant (though strictly speaking they are so only if $\varrho_g(z) \equiv 1$ identically).

It is well known that invariant operators exist for all the bounded symmetric domains B and for all multipliers $\varrho_g(z)$, for given ϱ the differential operators form a finitely generated ring, where the number of independent generators equals the rank of the group G (or of the symmetric space). In this ring all elements, except the constant, contain differentiations both with respect to z and $\bar{z}$. The form of integral operators is easily given explicitly for B irreducible, if $\varrho_g(z) = (j_g(z))^{-r}$ where $j_g(z)$ denotes the jacobian of the mapping g, then

$$Lf = \int_B x(z,\zeta)\left(\frac{k(z,\bar{\zeta}}{k(\zeta,\bar{\zeta})}\right)^r f(\zeta)dw_\zeta\,, \tag{2.1}$$

where dw_ζ is the invariant element of volume, $k(z,\bar{\zeta})$ is the Bergmann kernel function, and $x(z,\zeta)$ is a "point pair invariant" satisfying $x(gz, g\zeta) = x(z,\zeta)$ for all z and ζ in B and g in G.

In particular for analytic functions $f(z)$ we have the reproducing operator

$$f(z) = c_r \int_B \left(\frac{k(z,\bar{\zeta}}{k(\zeta,\bar{\zeta})}\right)^r f(\zeta)dw_\zeta\,, \tag{2.2}$$

where c_r is a certain polynomial in r. (2.2) is valid for a certain hilbertspace of analytic functions if $r > r_0$, the largest zero of the polynomial c_r.[4]
(b) Operators that change the multiplier, that is: which transform in the way (1.4), but with $\varrho_g(z) \not\equiv \sigma_g(z)$.

(b) is a more complex question than (a), but it is not difficult to establish that linear operators that change the multiplier do not exist for all the irreducible domains, but only for a certain subclass.

To see this we may look at the compact subgroup of G which leaves some point z_0 in B fixed, the socalled stability group or isotropy group of z_0. It is simplest to choose the point O where all coordinates are zero, and the compact subgroup K_0 which keeps O fixed. For all the six types of bounded symmetric domains, the way they are usually defined, the elements of K_0

[4] See Selberg [2].

are linear transformations, and K_0 is essentially (sometimes a slight change of variables is necessary, as in type III where we would put a factor $\frac{1}{\sqrt{2}}$ in the elements of the symmetric matrix which are off the main diagonal) a subgroup of the unitary group $U(N)$ where N is the complex dimension of B, and $j_k(z) = j_k(0)$, where $k \in K_0$, is a one-dimensional representation of K_0.

It is clear that if there exists a linear operator satisfying (1.4), then in particular (1.4) must hold if g is restricted to K_0 and we consider the functional $\mathcal{L}$ that L_z represents at $z = 0$.

On the other hand, it is not hard to show that if we have a linear functional $\mathcal{L}$ which has the required property (1.4) for g in K_0 then it can be extended to a linear operator L_z by means of the relation (1.4) with $z = gO$, but the general form of this operator seems awkward to obtain in this way, particularly if it is a differential operator.

It is easily seen that an integral operator

$$L_z f = \int_B h(z,\zeta) f(\zeta) dw_\zeta$$

where $h(z,\zeta)$ is short for $h(z,\bar{z};\zeta,\bar{\zeta})$, in order to satisfy (1.4) must have a kernel $h(z,\zeta)$ which satisfies

$$h(gz, g\zeta) = \varrho_g(z)\sigma_g^{-1}(\zeta)h(z,\zeta) , \tag{2.3}$$

and in particular for $g = k \in K_0$, if we put $\zeta = 0$, we get

$$h(kz, 0) = \varrho_k(z)\sigma_k^{-1}(0)h(z,0) ,$$

or since $\varrho_k(z) = \varrho_k(0)$,

$$h(kz, 0) = \varrho_k(0)\sigma_k^{-1}(0)h(z) . \tag{2.4}$$

Since we may assume that $h(z,\zeta)$ is analytic in z and $\bar{z}$,[5] it is clear that the expansion of $h(z,0)$ in terms of powers of the z and $\bar{z}$ for z near 0, must start with a homogeneous polynomial $p(z,\bar{z})$ which also transforms by the factor $\varrho_k(0)\sigma_k^{-1}(0)$ when we replace z by kz.

Similarly if D_z is a differential operator which obeys the transformation rule (1.4), at $z = 0$ it takes the form of a polynomial in the $\frac{\partial}{\partial z}$ and $\frac{\partial}{\partial \bar{z}}$

$$D_0 = P\left(\frac{\partial}{\partial z}, \frac{\partial}{\partial z}\right) .$$

When z and dz undergo a unitary transformation from K_0, the $\frac{\partial}{\partial z}$ undergoes the contragredient transformation so we are again led to a polynomial (which we may assume to be homogeneous, otherwise we take the homogeneous

[5] If our original $h(z,\zeta)$ is not so, we may form the convolution of L_z with a suitable operator of the form (2.1) on the left which preserves the multiplier $\varrho_g(z)$.

part of lowest degree that is not identically zero) which transforms in the way (2.4) when the variables undergo the contragredient transformation to k (actually if we interchange $\frac{\partial}{\partial z}$ and $\frac{\partial}{\partial \bar{z}}$ the vector $(\frac{\partial}{\partial \bar{z}}, \frac{\partial}{\partial z})$ undergoes the same transformation as $(z, \bar{z})$).

It is now easy to see for the various types of B whether such polynomials exist when $\varrho_g(z) \not\equiv \sigma_g(z)$.

We find that for type I they exist only if $m = n$ and are then of the form

$$|z|^r P(z, \bar{z}) \quad \text{or} \quad |\bar{z}|^r P(z, \bar{z}) ,$$

where $|z|$ is the determinant of z, r some positive integer and $P(z, \bar{z})$ some homogeneous polynomial which is invariant under K_0.

For type II they exist only if n is even and are then of the form

$$P_f^r(z) P(z, \bar{z}) \quad \text{or} \quad P_f^r(\bar{z}) P(z, \bar{z})$$

where $P_f(z)$ is the polynomial called the Pfaffian of z (actually $|z|^{\frac{1}{2}}$, since the determinant is a square in this case), r again is a positive integer, and P a homogeneous polynomial which is invariant under K_0.

For type III they exist for all n and are of the form

$$|z|^r P(z, \bar{z}) \quad \text{or} \quad |\bar{z}|^r P(z, \bar{z})$$

where r is a positive integer, and $P(z, \bar{z})$ homogeneous and invariant under K_0.

For type IV, they again exist and are of the form

$$(z'z)^r P(z, \bar{z}) \quad \text{or} \quad (\bar{z}'\bar{z})^r P(z, \bar{z})$$

with r a positive integer and $P(z, \bar{z})$ is again homogeneous and invariant under K_0.

For the types V and VI which (for good reason!) we have not exhibited explicitly, we find they do not exist for type V, but *for type VI they exist and are given by the form*

$$(p_3(z))^r \, P(z, \bar{z}) \quad \text{or} \quad (p_3(\bar{z}))^r \, P(z, \bar{z})$$

where r and P are as before and p_3 is a certain cubic polynomial in 27 variables.

3. In order to derive more explicitly the form of the linear operators that transform according to (1.4) in the cases when we have seen they can exist, we note that the cases we have listed in the previous section are precisely the cases when the bounded domain B by a suitable analytic mapping becomes a socalled "positive half-space",[6] and when the group G by this mapping

[6] M. Koecher [2] writes "half-space", I prefer "positive half-space" since it indicates the connection with a positivity domain.

becomes a real group (by which we mean that in this new unbounded version of our domain we have: $\overline{gz} = g\bar{z}$).

By a positive half-space we understand a domain $z = x + iy$, where the column vector x is unrestricted, whereas the vector y is required to lie in a homogeneous positivity-domain Y in the sense defined by Koecher.[7] As before we shall use N to denote the complex dimension.

We recall some of the properties of a homogeneous positivity domain Y : It is a cone such that for any two vectors $y^{(1)}$ and $y^{(2)}$ in Y, we have always

$$y^{(1)'}y^{(2)} > 0 \,, \tag{3.1}$$

(see footnote[8]), and so that if for some vector $y^{(1)}$ (3.1) holds for all $y^{(2)}$ in Y, then $y^{(1)}$ also lies in Y.

There also exists a group G_Y of real matrices A such that $y \to Ay$ maps Y onto itself, this group is transitive on Y. In particular, for any scalar $\lambda > 0$ we have λy is in Y if y is in Y so Y is a cone. It is seen from (3.1) that if A is in G_Y then $y \to A'^{-1}y$ also maps Y onto itself so we may without restriction assume that with A always also A'^{-1} lies in G_Y.

There exists a homogeneous polynomial $Q(y)$, which we choose to be of minimal degree $q > 0$, such that $Q(y)$ is positive in Y and

$$Q(Ay) = |A|^{\frac{q}{N}} Q(y) \,, \tag{3.2}$$

(see footnote[9]). If we define for $i = 1, \ldots, N$,

$$y_i^* = \frac{\partial \log Q(y)}{\partial y_i} \,, \tag{3.3}$$

then $y \to y^*$ is an involution which carries Y into itself. We have

$$Q(y^*)Q(y) = \text{const} \,, \tag{3.4}$$

and by a suitable choice of Q (which by (3.2) is only determined up to a constant factor) we get

$$Q(y^*)Q(y) = 1 \,. \tag{3.4'}$$

(3.2) also gives $y^{*'}y = q$ and $(Ay)^* = A'^{-1}y^*$.

On Y we have an invariant volume element

$$dV_y = \frac{dy}{(Q(y))^{\frac{N}{q}}} \,, \tag{3.5}$$

---

[7] M. Koecher [1]

[8] Koecher's definition is more general, he has (3.1) in the form $y^{(1)'}Sy^{(2)} > 0$, where S is a nonsingular symmetric real matrix, but (3.1) covers the cases we consider.

[9] Our $Q(y) = (N(y))^{\frac{q}{N}}$, where $N(y)$ is Koecher's "Norm-function".

where we have written dy for the euclidean volume element. We also have an invariant metric

$$ds^2 = -\sum_{1\le i,j\le N} \frac{\partial^2 \log Q(y)}{\partial y_i \partial y_j} dy_i dy_j \,. \tag{3.6}$$

The involution $y \to y^*$ (which is actually a symmetry) has a fixpoint e and we have $Q(e) = 1$.

Now consider the positive halfspace $z = x + iy$ where x is unrestricted and y is in Y, and the group generated by transformations of the form $z \to z + a$, where a is a real vector, $z \to Az$, where A is in G_Y, and $z \to z^*$, where

$$z_i^* = -\frac{\log Q(z)}{\partial z_i}$$

for $i = 1, \ldots, N$. We call this group G.

If we write

$$D_z = Q\left(\frac{\partial}{\partial z}\right) \tag{3.7}$$

we shall show that for g in G

$$D^r_{gz} = (j_g(z))^{-\frac{rg+N}{2N}} D^r_z \, (j_g(z))^{-\frac{rg-N}{2N}} \,, \tag{3.8}$$

where r is any positive integer.

(3.8) is obvious if $gz = z + a$, and also for $gz = Az$ with A in G_Y; so we really need to prove (3.8) only for $gz = z^*$.

To do this we look first at the Y space. Y is actually a symmetric space, for any two points $y^{(1)}$ and $y^{(2)}$ in Y there exists an A in G_Y such that $Ay^{(1)} = y^{(2)*}$, $Ay^{(2)} = y^{(1)*}$. Thus G_Y and the $*$ operation satisfy the conditions for G and μ in Selberg [1].[10]

Also we see that if r is a positive integer then

$$L_y = Q^r(y) Q^r\left(\frac{\partial}{\partial y}\right) \tag{3.9}$$

is an invariant operator under the group G_Y.

It follows from a general result[11] that under the $*$ operation the operator L given by (3.9) goes into its formal adjoint L^* with respect to the invariant measure dV_y or otherwise expressed

$$L_{y^*} = L_y^* \,.$$

Thus for two suitable functions f and g we have

[10] See Selberg [1], p. 51.

[11] See Selberg [1], top of p. 53. In the context given there, the proof is obvious.

$$\int_Y f(y)L_y g(y)\frac{dy}{(Q(y))^{\frac{N}{q}}} = \int_Y g(y)L_y^* f(y)\frac{dy}{(Q(y))^{\frac{N}{q}}} \,.$$

Inserting the expression for L we see that it is easy to find the formal adjoint L^* since the formal adjoint of $Q^r(\frac{\partial}{\partial y})$ with respect to the euclidean measure is $Q^r(-\frac{\partial}{\partial y}) = (-1)^{rq}Q^r(\frac{\partial}{\partial y})$.

We get:

$$\begin{aligned}\int_Y f(y)Lg(y)\frac{dy}{(Q(y))^{\frac{N}{q}}} &= \int_Y f(y)Q(y)^{r-\frac{N}{q}}Q^r\left(\frac{\partial}{\partial y}\right)g(y)dy \\ &= \int_Y g(y)Q^r\left(-\frac{\partial}{\partial y}\right)\left(Q(y)^{r-\frac{N}{q}}f(y)\right)dy \\ &= \int_y g(y)\left(Q(y)^{\frac{N}{q}}Q^r\left(-\frac{\partial}{\partial y}\right)Q(y)^{r-\frac{N}{q}}f(y)\right)\frac{dy}{(Q(y))^{\frac{N}{q}}}\end{aligned}$$

Thus

$$L^* = Q^{\frac{N}{q}}(y)Q^r\left(-\frac{\partial}{\partial y}\right)Q^{r-\frac{N}{q}}(y)\,.$$

Also

$$\begin{aligned}L^* &= Q^r(y^*)Q^r\left(\frac{\partial}{\partial y^*}\right) \\ &= Q^{-r}(y)Q^r\left(\frac{\partial}{\partial y^*}\right)\,.\end{aligned}$$

Comparing these two expressions for L^*, we get

$$\begin{aligned}(3.10)\quad Q^r\left(\frac{\partial}{\partial y^*}\right) &= Q^{r+\frac{N}{q}}(y)Q^r\left(-\frac{\partial}{\partial y}\right)Q^{r-\frac{N}{q}}(y) \\ &= (-1)^{rq}Q^{r+\frac{N}{q}}(y)Q^r\left(\frac{\partial}{\partial y}\right)Q^{r-\frac{N}{q}}(y)\,.\end{aligned}$$

But from (3.10) follows immediately

$$(3.11)\qquad Q^r\left(\frac{\partial}{\partial z^*}\right) = Q^{r+\frac{N}{q}}(z)Q^r\left(\frac{\partial}{\partial z}\right)Q^{r-\frac{N}{q}}(z)\,.$$

It remains to determine the jacobian of the mapping $z \to z^*$ or $j_*(z)$. We have

$$\frac{\partial z_i^*}{\partial z_j} = -\frac{\partial^2 \log Q(z)}{\partial z_i \partial z_j}$$

so that

$$j_*(z) = \left|-\frac{\partial^2 \log Q(z)}{\partial z_i \partial z_j}\right|\,.$$

If, as before, dy denotes the euclidean volume element we have for the invariant volume element in Y

$$\frac{dy^*}{(Q(y^*))^{\frac{N}{q}}} = \frac{dy}{(Q(y))^{\frac{N}{q}}}$$

or using (3.4′),

$$dy^* = \frac{dy}{(Q(y))^{\frac{2N}{q}}} \; .$$

Since the symmetry $y \to y^*$ preserves orientation or not according as N is even or odd, we get

$$\left| \frac{\partial y_i^*}{\partial y_j} \right| = (-1)^N \, (Q(y))^{-\frac{2N}{q}} \; ,$$

or

$$\left| \frac{\partial^2 \log Q(y)}{\partial y_i \partial y_j} \right| = (-1)^N \, (Q(y))^{-\frac{2N}{q}} \; .$$

It is therefore obvious that

$$j_*(z) = \left| - \frac{\partial^2 \log Q(z)}{\partial z_i \partial z_j} \right| = (Q(z))^{-\frac{2N}{q}} \; .$$

Combining this with (3.11), we get that (3.8) holds also for $gz = z^*$, thus (3.8) holds for all g in G.

It is however clear that (3.8), which is really an algebraic identity, holds in a much larger group than G. Let us define $\tilde{G}_Y$ as the group of complex matrices, whose entries satisfy the algebraic relations which define G_Y, and consider the group $\tilde{G}$ generated by translations $z \to z + a$ where a now may be a complex vector, $z \to Az$ where A is in $\tilde{G}_Y$ and $z \to z^*$. *Clearly (3.8) as an algebraic identity holds for any transformation g in $\tilde{G}$.*

The transformations of $\tilde{G}$ do not in general map the positive half-space onto itself. $\tilde{G}$ is actually large enough to map the positive half-space back into a bounded symmetric domain, in most cases also the original one (this is for instance true for the first three types listed at the end of Section 2), or one may have to add a final unitary transformation which does not lie in K_0 (this is the case for type IV where the transformation $z_1 \to z_1$, $z_j \to iz_j$ for $1 < j \leq n$, would be needed at the end. For type IV the positivity domain can be defined as $y_1 > 0$, $y_1^2 - y_2^2 - \ldots - y_n^2 > 0$ and we have $Q(y) = \frac{1}{2}(y_1^2 - y_2^2 - \ldots - y_n^2)$, so the last transformation is needed to transform $z_1^2 - z_2^2 \ldots - z_n^2$ into $z'z = z_1^2 + \ldots + z_n^2$).

At any rate we get in each case the form of the differential operator and its transformation formula for the original bounded domain.

In the case of type I with $m = n$ we get, writing $|\frac{\partial}{\partial z}|$ for $|\frac{\partial}{\partial z_{ij}}|$, that *if*

$$gz = (Az + B)(Cz + D)^{-1}$$

where A, B, C and D are complex $n \times n$ matrices such that

$$\begin{vmatrix} A & B \\ C & D \end{vmatrix} = 1 \,,$$

then

$$\left| \frac{\partial}{\partial gz} \right|^r = |Cz + D|^{r+n} \left| \frac{\partial}{\partial z} \right|^r |Cz + D|^{r-n} \,. \tag{3.12}$$

In the case of type II, for $z = z^{(2n,2n)}$ and $z' = -z$, *if g is the transformation*

$$gz = (Az + B)(Cz + D)^{-1}$$

where A, B, C and D are $2n \times 2n$ complex matrices with the property that for $M = \left(\begin{smallmatrix} A & B \\ C & D \end{smallmatrix}\right)$, $J = \left(\begin{smallmatrix} 0 & E \\ E & O \end{smallmatrix}\right)$, we have

$$M'JM = J$$

and $|M| = 1$. Then writing

$$P_f\left(\frac{\partial}{\partial z}\right) \quad for \quad P_f\left(\frac{\partial}{\partial z_{ij}}\right)$$

where P_f is the Pfaffian, we have

$$\left(P_f\left(\frac{\partial}{\partial gz}\right)\right)^r = |Cz + D|^{\frac{r+2n-1}{2}} \left(P_f\left(\frac{\partial}{\partial z}\right)\right)^r |Cz + D|^{\frac{r-2n+1}{2}} \,. \tag{3.13}$$

For type III, *if we define*

$$\left| \frac{\partial}{\partial z} \right| = \left| \frac{1 + \delta_{ij}}{2} \, \frac{\partial}{\partial z_{ij}} \right| \,,$$

where δ_{ij} is the Kronecker symbol (1 on the main diagonal, 0 off it) and

$$gz = (Az + B)(Cz + D)^{-1}$$

where for

$$M = \begin{pmatrix} A & B \\ C & D \end{pmatrix} \,, \quad I = \begin{pmatrix} O & -E \\ E & O \end{pmatrix}$$

we have

$$M'IM = I$$

and $|M| = 1$, then again

$$\left| \frac{\partial}{\partial gz} \right|^r = |Cz + D|^{r+\frac{n+1}{2}} \left| \frac{\partial}{\partial z} \right|^r |Cz + D|^{r-\frac{n+1}{2}} \,. \tag{3.14}$$

In this case of type IV, we will confine ourselves to stating the form for the original bounded domain without defining the more general group $\tilde{G}$ or giving the explicit forms of g or the jacobian $j_g(z)$.

If we define

$$D_z = \sum_{i=1}^{n} \frac{\partial^2}{\partial z_i^2} ,$$

then

$$(3.15) \qquad D_{gz}^r = (j_g(z))^{-(\frac{r}{n}+\frac{1}{2})} D_z^r \, (j_g(z))^{-(\frac{r}{n}-\frac{1}{2})} .$$

For the type VI we do not give explicit formulas.

Since these differential operators only contain differentiations with respect to z and not $\bar{z}$, we see that if we for real α put in a factor $(k(z,\bar{z}))^{-\alpha}$ on the left side and a factor $(k(z,\bar{z}))^{\alpha}$ on the right (k being again the Bergmann kernel function), we again get an operator which satisfies (1.4) but the two multipliers ϱ_g and σ_g have each been multiplied by $(j_g(z))^{\alpha}$.

Beside the operators D so constructed we may of course also consider their complex conjugates $\bar{D}$. These do not satisfy (1.4) since we required our multipliers to be analytic in z. $\bar{D}$ would transform in the way

$$\bar{D}_{gz} = (\overline{j_g(z)})^{-\alpha} \bar{D}_z (\overline{j_g(z)})^{\beta} ,$$

where α and β depend on D. If we now define

$$\tilde{D}_z = (k(z,\bar{z}))^{-\alpha} \, \bar{D}_z \, (k(z,\bar{z}))^{\beta} ,$$

we see that

$$\tilde{D}_{gz} = (j_g(z))^{\alpha} \, \tilde{D}_z \, (j_g(z))^{-\beta} ,$$

so this operator has the behavior required. Here, since the differentiations in $\tilde{D}_z$ are with respect to $\bar{z}$ and not z, we can clearly replace the pair (α,β) by any other pair of real numbers (α',β') as long as

$$\alpha' - \beta' = \alpha - \beta .$$

It can be shown that all differential operators which satisfy (1.4) can be obtained by combining the operators D or $\tilde{D}$ with suitable invariant differential operators of the kind mentioned under (a) at the beginning of Section 2.

4. To find the general form of integral operators that transform in the required way, we may again look at the representation of the domain B as a positive half-space where the analytic mappings gz are real, which is to say: $g\bar{z} = \overline{gz}$. We have of course also that $j_g(\bar{z}) = \overline{j_g(z)}$.

Considering the Bergmann kernel function of this halfspace, we get thus

$$\begin{aligned} k(gz, g\bar{\zeta}) &= k(gz, \overline{g\zeta}) \\ &= \left(j_g(z)\overline{j_g(\zeta)}\right)^{-1} k(z,\bar{\zeta}) \\ &= \left(j_g(z) j_g(\bar{\zeta})\right)^{-1} k(z,\bar{\zeta}) . \end{aligned}$$

If we now write ζ instead of $\bar{\zeta}$, this becomes

$$k(gz, g\zeta) = (j_g(z)j_g(\zeta))^{-1}\, k(z,\zeta)\,.$$

From this we see that if we put

$$h_{a,b}(z,\zeta) = \frac{(k(z,\bar{\zeta}))^{\frac{a+b}{2}}(k(z,\zeta))^{\frac{a-b}{2}}}{(k(\zeta,\bar{\zeta}))^{\frac{a+b}{2}}}\,, \tag{4.1}$$

where $b > a$ and $b - a$ is such that $(k(z,\zeta))^{\frac{a-b}{2}}$ is single valued for z and ζ in the positive halfspace,[12] then (4.1) transforms in the way given by (2.3), with $\varrho_g(z) = (j_g(z))^{-a}$, $\sigma_g(\zeta) = (j_g(\zeta))^{-b}$.

If $b < a$, we write

$$h_{a,b}(z,\zeta) = \frac{k^a(z,\bar{z})}{k^b(\zeta,\bar{\zeta})}\overline{h_{b,a}(\zeta,z)}\,. \tag{4.1'}$$

The most general form of a kernel which transforms in the way given by (2.3) is of the form

$$h(z,\zeta) = x(z,\zeta)h_{a,b}(z,\zeta)\,, \tag{4.2}$$

where x is an invariant of the point pair z and ζ while $h_{a,b}(z,\zeta)$ is given by (4.1) or (4.1′) according to the sign of $b - a$.

For the bounded domains the form of the kernels is more complicated than for the positive half-spaces.

5. For the reducible bounded symmetric domains these same questions can be answered by using our results for the irreducible factors.

It is possible to generalize the problem we considered and ask similar questions for, say, bilinear operators operating on two functions, or more generally q-linear operators acting on q functions; for instance to be able to produce from two automorphic forms a new one which depends linearly on these two, but whose multiplier is not the product of the multipliers of these two forms. Again one would begin by looking at the stability group of the point O in B. Thus, for instance, it is easy to show that such bilinear operators exist for type I if $Z^{(m,n)} = Z^{(2n,n)}$, whereas for $Z^{n,1}$ there are no such q-linear operators for $q < n$.[13] Whether such multi-linear operators are of much interest is doubtful.

I originally determined the explicit transformation formulas for the differential operators considered in Section 3 in the year 1960. My first aim was to construct operators that effected the shift in automorphy factors in

[12] This is true if $b - a$ is an integral multiple of $\frac{q}{N}$, since it is not hard to show that apart from a constant factor $k(z,\zeta)$ is equal to $(Q(\frac{z-\zeta}{2i}))^{-\frac{2N}{q}}$.

[13] More generally, for $m \geq n$, such q-linear operators exist iff $q \geq \frac{m}{n}$.

the same way as the operators $y^\alpha \frac{d^k}{dz^k} y^{-\alpha}$ and $y^{k+1} \frac{d^k}{d\bar{z}^k} y^{k-1}$ do in the case of the upper halfplane.

Later I used them to effect analytic continuation of dirichlet series associated with the fourier expansions of modular forms in positive halfspaces where the fourier expansion contains singular terms, and also to get the analytic continuation for the dirichlet series associated with two such modular forms in the case when singular terms are present.

I lectured off and on on these matters, the first time in Hamburg in the summer of 1961; later at various conferences, Copenhagen 1964, Jyväskylä 1970, Bar Ilan 1981 and other places abroad.

In the sixties some of the applications were privately communicated to Hans Maass, Howard Resnikoff and Audrey Terras, all of whom (with my permission) utilized some of this material in their publications.

References

Koecher, Max [1]: *Positivitätsbereiche im R^n*, American Journal of Mathematics, vol. 79, (1957), pp. 575–596

Koecher, Max [2]: *Automorphic forms in half-spaces,* Seminars on Analytic Functions, Institute for Advanced Study, Princeton, New Jersey, vol. 2 (1957), pp. 105–119

Selberg, A [1]: *Harmonic analysis and discontinuous groups,* Journal of Indian Math. Soc., vol. 20, (1956), pp. 47–87

Selberg, A. [2]: *Automorphic forms and integral operators,* Seminars on Analytic Functions, Institute for Advanced Study, Princeton, New Jersey, vol. 2, (1957), pp. 152–161

Siegel, C.L. [1]: *Analytic Functions of Several Complex Variables,* Institute for Advanced Study Lecture Notes, revised edition 1962

43.
Remarks on the distribution of poles of Eisenstein series

Festschrift in honor of I.I. Piatetski-Shapiro,
Weizmann Science Press, Israel, 1990. Vol. 2, pp. 251–278

0. Introduction Let Γ be a discrete group of motions of the hyperbolic plane. We denote a fundamental domain of Γ by $\mathcal{D} = \mathcal{D}_\Gamma$, and its area, assumed to be finite, by $A = A(\mathcal{D}) = 4\pi\mu(\mathcal{D})$. χ denotes a one-dimensional representation of Γ.

In the case when $\mathcal{D}$ is compact or when χ is nonsingular[1] with respect to all cusps of Γ, the trace formula allows us to estimate the number of eigenvalues $< \frac{1}{4} + T^2$ in the discrete spectrum of the hyperbolic Laplacian acting on functions defined on the hyperbolic plane which transform the multiplier χ under the action of Γ. The asymptotic formula has the form

$$N_\chi(T) = \mu(\mathcal{D})T^2 + B(\chi)T + O\left(\frac{T}{\log T}\right), \tag{0.1}$$

where

$$B(\chi) = -\frac{1}{\pi} \sum_{1 \le i \le \kappa} \log|1 - \chi(S_i)| .$$

Here the sum on the right hand side is taken over a complete set of inequivalent cusps ξ_i and S_i denotes the primitive parabolic element in Γ which has ξ_i as fixpoint.

If, on the other hand, χ is singular with respect to $\kappa_1 > 0$ cusps, say ξ_i for $1 \le i \le \kappa_1$, we have instead of (0.1) the asymptotic formula

$$\begin{aligned} N_\chi(T) - \frac{1}{2\pi}\int_0^T \frac{\varphi'}{\varphi}\left(\frac{1}{2} + it, \chi\right) dt \\ = \mu(\mathcal{D})T^2 + B_1 T \log T + B_2(\chi)T + O\left(\frac{T}{\log T}\right) . \end{aligned} \tag{0.2}$$

Here

[1] We shall be using the terminology and notations of my Göttingen and Bombay lectures, Selberg [1] pp. 632–672, and pp. 423–463, except that we here use t and T where there r and R were used, also $\mu(\mathcal{D})$ here is one half of $\mu(\mathcal{D})$ there. Further $\varrho = \beta + i\gamma$ will here denote the zeros of $\varphi(s,\chi)$ rather than the poles. Finally, we write here $L_{i,j}(s,\chi)$ instead of the $L_0^{(i,j)}(s,\chi)$ used there.

$$B_1 = -\frac{\kappa_1}{\pi} ,$$

and

$$B_2(\chi) = \frac{\kappa_1(1-\log 2)}{\pi} - \frac{1}{\pi} \sum_{\kappa_1 \le i \le \kappa} \log |1-\chi(S_i)| \ .$$

The second term on the left hand side of (0.2) measures the contribution of the continuous spectrum which thus is connected with the variation of the argument of $\varphi(\frac{1}{2}+it,\chi)$ on the interval $0 \le t \le T$. The function $\varphi(s,\chi)$ is defined as the determinant of a κ_1 by κ_1 matrix

$$\phi(s,\chi) = (\varphi_{i,j}(s,\chi)) \ , \tag{0.3}$$

where the $\varphi_{i,j}(s,\chi)$ arises as the coefficient of y_j^{1-s} in the "constant term" of the Fourier expansion of the Eisenstein series $E_i(z,s;\chi)$ for $1 \le i \le \kappa_1$ at the cusp ξ_j for $1 \le j \le \kappa_1$. This coefficient can be expressed as

$$\varphi_{i,j}(s,\chi) = \frac{\sqrt{\pi}\Gamma(s-\frac{1}{2})}{\Gamma(s)} L_{i,j}(s,\chi) \ , \tag{0.4}$$

where $L_{i,j}$ is a dirichlet series of the form

$$L_{i,j}(s,\chi) = \sum_{n=1}^{\infty} \frac{\alpha_{i,j}(s,\chi)}{\lambda_{i,j}(n)^s} \ , \tag{0.5}$$

where the $\lambda_{i,j}$ are real and positive, and the series converges absolutely for $\sigma > 1$. If we denote by χ_0 the identity representation, we have

$$|\alpha_{i,j}(n,\chi)| \le \alpha_{i,j}(n,\chi_0) \ . \tag{0.6}$$

From (0.4) and (0.5) it follows that

$$\varphi(s,\chi) = \operatorname{Det} \phi(s,\chi) = \left(\frac{\sqrt{\pi}\Gamma(s-\frac{1}{2})}{\Gamma(s)}\right)^{\kappa_1} L(s,\chi) \ , \tag{0.7}$$

where $L(s,\chi)$ is a dirichlet series. Since

$$\varphi_{i,j}(s,\chi) = \varphi_{j,i}(s,\bar{\chi})$$

we can conclude that $L(s,\chi)$ has real coefficients. Like the series $L_{i,j}(s,\chi)$ it converges absolutely for $\sigma > 1$. We shall write[2]

$$L(s,\chi) = ab^{1-2s} L^*(s,\chi) \ , \tag{0.8}$$

where $a \ne 0$ and real, b real and positive and

[2] a depends on χ, as will in general b. If the set of χ depends on continuous parameters we will have $b(\chi) = b(\chi_0)$ except possibly for a set of χ of lower dimension. It is easy to see that we always have $b \ge 1$.

$$L^*(s,\chi) = 1 + \sum_{n=1}^{\infty} \frac{\alpha_n(\chi)}{\lambda_n^s}, \tag{0.9}$$

where the λ_n are > 1. Furthermore, we have

$$\varphi(s,\chi)\varphi(1-s,\chi) = 1, \tag{0.10}$$

and in particular

$$\left|\varphi\left(\frac{1}{2}+it,\chi\right)\right| = 1. \tag{0.11}$$

$\varphi(s,\chi)$ is holomorphic for $\sigma \geq \frac{1}{2}$, except possibly for a finite number of poles on the stretch $\frac{1}{2} < s \leq 1$ of the real line. (0.10) shows $\varphi(s,\chi)$ to be meromorphic in the whole complex plane. The poles of $\varphi(s,\chi)$ correspond to poles of the Eisenstein series $E_i(z,s;\chi)$, and it is easily seen that there are infinitely many such poles in the half plane $\sigma < \frac{1}{2}$. From (0.10) and the fact that $\varphi(s,\chi)$ is real for real s, we observe that if we denote the zeros of $\varphi(s,\chi)$ in $\sigma < \frac{1}{2}$ by $\varrho = \beta + i\gamma$, then $1 - \bar{\varrho} = 1 - \beta + i\gamma$ are the poles of $\varphi(s,\chi)$ in the region $\sigma < \frac{1}{2}$. Studying the zeros in $\sigma > \frac{1}{2}$ is thus equivalent to studying the poles in $\sigma < \frac{1}{2}$, and the zeros of $\varphi(s,\chi)$ in $\sigma > \frac{1}{2}$ are exactly the zeros of $L^*(s,\chi)$ in $\sigma > \frac{1}{2}$.

The second term on the left hand side of (0.2) differs from

(0.12)

$$-\frac{1}{2\pi}\int_0^T \frac{L^{*\prime}}{L^*}\left(\frac{1}{2}+it,\chi\right)dt$$
$$= N_\chi\left(\frac{1}{2},T\right) - \frac{\nu}{2} + \frac{1}{2\pi}\arg L^*\left(\frac{1}{2},\chi\right) - \frac{1}{2\pi}\arg L^*\left(\frac{1}{2}+iT,\chi\right),$$

by an amount which is seen to be

$$\frac{T}{\pi}\log b + O(1) = O(T). \tag{0.13}$$

On the right hand side of (0.12) $N_\chi(\sigma_0,T)$ denotes the number of $\varrho = \beta + i\gamma$ in the region $\frac{1}{2} \leq \sigma_0 \leq \beta$, $0 \leq \gamma \leq T$ with the usual convention of counting zeros on the boundary with half multiplicity.[3] ν is the number of poles of $\varphi(s,\chi)$ or $L^*(s,\chi)$ on the stretch $\frac{1}{2} < s \leq 1$. The arguments on the right hand side of (0.12) are as usual determined by continuity letting σ go from $+\infty$ to $\frac{1}{2}$ starting with the argument zero at $\sigma = +\infty$, and observing the usual conventions if the path passes through a zero or a pole. We get thus

[3] And the zero of order κ_1 at $s = \frac{1}{2}$, which $L^*(s,\chi)$ has, but $\varphi(s,\chi)$ has not, is counted with $\frac{1}{4}$ multiplicity.

$$(0.14)\quad -\frac{1}{2\pi}\int_0^T \frac{\varphi'}{\varphi}\left(\frac{1}{2}+it,\chi\right)dt = N_\chi\left(\frac{1}{2},T\right) + \frac{\log b}{\pi}T - \frac{1}{2\pi}\arg L^*\left(\frac{1}{2}+iT,\chi\right) + O(1)\,.$$

A quick estimation of the third term on the right hand side of (0.14) can be given by noting that both

$$Z_\Gamma(s,\chi) \text{ and } L^*(s,\chi)Z_\Gamma(s,\chi)$$

are regular for $t > t_0$, and in the region $t > t_0$, $\sigma \geq 0$ they are both bounded by the expression

$$O(e^{ct})\,,$$

with some positive constant c. A standard argument then permits us to estimate the argument of both $Z_\Gamma(s,\chi)$ and $L^*(s,\chi)Z_\Gamma(s,\chi)$ at $\frac{1}{2}+iT$ as $O(T)$, and so by taking the difference we get the same estimate for the term

$$\arg L^*\left(\frac{1}{2}+iT,\chi\right)\,.$$

(0.14) thus gives

$$(0.15)\quad -\frac{1}{2\pi}\int_0^T \frac{\varphi'}{\varphi}\left(\frac{1}{2}+it,\chi\right)dt = N_\chi\left(\frac{1}{2},T\right) + O(T)\,.$$

We may therefore rewrite (0.2) as

$$(0.16)\quad N_\chi(T) + N_\chi\left(\frac{1}{2},T\right) = \mu(\mathcal{D})T^2 + B_1 T\log T + O(T)\,.$$

In a later section we shall give a somewhat sharper form of (0.15) and (0.16), but already in the present cruder form (0.16) shows that getting additional information about the zeros $\beta + i\gamma$ of $L^*(s,\chi)$ and so about $N_\chi(\frac{1}{2},T)$ could throw additional light on the behavior of $N_\chi(T)$. While we are unable directly to count the zeros of $L^*(s,\chi)$, there are however expressions counting them with certain weights, which can be very precisely estimated.

1. We first prove some lemmas, which we shall give in a rather more general form than we actually need here.

Lemma 1. *Let $f(s)$ be holomorphic for $\sigma > \alpha$, except for at most a finite number of poles in this region, and let $f(s)$ have continuous boundary values on $\sigma = \alpha$. Further, assume that we have*

$$\sigma\,(f(s)-1) \to 0$$

as $\sigma \to \infty$. *Denoting the zeros of* $f(s)$ *in* $\sigma > \alpha$ *by* $\beta + i\gamma$ *and the poles by* $\sigma_j + it_j$, $1 \le j \le r$, *we have for* $T > \max_j |t_j|$, *that*

(1.1)

$$\sum_{\substack{|\gamma|<T \\ \beta>\alpha}} (T - |\gamma|)(\beta - \alpha) = \frac{1}{2\pi} \int_{-T}^{T} (T - |t|) \log |f(\alpha + it)|\, dt$$

$$+ \sum_j (T - |t_j|)(\sigma_j - \alpha) + \frac{1}{2\pi} \int_{\alpha}^{\infty} (\sigma - \alpha) \log \frac{|f(\sigma + iT) f(\sigma - iT)|}{|f(\sigma)|^2} d\sigma .$$

To prove this, we first assume that $f(s)$ is analytic also on $\sigma = \alpha$, and write

$$F(s) = f(s)\overline{f(\bar{s})} .$$

Let $\sigma_0 > \max_{1 \le j \le r} \sigma_j$, and consider the rectangle R with corners σ_0, α, $\alpha + iT$, $\sigma_0 + iT$. We remove from R the points that lie within the distance δ from a zero or a pole of $F(s)$ in R or on its boundary, δ is here chosen so small that there is no overlapping of the discs or halfdiscs removed. We call the resulting region R_δ, and write $u(s) = (T-t)(\sigma-\alpha)$ and $v(s) = \log |F(s)|$. Both u and v are harmonic in R_δ, so we have

$$\frac{1}{2\pi} \int\!\!\int_{R_\delta} (u\Delta v - v\Delta u) d\sigma\, dt = 0 .$$

We now use Green's theorem, and in the integral over the boundary ∂R_δ that arises, we observe that $u = 0$ for $t = T$ and also for $\sigma = \alpha$, and that $\frac{\partial v}{\partial n} = 0$ on the part of ∂R_δ where $t = 0$. In the resulting formula we let $\delta \to 0$ and $\sigma_0 \to \infty$. We observe that the integral over the boundary where $\sigma = \sigma_0$ tends to zero, while the contribution from the boundaries of the small discs around the zeros tends to the lefthand side of (1.1). The contribution from the poles, as well as that of the three remaining sides of R, we move to the righthand side. It becomes

$$\frac{1}{2\pi} \int_0^T (T - t) \log |F(\alpha + it)|\, dt + \sum_j (T - |t_j|)(\sigma_j - \alpha)$$

$$+ \frac{1}{2\pi} \int_{\alpha}^{\infty} (\sigma - \alpha) \log \frac{|F(\sigma + iT)|}{|F(\sigma)|} d\sigma ,$$

which, remembering the definition of F, we see to equal the righthand side of (1.1). This proves (1.1) if $f(s)$ is regular analytic also on $\sigma = \alpha$. If it is not, we write down (1.1) with $\alpha + \varepsilon$, $\varepsilon > 0$ instead of α and now let

$\varepsilon \to 0$. The assumption of continuous boundary values on $\sigma = \alpha$ justifies the passage to the limit.[4]

Lemma 2. *If, in addition to the assumptions of Lemma 1, we assume that*

$$F(s) = O(|t|^c)\ ,$$

for $|t| > t_0$ *and* $\sigma \geq \alpha$, *with some positive constant* c, *and that*

$$f(s) = 1 + O(e^{-c'\sigma})$$

with some constant $c' > 0$, *uniformly for* $\sigma \geq \sigma_0$, *then, for* $T \geq 2$,

$$\sum_{\substack{|\gamma|<T\\ \beta>\alpha}} (T - |\gamma|)\,(\beta - \alpha) = \frac{1}{2\pi} \int_{-T}^{T} (T - |t|) \log|f(\alpha + it)|\, dt + T \sum_{\sigma_j > \alpha} (\sigma_j - \alpha) + O(\log T)\ . \tag{1.2}$$

From (1.1) it is clear that (1.2) follows if we can prove that

$$\int_\alpha^\infty (\sigma - \alpha) \log|f(\sigma \pm iT)|\, d\sigma = O(\log T)\ . \tag{1.3}$$

It is of course enough to consider (1.3) with the upper sign since $\overline{f(\bar{s})}$ satisfies the same conditions as $f(s)$.

For $\sigma \geq \sigma'$ where $\sigma' > \sigma_0$ is sufficiently large, we have

$$\log|f(\sigma + iT)| = O(e^{-c'\sigma})\ ,$$

and so

$$\int_{\sigma'}^\infty (\sigma - \alpha) \log|f(\sigma + iT)|\, d\sigma = O(1)\ ,$$

also for $T > T_0$, we have $|f(\sigma + iT)| < T^{1+c}$, so

$$\int_\alpha^{\sigma'} (\sigma - \alpha) \log|f(\sigma + iT)|\, d\sigma < (\sigma' - \alpha)^2 (1 + c) \log T\ .$$

Thus for $T > 2$ we have

[4] An alternative proof of the lemma one gets by considering $\frac{1}{2\pi}\Re \int_{\bar{R}} (s - \alpha - iT) \log F(s)\, ds = 0$, where $\bar{R}$ is the boundary of the rectangle R, but with added horizontal slits joining the zeros and poles of $F(s)$ in R to the line $\sigma = \alpha$. Evaluation of the contributions from the slits, as well as the boundaries of R, again gives (1.1) as $\sigma \to \infty$.

$$\int_{\alpha}^{\infty} (\sigma - \alpha) \log |f(\sigma + iT)| \, d\sigma < c'' \log T \,, \tag{1.4}$$

with some positive constant c''.

We need to show that we have a similar lower bound for the lefthand side of (1.3). To this end we use Lemma 14 of Selberg [1] p. 319, taking here $t_2 = T$ and $t_1 = T - \frac{2\pi}{c'}$ in that lemma, which then gives for $T > T_0$,

$$\frac{\pi}{c'} \sum_{\substack{\beta > \alpha \\ T - \frac{2\pi}{c'} < \gamma < T}} \sin \frac{c'(T-\gamma)}{2} \sinh \frac{c'}{2}(\beta - \alpha)$$

$$= \int_{T - \frac{2\pi}{c'}}^{T} \sin \frac{c'}{2}(T - t) \log |f(\alpha + it)| \, dt +$$

$$+ \int_{\alpha}^{\infty} \sinh \frac{c'}{2}(\sigma - \alpha) \left\{ \log |f(\sigma + iT)| + \log \left| f\left(\sigma + i\left(T - \frac{2\pi}{c'}\right)\right)\right| \right\} d\sigma.$$

Here the lefthand side is ≥ 0 and in the last integral on the righthand side the part over (σ', ∞) is seen to be $O(1)$.

Since we have

$$|f(s)| < |t|^{1+c} \,,$$

for $|t| > T_0$, and $\sigma \geq \alpha$, we get from the above

$$\int_{\alpha}^{\sigma'} \sinh \frac{c'}{2}(\sigma - \alpha) \log |f(\sigma + iT)| \, d\sigma$$

$$> - \int_{T - \frac{2\pi}{c'}} (1 + c) \log T \, dt - \int_{\alpha}^{c'} (1 + c) \log T \sinh \frac{c'}{2}(\sigma - \alpha) d\sigma$$

$$- O(1) > -c''' \log T \,,$$

with some positive constant x'''. From this we derive

$$\int_{\alpha}^{\sigma'} \sinh \frac{c'}{2}(\sigma - \alpha) \log \frac{|f(\sigma + iT)|}{T^{1+c}} d\sigma > -c'''' \log T \,,$$

where again c'''' is some positive constant. Since for $T > T_0$ the logarithm in the integral is negative, we may in this inequality replace

$$\sinh \frac{c'}{2}(\sigma - \alpha)$$

by its lower bound

$$\frac{c'}{2}(\sigma - \alpha) \,,$$

this gives

$$\frac{c'}{2} \int_{\alpha}^{\sigma'} (\sigma - \alpha) \log \frac{|f(\sigma + iT)|}{T^{1+c}} d\sigma > -c'''' \log T \,.$$

Combining this with the estimate for the integral over (σ',∞), we get the desired lower bound for the lefthand side of (1.3). This proves the lemma.

We now return to the function $L^*(s,\chi)$, which clearly satisfies the assumptions of Lemma 1 and Lemma 2 with some $c'>0$. It follows from (0.7), (0.8) and (0.11) that

$$\left|L^*\left(\frac{1}{2}+it,\chi\right)\right| = \frac{1}{|a|}\left|\frac{\Gamma(\frac{1}{2}+it)}{\Gamma(it)}\right|^{\kappa_1} = \frac{1}{|a|}\left(\frac{|t|}{\pi}\right)^{\frac{\kappa_1}{2}}\left(\frac{1-e^{-2\pi|t|}}{1+e^{-2\pi|t|}}\right)^{\frac{\kappa_1}{2}} . \tag{1.5}$$

Also for $\sigma>\frac{1}{2}$, $t\neq 0$, we have

$$|\varphi(\sigma+it,\chi)| < \left(\sqrt{1+\left(\frac{\sigma-\frac{1}{2}}{t}\right)^2}+\frac{\sigma-\frac{1}{2}}{|t|}\right)^{\kappa_1} , \tag{1.6}$$

(see footnote[5]); this shows

$$|L^*(\sigma+it,\chi)| = O\left(|t|^{\frac{\kappa_1}{2}}\right) ,$$

for $|t|>2$ and $\frac{1}{2}\le\sigma\le 2$, while for $\sigma\ge 2$ we have

$$L^*(\sigma+it,\chi) = 1+O(\lambda_1^{-\sigma}) .$$

From (1.5) we get

$$\begin{aligned}
&\int_{-T}^{T}(T-|t|)\log\left|L^*\left(\frac{1}{2}+it,\chi\right)\right| dt\\
&= \kappa_1\int_0^T (T-t)\log\frac{t}{\pi}dt - 2\log|a|\int_0^T(T-t)dt\\
&\quad + \kappa_1\int_0^T (T-t)\log\frac{1-e^{-2\pi t}}{1+e^{-2\pi t}}dt\\
&= \frac{\kappa_1}{2}T^2\log\frac{T}{\pi} - \left(\frac{3}{4}\kappa_1+\log|a|\right)T^2 - \frac{\pi}{8}\kappa_1 T + O(1) .
\end{aligned}$$

Lemma 2 now gives

[5] This follows from (7.44), Selberg [1] p. 520, if we note that (as can easily be shown) there A may always be taken as equal to 1, and that the Hermitian matrix on the righthand side of (7.44) is positive. From this we may deduce

$$\phi\bar{\phi}' < \left(\sqrt{1+\left(\frac{\sigma-\frac{1}{2}}{t}\right)^2}+\frac{\sigma-\frac{1}{2}}{|t|}\right)^2 E ,$$

from which (1.6) follows immediately.

Theorem 1. *We have*

$$\sum_{\substack{|\gamma|<T\\ \beta>\frac{1}{2}}} (T-|\gamma|)\left(\beta-\frac{1}{2}\right) = \frac{\kappa_1}{4\pi}T^2\log\frac{T}{\pi} - \frac{1}{2\pi}\left(\frac{3}{4}\kappa_1+\log|a|\right)T^2 - \frac{1}{16}\kappa_1 T + T\sum_{\frac{1}{2}<\sigma_j\le 1}\left(\sigma_j-\frac{1}{2}\right) + O(\log T)\,, \tag{1.7}$$

or, since the zeros are symmetric around the real axis,

$$\sum_{\substack{0\le\gamma<T\\ \beta<\frac{1}{2}}} (T-\gamma)\left(\beta-\frac{1}{2}\right) = \frac{\kappa_1}{8\pi}T^2\log\frac{T}{\pi} - \frac{1}{4\pi}\left(\frac{3}{4}\kappa_1+\log|a|\right)T^2 - \frac{1}{32}\kappa_1 T + \frac{1}{2}T\sum_{\frac{1}{2}<\sigma_j\le 1}\left(\sigma_j-\frac{1}{2}\right) + O(\log T)\,. \tag{1.7'}$$

We also have

$$\sum_{\substack{0\le\gamma\le T\\ \beta>\frac{1}{2}}}\left(\beta-\frac{1}{2}\right) = \frac{\kappa_1}{4\pi}T\log\frac{T}{\pi} - \frac{1}{2\pi}\left(\frac{\kappa_1}{2}+\log|a|\right)T + O(\log T)\,. \tag{1.8}$$

While (1.7) and (1.7′) are evident from the above, (1.8) follows by a simple differencing argument. If we denote the lefthand side of (1.7′) by $A(T)$ and the lefthand side of (1.8) by $B(T)$, we have

$$A(T)-A(T-1)\le B(T)\le A(T+1)-A(T)\,.$$

Thus (1.8) follows immediately from (1.7′).

While (1.7′) and (1.8) do occur in the literature[6] the above proof is simpler and based on more general principles than the only published proof. Also the form of these statements given here is rather more explicit.

If Γ is the modular group and $\chi=\chi_0$ (which is the only singular χ in this case), we get

$$L^*(s,\chi_0)=\frac{\zeta(2s-1)}{\zeta(2s)}\,.$$

It is easily seen that (1.7′) and (1.8) in this case are equivalent to the von Mangoldt formula for the number of zeros of $\zeta(s)$ in $0<t\le T$, and

[6] Selberg [1] p. 674 and Hejhal [1] p. 456, Theorem 2.22.

its integrated form, both with the best remainder terms we can get today without assuming any hypothesis.

2. If we insert the full expression (0.14) into (0.2), we get instead of (0.16),

$$(2.1)\qquad N_\chi(T) + N_\chi\left(\frac{1}{2}, T\right) = \mu(\mathcal{D})T^2 + B_1 T\log T + B_2'(\chi)T + \frac{1}{2\pi}\arg L^*\left(\frac{1}{2}+iT,\chi\right) + O\left(\frac{T}{\log T}\right),$$

where

$$B_2'(\chi) = B_2(\chi) - \frac{\log b}{\pi}\,.$$

We wish now to improve our estimation of

$$\arg L^*\left(\frac{1}{2}+iT,\chi\right)$$

to that of the O-term in (0.2) or (2.1). Though there does not at present seem much hope of improving our estimation of this O-term beyond $O(\frac{T}{\log T})$, we shall show also that, up to a point, for any such improvement, we can make the same improvement in our estimation of

$$\arg L^*\left(\frac{1}{2}+iT,\chi\right)\,.$$

Theorem 2. *Assume that*

$$(2.2)\qquad N_\chi(T) + N_\chi\left(\frac{1}{2}, T\right) = \mu(\mathcal{D})T^2 + B_1 T\log T + B_2'(\chi)T + \frac{1}{2\pi}\arg L^*\left(\frac{1}{2}+iT,\chi\right) + O\left(R_1(T)\right)$$

holds, where $R_1(T)$ is an increasing function with the property that

$$R_1(T+1) < 2R_1(T)\,,\quad R_1(T) = O(T)\,,$$

then we have

$$(2.3)\qquad N_\chi(T) + N_\chi\left(\frac{1}{2}, T\right) = \mu(\mathcal{D})T^2 + B_1 T\log T + B_2'(\chi)T + O\left(R_2(T)\right)$$

where

$$R_2(T) = \max\left(\sqrt{T}\log T, R_1(T)\right)\,.$$

Corollary. *We have unconditionally*

$$N_\chi(T) + N_\chi\left(\frac{1}{2}, T\right) = \mu(\mathcal{D})T^2 + B_1 T \log T + B_2'(\chi)T + O\left(\frac{T}{\log T}\right) . \tag{2.4}$$

Proof. We have[7]

$$\varphi(s,\chi) = \operatorname{sgn} a(\chi) b^{1-2s} \prod_j \frac{s-1+\sigma_j}{s-\sigma_j} \prod_\varrho \frac{s-\varrho}{s-1+\bar\varrho} \tag{2.5}$$

which gives

$$L^*(s,\chi) = \frac{1}{|a(\chi)|}\left(\frac{\Gamma(s)}{\sqrt{\pi}\Gamma(s-\frac{1}{2})}\right)^{\kappa_1} \prod_j \frac{s-1+\sigma_j}{s-\sigma_j} \prod_\varrho \frac{s-\varrho}{s-1+\bar\varrho} . \tag{2.6}$$

Forming the logarithmic derivative and taking imaginary parts, we get for $\sigma \geq \frac{1}{2}$, $|t| > 2$

$$\operatorname{Im} \frac{L^{*\prime}}{L^*}(\sigma+it,\chi) = -\sum_\varrho \frac{4(\sigma-\frac{1}{2})(\beta-\frac{1}{2})(t-\gamma)}{((\sigma-\beta)^2+(t-\gamma)^2)((\sigma+\beta-1)^2+(t-\gamma)^2)} + O(1) . \tag{2.7}$$

For $H \geq 1$, we obtain, using (1.8), that

$$\begin{aligned}&\left|\sum_{H\leq|t-\gamma|\leq 2H} \frac{4(\sigma-\frac{1}{2})(\beta-\frac{1}{2})(t-\gamma)}{((\sigma-\beta)^2+(t-\gamma)^2)((\sigma+\beta-1)^2+(t-\gamma)^2)}\right| \\ &\quad < \frac{4}{H^2}\sum_{|t-\gamma|\leq 2H}\left(\beta-\frac{1}{2}\right) = O\left(\frac{\log(|t|+2H)}{H}\right) \\ &\quad = O\left(\frac{\log|t|+\log 2H}{H}\right) .\end{aligned}$$

From this we see that

$$\sum_{|t-\gamma|\geq 1} \frac{4(\sigma-\frac{1}{2})(\beta-\frac{1}{2})(t-\gamma)}{((\sigma-\beta)^2+(t-\gamma)^2)((\sigma+\beta-1)^2+(t-\gamma)^2)} = O(\log|t|) ,$$

and so

[7] See Selberg [1] p. 656, equation (8.8). In the product $\prod_\varrho$ we combine terms with ϱ and $\bar\varrho$ to $\frac{(s-\varrho)(s-\bar\varrho)}{(s-1+\bar\varrho)(s-1+\varrho)}$ in order to get absolute convergence.

(2.8) $$\operatorname{Im}\frac{L^{*'}}{L^*}(\sigma+it,\chi)=$$
$$-\sum_{|t-\gamma|<1}\frac{4(\sigma-\frac{1}{2})(\beta-\frac{1}{2})(t-\gamma)}{((\sigma-\beta)^2+(t-\gamma)^2)((\sigma+\beta-1)^2+(t-\gamma)^2)}+O(\log|t|)$$

We now assume that we have

(2.9) $$N_\chi(T)+N_\chi\left(\frac{1}{2},T\right)=\mu(\mathcal{D})T^2+B_1T\log T + B_2'(\chi)T+\theta(T)R_2(T)\,,$$

with $|\theta(T)|\le 1$ and $R_2(T)$ some increasing function with $\sqrt{T}\log T\le R_2(T)=O(T)$, and such that

$$R_2(T+1)\le 2R_2(T).$$

For $T\ge 1000$, we define

(2.10) $$h=(TR_2(T))^{-\frac{1}{3}}(\log T)^{\frac{4}{3}}\,,$$

(2.10′) $$l=\left[\left(\frac{R_2^2(T)}{T\log^2T}\right)^{\frac{1}{3}}\right]\ge 1\,,$$

(see footnote[8]), and

(2.10″) $$h^*=2lh\le 2T^{-\frac{2}{3}}(R_2(T))^{\frac{1}{3}}(\log T)^{\frac{2}{3}}<1\,.$$

From (2.9) we derive that the number of ϱ with $T-h^*\le\gamma\le T+h^*$ is

$$O(R_2(T))+O(Th^*)=O(R_2(T))\,.$$

Looking at the l intervals

$$T+2\nu h\le t\le T+2(\nu+1)h$$

for $\nu=0,1,\ldots,l-1$, we see that at least one has at most

$$O\left(\frac{R_2(T)}{l}\right)\,,$$

γ's in it. We denote its midpoint by T', we have then

$$0<T'-T<h^*$$

and that

[8] [] here means greatest integral part.

$$\sum_{|T'-\gamma|\le h} 1 = O\left(\frac{R_2(T)}{l}\right) . \tag{2.11}$$

In the same way we conclude there is a T'' with

$$0 < T - T'' < h^*$$

and so that

$$\sum_{|T''-\gamma|\le h} 1 = O\left(\frac{R_2(T)}{l}\right) . \tag{2.11'}$$

We now proceed to estimate

$$\arg L^*\left(\frac{1}{2} + iT', \chi\right)$$

and

$$\arg L^*\left(\frac{1}{2} + iT'', \chi\right) .$$

Clearly it is enough to consider the case of T', since T'' is handled the same way.

We have

$$\begin{aligned} \left|\arg L^*\left(\frac{1}{2} + it, \chi\right)\right| &\le \int_{\frac{1}{2}}^{\sigma_0} \left|\operatorname{Im} \frac{L^{*'}}{L^*}(\sigma + it, \chi)\right| d\sigma \\ &= \int_{\frac{1}{2}}^{\sigma_0} \left|\operatorname{Im} \frac{L^{*'}}{L^*}(\sigma + it, \chi)\right| d\sigma + O(1) , \end{aligned} \tag{2.12}$$

where we have chosen σ_0 so large that, according to (0.9), we have for $\sigma \ge \sigma_0$, uniformly in σ and t,

$$\frac{L^{*'}}{L^*}(\sigma + it, \chi) = O(\lambda_1^{-\sigma})$$

where $\lambda_1 > 1$.

For $\frac{1}{2} \le \sigma \le \sigma_0$, $t = T'$, we go back to (2.8) and split the sum on the righthand side in two parts, one where $h \le |T' - \gamma| < 1$, and another where $|T' - \gamma| < h$.

For the first part we get, using again (1.8), that

(2.13)

$$\begin{aligned} &\sum_{h\le|T'-\gamma|<1} \frac{4(\sigma - \frac{1}{2})(\beta - \frac{1}{2})|T' - \gamma|}{((\sigma - \beta)^2 + (T' - \gamma)^2)((\sigma + \beta - 1)^2 + (T' - \gamma)^2)} \\ &< \frac{4(\sigma - \frac{1}{2})}{h((\sigma - \frac{1}{2})^2 + h^2)} \sum_{|T'-\gamma|<1} \left(\beta - \frac{1}{2}\right) = O\left(\frac{\sigma - \frac{1}{2}}{h((\sigma - \frac{1}{2})^2 + h^2)} \log T\right) , \end{aligned}$$

while for the second part we have

$$(2.14)\qquad \sum_{|T'-\gamma|<h} \frac{4(\sigma-\frac{1}{2})(\beta-\frac{1}{2})|T'-\gamma|}{((\sigma-\beta)^2+(T'-\gamma)^2)((\sigma+\beta-1)^2+(T'-\gamma)^2)}$$
$$< \sum_{|T'-\gamma|<h} \frac{|T'-\gamma|}{(\sigma-\beta)^2+(T'-\gamma)^2} .$$

We now get from (2.12), using (2.8), (2.13), (2.14) and (2.11), that

$$(2.15)\qquad \int_{\frac{1}{2}}^{\sigma_0} \left| \operatorname{Im} \frac{L^{*\prime}}{L^*}(\sigma+iT',\chi)\right| d\sigma$$
$$< \int_{\frac{1}{2}}^{\infty} \sum_{|T'-\gamma|<h} \frac{|T'-\gamma|}{(\sigma-\beta)^2+(T'-\gamma)^2} d\sigma + O\left(\frac{\log T}{h}\int_{\frac{1}{2}}^{\sigma_0} \frac{(\sigma-\frac{1}{2})d\sigma}{(\sigma-\frac{1}{2})^2+h^2}\right)$$
$$+ O(\log T) = O\left(\frac{R_2(T)}{l}\right) + O\left(\frac{\log T\cdot \log(1+\frac{\sigma_0^2}{h^2})}{h}\right) + O(\log T)$$
$$= O\left(\frac{R_2(T)}{l}\right) + O\left(\frac{\log^2 T}{h}\right)$$
$$= O\left((TR_2(T))^{\frac{1}{3}}\log^{\frac{2}{3}} T\right) = O\left(\sqrt{T}\log T\left(\frac{R_2(T)}{\sqrt{T}\log T}\right)^{\frac{1}{3}}\right) ,$$

and so altogether

$$(2.16)\qquad \arg L^*\left(\frac{1}{2}+iT',\chi\right) = O\left(\sqrt{T}\log T\left(\frac{R_2(T)}{\sqrt{T}\log T}\right)^{\frac{1}{3}}\right) ,$$

and in the same way

$$(2.16')\qquad \arg L^*\left(\frac{1}{2}+iT'',\chi\right) = O\left(\sqrt{T}\log T\left(\frac{R_2(T)}{\sqrt{T}\log T}\right)^{\frac{1}{3}}\right) .$$

Going now back to (2.2), the lefthand side is a monotonically increasing function of T, call it $M(T)$, then clearly

$$M(T'') \le M(T) \le M(T') .$$

Here $M(T')$ and $M(T'')$ can now be estimated by (2.2) using (2.16) and (2.16′). Since the three first terms on the righthand side of (2.2) change at most by an amount

$$O(Th^*) = O\left(\sqrt{T}\log T\left(\frac{R_2(T)}{\sqrt{T}\log T}\right)^{\frac{1}{3}}\right) ,$$

when we replace T by T' or T'', we get finally

$$(2.17)\ M(T) = N_\chi(T) + N_\chi\left(\frac{1}{2}, T\right) = \mu(\mathcal{D})T^2 + B_1 T \log T + B_2'(\chi)T$$
$$+ O(R_1(T)) + O\left(\sqrt{T}\log T\left(\frac{R_2(T)}{\sqrt{T}\log T}\right)^{\frac{1}{3}}\right).$$

Going back to (2.9) we see that this means we can replace our function $R_2(T)$ by

$$c_1 R_1(T) + c_2\sqrt{T}\log T\left(\frac{R_2(T)}{\sqrt{T}\log T}\right)^{\frac{1}{3}},$$

where c_1 and c_2 are constants which do not depend on R_2. From this we see that the conclusion of Theorem 2 follows by iteration since by (0.16), we know that (2.9) holds if $R_2(T)$ is a sufficiently large multiple of T.[9]

The corollary follows immediately from (2.1).

The term $R_1(T)$ in (2.2) arises essentially from the estimation of $\arg Z_\Gamma(s,\chi)$ for $s = \frac{1}{2} + iT$, while the best estimate so far proved is

$$O\left(\frac{T}{\log T}\right),$$

it is conceivable that $R_1(T)$ might be of smaller order than $\sqrt{T}\log T$, but it could not in general be of much smaller order, since one can show there are Γ and χ for which

$$\arg Z_\Gamma\left(\frac{1}{2} + iT, \chi\right) = \Omega_\pm\left(\frac{\sqrt{T}}{\log T}\right).$$

3. It can be shown by examples that there is no limitation to how close to the line $\sigma = \frac{1}{2}$, or how far away from it the zeros of $L^*(s,\chi)$ may lie.

Consider the subgroup $\Gamma_0(4)$[10] of the modular group. This group is generated by two elements, say $A = \begin{pmatrix} 1 & 1 \\ 0 & 1 \end{pmatrix}$ and $B = \begin{pmatrix} 1 & 0 \\ -4 & 1 \end{pmatrix}$ and is a

---

[9] The iteration is simpler to follow if we note that

$$c_2\left(\frac{R_2(T)}{\sqrt{T}\log T}\right)^{\frac{1}{3}} < c_2^{\frac{2}{3}} + \frac{R_2(T)}{3\sqrt{T}\log T},$$

this gives that we can replace $R_2(T)$ by

$$c_1 R_1(T) + c_2^{\frac{3}{2}}\sqrt{T}\log T + \frac{R_2(T)}{3}$$

in (2.9), the conclusion is then obvious.

[10] As usual $\Gamma_0(N)$ denotes the subgroup of the modular group whose elements $\begin{pmatrix} a & b \\ c & d \end{pmatrix}$ satisfy $c \equiv 0 (\mathrm{mod}\, N)$.

free group. There are three inequivalent cusps, say the fixpoints of A, B and $B^{-1}A^{-1}$ respectively. If we define $\chi(A) = 1$ and $\chi(B) = e^{i\alpha}$, where we for the moment will assume $-\pi \le \alpha \le \pi$ and $\alpha \ne 0, \frac{\pi}{2}, -\frac{\pi}{2}$, then χ is singular with respect to the cusp at ∞ but not with respect to the two other cusps. Denoting by Γ_∞ the group generated by A, we have in this case

$$L(s,\chi) = \sum_{\Gamma_\infty \backslash \Gamma_0(4)/\Gamma_\infty} \frac{\chi(M)}{|c|^{2s}}, \tag{3.1}$$

where $M = \begin{pmatrix} a & b \\ c & d \end{pmatrix}$ runs over a set of representatives of the double cosets. We have

$$L^*(s,\chi) = \frac{4^{2s}}{a_1(\chi)} L(s,\chi) = \frac{1}{a_1(\chi)} \sum_{n=1}^{\infty} \frac{a_n(\chi)}{n^{2s}}, \tag{3.2}$$

where

$$a_1(\chi) = 2\cos\alpha\,, \quad a_2(\chi) = 4\cos 2\alpha\,.$$

If we let α tend to $\frac{\pi}{2}$ or $-\frac{\pi}{2}$, we see that $a_1(\chi) \to 0$ while $a_2(\chi) \to -4$, and we easily conclude that there is a family of zeros moving far out to the right, these have real parts

$$\beta \approx \frac{\log|\frac{a_2(\chi)}{a_1(\chi)}|}{\log 4} = \frac{\log 2|\frac{\cos 2\alpha}{\cos\alpha}|}{\log 4}, \tag{3.3}$$

and approximately

$$\frac{\log 4}{\pi} T \tag{3.4}$$

of them have imaginary parts $|\gamma| < T$ for large T. For $\alpha = \frac{\pi}{2}$ or $\alpha = -\frac{\pi}{2}$, this family of zeros have disappeared, and for $\chi(B) = \pm i$, we have that the form of $L^*(s,\chi)$ has changed to

$$-\frac{8^{2s}}{4} L(s,\chi)\,.$$

Next, let us consider what happens when $\alpha \to 0$. For $\alpha = 0$, we have $\chi = \chi_0$ which is singular with respect to all three cusps, so $\kappa_1 = 3$ in this case and the dirichlet series (3.1) becomes

$$\begin{aligned} L_{1,1}(s,\chi_0) &= \sum_{\Gamma_\infty \backslash \Gamma_0(4)/\Gamma_\infty} \frac{1}{|c|^{2s}} \\ &= \sum_{n=1}^{\infty} \frac{\varphi(4n)}{(4n)^{2s}} = 2.4^{-2s}(1-2^{-2s})^{-1} \frac{\zeta(2s-1)}{\zeta(2s)}, \end{aligned}$$

(see footnote[11]). Denoting by $\varphi_0(s)$ the $\varphi(s,\chi)$ corresponding to the full modular group and $\chi = \chi_0$, we get

$$\varphi_{1,1}(s,\chi_0) = \sqrt{\pi}\frac{\Gamma(s-\frac{1}{2})}{\Gamma(s)}L_{1,1}(s,\chi_0) = 2.4^{-2s}(1-2^{-2s})^{-1}\varphi_0(s) .$$

since $|\varphi_0(\frac{1}{2}+it)| = 1$, we see that

$$\left|\varphi_{1,1}\left(\frac{1}{2}+it,\chi_0\right)\right| \leq 1 ,$$

with equality only where $2^{2it} = 1$.

Of course that this inequality holds, with equality at most only at isolated points, follows also from the fact that

$$\phi\left(\frac{1}{2}+it,\chi_0\right)$$

is a unitary 3 by 3 matrix so that

$$\sum_{j=1}^{3}\left|\varphi_{1,j}\left(\frac{1}{2}+it,\chi_0\right)\right|^2 = 1 ,$$

and neither of the two functions $\varphi_{1,2}(s,\chi_0)$ and $\varphi_{1,3}(s,\chi_0)$ can vanish identically.

On the other hand, for $\chi \neq \chi_0$, we have always

$$\left|\varphi\left(\frac{1}{2}+it,\chi_0\right)\right| = 1 .$$

Given now a t_0 and a positive $\delta < \frac{1}{2}$, let us assume that there is a sequence of $\chi_j \neq \chi_0$ such that $\chi_j \to \chi_0$ as $j \to \infty$, and that

$$\varphi(s,\chi_j)$$

has no zero in $|s-s_0| < \delta$, where $s_0 = \frac{1}{2}+it_0$, and seek to get a contradiction.

In the case we are considering, we can show that there is at most one pole of $\varphi(s,\chi)$ on $\frac{1}{2} < s \leq 1$,[12] and if χ is close enough to χ_0 (or α close enough to zero) there is definitely one pole there. Since $a(\chi)$ is positive for χ close enough to χ_0, (2.5) in this case takes the form

---

[11] $\varphi(n)$ here is Eulers function $n\prod_{p|n}(1-\frac{1}{p})$.

[12] It can be shown that we have at most one eigenvalue on $0 \leq \lambda < \frac{1}{4}$, if the corresponding Riemann surface is either: (a) a sphere with 3 branchpoints, (b) a sphere with 4 branchpoints of which 3 are of order 2, (c) a sphere with 5 branchpoints of order 2, (d) a torus with one branchpoint. $\Gamma_0(4)$ corresponds to a sphere with 3 branchpoints of infinite order.

$$\varphi(s,\chi) = 4^{1-2s}\frac{s-1+\sigma_1}{s-\sigma_1}\prod_{\varrho}\frac{s-\varrho}{s-1+\bar{\varrho}}\,,$$

from this we get, taking the square and absolute values, for $\sigma \geq \frac{1}{2}$

$$|\varphi(s,\chi)|^2 \geq 4^{2-4\sigma}\prod_{\varrho}\left(1-\frac{4(\sigma-\frac{1}{2})(\beta-\frac{1}{2})}{(\sigma-1+\beta)^2+(t-\gamma)^2}\right)$$

$$\geq 4^{2-4\sigma}\left(1-4\left(\sigma-\frac{1}{2}\right)\sum_{\varrho}\frac{\beta-\frac{1}{2}}{(\sigma-1+\beta)^2+(t-\gamma)^2}\right)\,. \tag{3.5}$$

From (1.8) we get, observing that our proof of (1.8) actually shows that (1.8) holds uniformly for a fixed Γ as long as $a(\chi)$ is bounded away from zero, and for a sequence $\Gamma_j, \chi_j \to \Gamma, \chi$ if $|a(\chi_j)| > a_0 > 0$, that

$$\sum_{|t_0-\gamma|\leq 1}\left(\beta-\frac{1}{2}\right) = O\left(\log\left(2+|t_0|\right)\right)\,, \tag{3.6}$$

and that if $H \geq 1$, then

$$\sum_{H\leq|t_0-\gamma|<2H}\left(\beta-\frac{1}{2}\right) = O\left(H\log\left(2+H+|t_0|\right)\right)\,, \tag{3.7}$$

uniformly if $|\alpha| < \alpha_0 < \frac{\pi}{2}$, and thus for the χ_j for $j > j_0$. For $j > j_0$ and $|s-s_0| \leq \frac{\delta}{2}$ we thus have

$$\sum_{\varrho}\frac{\beta-\frac{1}{2}}{(\sigma-1+\beta)^2+(t-\gamma)^2} \leq \frac{4}{\delta^2}\sum_{|t_0-\gamma|<1}\left(\beta-\frac{1}{2}\right)$$

$$+\sum_{|t_0-\gamma|\geq 1}\frac{\beta-\frac{1}{2}}{(t-\gamma)^2} = O\left(\frac{\log(2+|t_0|)}{\delta^2}\right)\,,$$

since there are no zeros in $|s-s_0| < \delta$. Inserting this in (3.5), we get for $|s-s_0| < \frac{\delta}{2}$, $\sigma \geq \frac{1}{2}$, that for $j > j_0$,

$$|\varphi(s,\chi_j)|^2 \geq 4^{2-4\sigma}\left(1-c\frac{\sigma-\frac{1}{2}}{\delta^2}\log\left(2+|t_0|\right)\right)\,,$$

where c is a positive constant. Hence, if

$$\eta = \frac{\delta^2}{2c\log(2+|t_0|)}\,,$$

we have for $j > j_0$, $|s-s_0| \leq \min(\eta, \frac{\delta}{2})$, $\sigma \geq \frac{1}{2}$ that

$$|\varphi(s,\chi_j)| > \frac{1}{2}\,.$$

Because of (0.10), we now see that $\varphi(s,\chi_j)$ is uniformly bounded for $j > j_0$, and

$$|s - s_0| \leq \eta' = \min\left(\eta, \frac{\delta}{2}\right) .$$

$\varphi(s,\chi_j)$ is also for $j > j_0$ uniformly bounded in the region $\sigma \geq \frac{1}{2}$ minus a disc with radius $\frac{1}{3}$ and center at 1 (since we may assume j_0 so large that $\sigma_1 \geq \frac{3}{4}$ for $j > j_0$), and therefore uniformly bounded in the union of these two regions which we may call Ω. This implies that given any compact subregion inside Ω we may select a subsequence j^* of the χ_j such that $\varphi(s,\chi_{j^*})$ converges uniformly to a limit function, which is holomorphic in Ω, as $j^* \to \infty$. In the region $\sigma > 1$ we clearly have

$$\varphi(s,\chi) \to \varphi_{1,1}(s,\chi_0)$$

as $\chi \to \chi_0$, therefore the limit function of $\varphi(s,\chi_{j^*})$ must be $\varphi_{1,1}(s,\chi_0)$, but on the stretch $s = \frac{1}{2} + it$, $|t - t_0| < \eta'$, $|\varphi(s,\chi_{j^*})|$ can *not* converge to $\varphi_{1,1}(s,\chi_0)$ since $|\varphi(s,\chi_{j^*})| = 1$ there while $|\varphi_{1,1}(s,\chi_0)| < 1$ except possibly for one point of this interval. Our original assumption is therefore false, and so we have

Theorem 3. *Given a* T_0 *and an* $\varepsilon > 0$, *there exists a* $\delta = \delta(\varepsilon, T_0) > 0$, *such that for* $0 < |\alpha| < \delta$, $\varphi(s,\chi)$ *always has a zero in* $|s - s_0| < \varepsilon$ *where* $s_0 = \frac{1}{2} + it_0$, $|t_0| \leq T_0$.

Essentially the same proof shows that if we consider the group Γ_q generated by $\begin{pmatrix} 0 & -1 \\ 1 & 0 \end{pmatrix}$ and

$$\begin{pmatrix} 1 & 2\cos\frac{\pi}{q} \\ 0 & 1 \end{pmatrix}$$

where $q > 2$ is an integer, and take $\chi = \chi_0$, then the same statement holds that given T_0 and $\varepsilon > 0$, there is a $q_0 = q_0(T_0, \varepsilon)$ such that for $q > q_0$, $\varphi_q(s,\chi_0)$ has a zero in $|s - s_0| < \varepsilon$, where $s_0 = \frac{1}{2} + it_0$ and $|t_0| < T_0$.

Here the Γ_q has just one cusp, while the limitgroup, generated by $\begin{pmatrix} 0 & -1 \\ 1 & 0 \end{pmatrix}$ and $\begin{pmatrix} 1 & 2 \\ 0 & 1 \end{pmatrix}$ has two cusps. Clearly the same argument applies whenever we consider a sequence Γ_j, χ_j which as $j \to \infty$ tends to a limitgroup Γ and a χ such that in the limitcase the number of Eisenstein series (that is, κ_1) has been increased.

4. For the modular group, the form of $\varphi(s,\chi_0) = \sqrt{\pi}\frac{\Gamma(s-\frac{1}{2})}{\Gamma(s)}\frac{\zeta(2s-1)}{\zeta(2s)}$, shows that the average value of $\beta - \frac{1}{2}$ is in this case $\frac{1}{4}$ (and if the Riemann hypothesis is true we have always $\beta = \frac{3}{4}$ in this case). We wish in this section to show that the average value of $\beta - \frac{1}{2}$ cannot be more than $\frac{1}{4}$ in any case.

Somewhat loosely expressed, this means that for the modular group and its congruence subgroups with $\chi = \chi_0$ in the equation (2.4) the term $N_\chi(\frac{1}{2}, T)$ takes about the minimal value it can have in relation to κ_1. Thus $N_\chi(T)$, which counts the discrete spectrum, in these cases takes about the maximal value it can take (given $A(\mathcal{D})$ and κ_1) for these groups.

What we essentially need to show is that for $\sigma > \frac{3}{4}$, and even for $\sigma = \frac{3}{4} + \frac{\omega(T)}{\log T}$, where $\omega(T)$ tends to infinity with T but much slower than $\log T$, there will be comparatively few zeros of $\varphi(s, \chi)$ with $\beta > \sigma$ and $|\gamma| < T$.

To this end, we shall prove some lemmas, which we give in a rather more general form than we actually need here.

Lemma 3. *We assume the series*

$$f(s) = \sum_{n=1}^{\infty} a_n \lambda_n^{-s}\,, \tag{4.1}$$

where $a_n > 0$ and $\lambda_n > 0$,[13] *is convergent for $\sigma > 1$, and holomorphic for $\sigma > 0$, except for a simple pole at $s = 1$, and possibly a finite number of poles in the region $0 < \sigma < 1$. We also assume that $f(s)$ has continuous boundary values on $\sigma = 0$, and that*

$$f(s) = O\left(\sqrt{|t|}\right)\,, \tag{4.2}$$

(see footnote[14]*), then, if c is larger than the imaginary parts of the poles of $f(s)$, we have for $\frac{1}{2} \le \sigma \le 2$*

$$\int_c^T |f(\sigma + it)|^2\, dt = O\left(T \min\left(\frac{1}{\sigma - \frac{1}{2}}, \log T\right)\right)\,. \tag{4.3}$$

To prove this we may, without loss of generality, assume that all $\lambda_n > 2$, since the part

$$\sum_{\lambda_n \le 2} a_n \lambda_n^{-s}$$

clearly is bounded for $0 \le \sigma \le 2$.

For $T > 2$ we have

$$\sum_{\lambda_n < T} \left(1 - \frac{\lambda_n}{T}\right) a_n \lambda_n^{-s} = \frac{1}{2\pi i} \int_{2-i\infty}^{2+i\infty} \frac{T^{z-s}}{(z-s)(z-s+1)} f(z)\, dz\,.$$

---

[13] We do not assume the λ_n are increasing with n, they could for that matter be everywhere dense on $(0, \infty)$.

[14] Less would do, for instance: $\int_T^{2T} |f(\sigma + it)|^2 dt = O(T^2)$, for $0 \le \sigma \le 2$.

Moving here the path of integration to the left to the imaginary axis, taking account of the residues at the poles at $z = s$ (and at $z = s - 1$ if $\sigma > 1$, if $\sigma = 1$ we avoid the pole at $s - 1$ on the imaginary axis by making a detour around it along a half circle with $s - 1$ as its centre and radius $\frac{1}{\log T}$), as well as the finite set of poles of $f(z)$. The contribution of these latter poles, as well as the remaining integral, we can estimate using (4.2).

Assuming $\sigma \geq \frac{1}{2}$ and $T^{\frac{1}{4}} \leq |t| \leq T$ we get, for $T > T_0$,

$$f(\sigma + it) = \sum_{\lambda_n < T} \left(1 - \frac{\lambda_n}{T}\right) a_n \lambda_n^{-\sigma - it} + O(1) . \tag{4.4}$$

From (4.2) and the fact that $f(s)$ is bounded on $\sigma = \frac{3}{2}$, we get, using the Phragmén-Lindelöf principle, that for $0 \leq \sigma \leq \frac{3}{2}$, $|t| > t_0$,

$$f(\sigma + it) = O\left(|t|^{\frac{1}{2} - \frac{\sigma}{3}}\right) . \tag{4.5}$$

We now write

$$A(x) = \sum_{\lambda_n \leq x} a_n , \tag{4.6}$$

and

$$A(x) = A_1(x) + R(x) , \tag{4.7}$$

where $A_1(x)$ denotes the sum of the residues of

$$\frac{x^s}{s} f(s)$$

at the poles in $\sigma > 0$, so that

$$A_1(x) = cx + \sum_i x^{\varrho_i} p_i(\log x) , \tag{4.8}$$

where the ϱ_i run over the finite set of poles of $f(s)$ in $0 < \sigma < 1$, and the p_i are certain polynomials. We also have

$$\begin{aligned} R(x) &= \frac{1}{2\pi i} \int_{\varepsilon - i\infty}^{\varepsilon + i\infty} \frac{x^s}{s} f(s)\, ds \\ &= \frac{x^\varepsilon}{2\pi} \int_{-\infty}^{\infty} \frac{x^{it}}{\varepsilon + it} f(\varepsilon + it)\, dt , \end{aligned} \tag{4.9}$$

where $\varepsilon > 0$ is chosen so small that all ϱ_i lie in the region $\varepsilon < \sigma < 1$.

From (4.9) we get the inequality

$$\int_1^\infty \frac{(R(x))^2}{x^{1+2\varepsilon}} dx < \int_{-\infty}^{\infty} \frac{|f(\varepsilon + it)|^2}{\varepsilon^2 + t^2} dt .$$

Using now (4.5), we get

$$\int_1^\infty \frac{(R(x))^2}{x^{1+2\varepsilon}}dx = O\left(\frac{1}{\varepsilon}\right) .$$

Taking here $\varepsilon = \frac{1}{\log x}$, we get for $x > x_0$

$$\int_1^x \frac{(R(u))^2}{u}du = O(\log x) . \tag{4.10}$$

Consider now the expression

(4.11)

$$\int_1^T \frac{(A(u+1)-A(u))^2}{u}du = \sum_{\substack{\lambda_n,\lambda_m<T+1\\ |\lambda_n-\lambda_m|<1}} a_n a_m \int_{\max(\lambda_n,\lambda_m)-1}^{\min(\lambda_n,\lambda_m)} \frac{du}{u}$$

$$> \sum_{\substack{\lambda_n,\lambda_m<T\\ |\lambda_n-\lambda_m|<1}} \frac{a_n a_m}{\sqrt{\lambda_n\lambda_m}}(1-|\lambda_n-\lambda_m|) .$$

By (4.6), (4.7), and (4.8),

$$A(u+1)-A(u) = A_1(u+1)-A_1(u)+R(u+1)-R(u) = R(u+1)-R(u)+O(1) ,$$

so that

$$(A(u+1)-A(u))^2 \le 3\,(R(u))^2 + 3\,(R(u+1))^2 + O(1) .$$

Combining this with (4.10), we get

$$\int_1^T \frac{(A(u+1)-A(u))^2}{u}du = O(\log T) ,$$

and so (4.11) gives

$$\sum_{\substack{\lambda_n,\lambda_m<T+1\\ |\lambda_n-\lambda_m|<1}} \frac{a_n a_m}{\sqrt{\lambda_n\lambda_m}}(1-|\lambda_n-\lambda_m|) = O(\log T) . \tag{4.12}$$

We now write

$$a_n^* = \left(1-\frac{\lambda_n}{T}\right)a_n ,$$

for $\lambda_n < T$, and

$$f^*(s) = \sum_{\lambda_n<T} a_n^* \lambda_n^{-s} .$$

Using, for $\sigma \ge \frac{1}{2}$, (4.2) to estimate the integral from T_0 to $T^{\frac{1}{4}}$, and (4.4) for the remaining part, we see that

$$\int_{T_0}^{T} |f(\sigma+it)|^2\,dt \le 2\int_0^T |f^*(\sigma+it)|^2\,dt + O(T)\,. \tag{4.13}$$

Here the first term on the righthand side equals

$$\begin{aligned}\int_{-T}^{T} |f^*(\sigma+it)|^2\,dt &< 2\int_{-\infty}^{\infty} |f^*(\sigma+it)|^2 \left(\frac{2T}{t}\sin\frac{t}{2T}\right)^2 dt\\ &= 4\pi T \sum_{\lambda_n,\lambda_m<T} \frac{a_n^* a_m^*}{(\lambda_n\lambda_m)^\sigma}\left(1 - T\left|\log\frac{\lambda_n}{\lambda_m}\right|\right)^+ ,\end{aligned} \tag{4.14}$$

where we here and later use the notation $(a)^+$ to denote $\max(a,0)$. Now, for $\lambda_n, \lambda_m < T$,

$$\left|\log\frac{\lambda_n}{\lambda_m}\right| > 2\frac{|\lambda_n-\lambda_m|}{\lambda_n+\lambda_m} > \frac{|\lambda_n-\lambda_m|}{T}\,,$$

and $a_n^* < a_n$, so (4.14) gives

$$\int_{-T}^{T} |f^*(\sigma+it)|^2\,dt < 4\pi T \sum_{\lambda_n,\lambda_m<T} \frac{a_n a_m}{(\lambda_n\lambda_m)^\sigma}\left(1-|\lambda_n-\lambda_m|\right)^+\,. \tag{4.15}$$

Observing that the double sum on the righthand side is less than or equal to the lefthand side of (4.12), we get from (4.13) and (4.15)

$$\int_{T_0}^{T} |f(\sigma+it)|^2\,dt = O(T\log T)\,. \tag{4.16}$$

For $\sigma > \frac{1}{2}$, we may sharpen this. We denote the lefthand side of (4.12) by $S(T)$ and can then bound the double sum on the righthand side of (4.15) by the Stieltjes-integral

$$\int_{\frac{3}{2}}^{T} \frac{dS(t)}{t^{\sigma-\frac{1}{2}}}\,.$$

By partial integration, using the fact that $S(t) = O(\log t)$, we easily obtain that the double sum in question is $O(\frac{1}{\sigma-\frac{1}{2}})$. This gives

$$\int_{T_0}^{T} |f(\sigma+it)|^2\,dt = O\left(\frac{T}{\sigma-\frac{1}{2}}\right)\,. \tag{4.17}$$

This, combined with (4.16), gives the result (4.3), so our lemma is proved.

A simple convexity argument now allows us to conclude that since

$$\int_{T_0}^{T} |f(\sigma+it)|^2\,dt = O(T^2)\,,$$

for $0 \le \sigma \le \frac{1}{2}$, and

$$\int_{T_0}^{T} \left| f\left(\frac{1}{2}+it\right)\right|^2 dt = O(T\log T)\ ,$$

we have, uniformly for $0 \le \sigma \le \frac{1}{2}$, that

$$\int_{T_0}^{T} |f(\sigma+it)|^2\, dt = O\left(T^{2-2\sigma}(\log T)^{2\sigma}\right)\ . \tag{4.18}$$

Theorem 4. *For $\frac{3}{4} \le \sigma \le 2$, $T > 2$, and $1 \le i,j \le \kappa_1$, we have*

$$\int_{1}^{T} |L_{i,j}(\sigma+it,\chi)|^2\, dt = O\left(T\min\left(\frac{1}{\sigma-\frac{3}{4}}, \log T\right)\right)\ , \tag{4.19}$$

where the constant implied by the O-symbol is an absolute constant.

From the inequality quoted in footnote 5, we see that

$$|\varphi_{i,j}(\sigma+it,\chi)| < \sqrt{1+\frac{(\sigma-\frac{1}{2})^2}{t^2}} + \frac{\sigma-\frac{1}{2}}{|t|} < 1 + \frac{2\sigma-1}{|t|}\ . \tag{4.20}$$

For $\chi = \chi_0$, this shows that

$$f_1(s) = L_{i,j}\left(\frac{s+1}{2}, \chi_0\right)$$

satisfies the conditions of Lemma 3 with an absolute constant in (4.2) for $|t| \ge 1$. In the case that $\chi \ne \chi_0$, we see that

$$f_2(s) = 2L_{i,j}\left(\frac{s+1}{2},\chi_0\right) + L_{i,j}\left(\frac{s+1}{2},\chi\right) + L_{i,j}\left(\frac{s+1}{2},\bar{\chi}\right)\ ,$$

and

$$f_3(s) = 2L_{i,j}\left(\frac{s+1}{2},\chi_0\right) + iL_{i,j}\left(\frac{s+1}{2},\chi\right) - iL_{i,j}\left(\frac{s+1}{2},\bar{\chi}\right)\ ,$$

also satisfy the conditions of Lemma 3, again with an absolute constant in (4.2) for $|t| \ge 1$. Since we have

$$L_{i,j}\left(\frac{s+1}{2},\chi\right) = \frac{1}{2}f_2(s) - \frac{i}{2}f_3(s) - (1-i)f_1(s)\ ,$$

and thus

$$\left|L_{i,j}\left(\frac{s+1}{2},\chi\right)\right|^2 \le \frac{5}{2}\left(|f_1(s)|^2 + |f_2(s)|^2 + |f_3(s)|^2\right)\ ,$$

(4.19) now follows directly from (4.3).

In a similar way, we can deduce from (4.18) that

$$\int_1^T |L_{i,j}(\sigma+it,\chi)|^2\,dt = O\left(T^{4-4\sigma}(\log T)^{4\sigma-2}\right) , \tag{4.21}$$

uniformly for $\frac{1}{2} \le \sigma \le \frac{3}{4}$.

The example of the full modular group with $\chi = \chi_0$ shows (as is easily seen) that (4.19) is sharp in the sense that $\log T$ cannot be replaced by any function of T of slower growth as $T \to \infty$, and $\frac{1}{\sigma - \frac{3}{4}}$ cannot be replaced by any function of σ of slower growth as $\sigma \to \frac{3}{4}$. However, for $\frac{1}{2} < \sigma < \frac{3}{4}$, (4.21) is *not* sharp for the modular group with $\chi = \chi_0$, since in this case the order of the lefthand side is easily seen actually to be

$$T^{4-4\sigma}\min\left(\frac{1}{\frac{3}{4}-\sigma}, \log T\right) .$$

Thus the exponent $4-4\sigma$ in (4.21) is best possible, but the factor $(\log T)^{4\sigma-2}$ probably does not represent the best possible, though the methods used here do not seem capable of yielding more.

We now turn to the function $L^*(s,\chi)$, by (0.8) we have

$$L^*(s,\chi) = \frac{b^{2s-1}}{a} L(s,\chi) ,$$

where

$$L(s,\chi) = \operatorname{Det} L_{i,j}(s,\chi) .$$

Thus

$$|L(s,\chi)| \le \left(\frac{1}{\kappa_1}\sum_{i,j}|L_{i,j}(s,\chi)|^2\right)^{\frac{\kappa_1}{2}} ,$$

and so

$$|L^*(s,\chi)| \le \frac{b^{2\sigma-1}}{|a|}\left(\frac{1}{\kappa_1}\sum_{i,j}|L_{i,j}(s,\chi)|^2\right)^{\frac{\kappa_1}{2}} .$$

The inequality between the geometric and the arithmetic mean then gives

$$\int_1^T \log|L^*(\sigma+it,\chi)|\,dt \le ((2\sigma-1)\log b - \log|a|)(T-1) \tag{4.22}$$
$$+\frac{\kappa_1}{2}T\log\left(\frac{1}{T}\int_1^T \frac{1}{\kappa_1}\sum_{i,j\le\kappa_1}|L_{i,j}(\sigma+it,\chi)|^2\,dt\right) .$$

By means of the estimate

$$|\varphi(\sigma+it,\chi)| < \left(\sqrt{1+\frac{(\sigma-\frac{1}{2})^2}{t^2}}+\frac{\sigma-\frac{1}{2}}{|t|}\right)^{\kappa_1} < \left(1+\frac{2\sigma-1}{|t|}\right)^{\kappa_1},$$

valid for $\sigma > \frac{1}{2}$, $t \neq 0$, which follows from the inequality given in footnote 5, since

$$|\varphi(\sigma+it,\chi)| = (\mathrm{Det}\,\phi\bar{\phi}')^{\frac{1}{2}},$$

we see that for $\sigma > \frac{1}{2}$

$$\int_0^1 \log|L^*(\sigma+it,\chi)|\,dt \le (2\sigma-1)\log b - \log|a| + O(\kappa_1).$$

Combining this with (4.22), we get for $\sigma > \frac{1}{2}$

$$\int_{-T}^{T} \log|L^*(\sigma+it,\chi)|\,dt \le 2\left((2\sigma-1)\log b - \log|a|\right)T$$
$$+ O(\kappa_1) + \kappa_1 T \log\left(\frac{1}{T}\int_1^T \frac{1}{\kappa_1}\sum_{i,j}|L_{i,j}(\sigma+it,\chi)|^2\,dt\right).$$

In the case that $2 \ge \sigma \ge \frac{3}{4}$, Theorem 4 now gives for $T > 1$,

$$\int_{-T}^{T} \log|L^*(\sigma+it,\chi)|\,dt < \kappa_1 T \log\left(\min\left(\frac{1}{\sigma-\frac{3}{4}}, \log T\right)\right) \tag{4.23}$$
$$+ 2\left((2\sigma-1)\log b - \log|a|\right)T + O(\kappa_1 T),$$

where the constants implied by the O-symbol are absolute.

Theorem 5. *We have for* $\frac{3}{4} \le \sigma \le 2$,

(4.24)

$$\sum_{\substack{|\gamma|<T\\ \beta>\sigma}} (\beta-\sigma) < \frac{\kappa_1}{2\pi} T \log\left(\min\left(\frac{1}{\sigma-\frac{3}{4}}, \log T\right)\right)$$
$$+ \frac{1}{\pi}\left((2\sigma-1)\log b - \log|a|\right)T + c\kappa_1 T + O(\log T),$$

here c is an absolute constant, while the constants implied in the O-symbol may depend on Γ and χ.

This follows fairly directly from (4.23) by applying a well-known formula due to Littlewood[15] which gives

[15] Littlewood [1].

$$\sum_{\substack{|\gamma|<T \\ \beta>\sigma}} (\beta - \sigma) = \frac{1}{2\pi} \int_{-T}^{T} \log |L^*(\sigma + it, \chi)| \, dt$$

$$+ \frac{1}{2\pi} \int_{\sigma}^{\infty} \arg L^*(\sigma' + iT, \chi) \, d\sigma'$$

$$- \frac{1}{2\pi} \int_{\sigma}^{\infty} \arg L^*(\sigma' - iT, \chi) \, d\sigma' + \sum_{\sigma_j > \sigma} (\sigma_j - \sigma)$$

$$= \frac{1}{2\pi} \int_{-T}^{T} \log |L^*(\sigma + it, \chi)| \, dt + \frac{1}{\pi} \int_{\sigma}^{\infty} \arg L^*(\sigma' + iT, \chi) d\sigma'$$

$$+ \sum_{\sigma_j > \sigma} (\sigma_j - \sigma) \, .$$

Here the first term on the righthand side is estimated from above by (4.23), the last term is constant and can be bounded by the area $A(\mathcal{D})$. The argument of $L^*(\sigma' + iT, \chi)$ for $\sigma' \geq \frac{3}{4}$ can be estimated by a standard method using the estimate

$$L^*(\sigma + iT, \chi) = O\left(\frac{\sqrt{T}}{|a|}\right)$$

valid for $T > 1$, $\sigma \geq \frac{1}{2}$, and is then found to be

$$O\left(\sigma_0 \log \frac{T}{|a|^2}\right)$$

with an absolute constant implied by the O-symbol. Here σ_0 is taken so large that

$$|L^*(s, \chi) - 1| \leq \frac{1}{2}$$

for $\sigma \geq \sigma_0$. However, it seems not possible to give any upper bound for σ_0 in terms of the other parameters $A(\mathcal{D})$ (or $\mu(\mathcal{D})$), κ_1, b and $|a|$, though it is obvious that if we keep Γ fixed and vary χ, a bound of the form $c_1 + c_2 \log \frac{1}{|a|}$ can be given where c_1 and c_2 are constants depending on Γ.

Combined, the above estimations give (4.24). It is clear that (4.21) would serve to estimate the lefthand side of (4.24) from above also for $\frac{1}{2} < \sigma < \frac{3}{4}$, but the resulting inequality is of somewhat less interest.

It is immediate from (4.24) that for any fixed $\sigma > \frac{3}{4}$, the number of zeros $\varrho = \beta + i\gamma$ with $\beta > \sigma$ and $0 \leq \gamma \leq T$, which we have called $N_\chi(\sigma, T) = O(T)$.

Taking in (4.24) $\sigma = \frac{3}{4}$, and considering the ϱ with $0 \leq \gamma \leq T$ only, which gives

$$\text{(4.24')} \quad \sum_{\substack{0 \leq \gamma \leq T \\ \beta > \frac{3}{4}}} \left(\beta - \frac{3}{4}\right) < \frac{\kappa_1}{4\pi} T \log \log T + \frac{1}{4\pi} (\log b - 2 \log |a|) \, T$$

$$+ \frac{c}{2} \kappa_1 T + O(\log T)$$

and subtracting this from (1.8), we get

$$\sum_{0\le\gamma\le T} \min\left(\frac{1}{4}, \beta-\frac{1}{2}\right) > \frac{\kappa_1}{4\pi} T \log \frac{T}{\pi \log T} - \frac{\log b}{4\pi} T - c'\kappa_1 T - O(\log T) , \tag{4.25}$$

which shows that

$$N_\chi\left(\frac{1}{2}, T\right) > \frac{\kappa_1}{\pi} T \log \frac{T}{\pi \log T} - \frac{1}{\pi} T \log b - c''\kappa_1 T - O(\log T) . \tag{4.26}$$

This shows that $N_\chi(\frac{1}{2}, T)$ is at least of the order $T \log T$, while (4.24′) shows that if $\frac{\omega(T)}{\log\log T} \to \infty$ as $T \to \infty$, then

$$N_\chi\left(\frac{3}{4} + \frac{\omega(T)}{\log T}, T\right) = o(T \log T) ,$$

so that *almost all* ϱ *have* $\beta < \frac{3}{4} + \frac{\omega(\gamma)}{\log \gamma}$. The inequality (4.26) is not so far from being an equality in the case of the modular group and χ_0, the same is the case for congruence subgroups of the modular group and χ_0. It is not unreasonable to conjecture that *for these groups* $N_\chi(\frac{1}{2}, T)$ *has the minimal growth that it can have given* κ_1, and so because of the relation between $N_\chi(T)$ and $N_\chi(\frac{1}{2}, T)$, *for these groups* $N_\chi(T)$ *would have the maximal growth that it can have given* $\mu(\mathcal{D})$ *and* κ_1, *or otherwise expressed, the largest discrete spectrum.*

For the modular group with χ_0, we can evaluate the lefthand side of (4.23) rather more accurately than what is given by the righthand side of (4.23) in that case. This is done by multiplying $L^*(s, \chi)$ in that case with a suitable dirichlet polynomial which acts as a mollifier. There seems to be no possibility of applying any similar device in the general case though the $\log\log T$ factor occurring in (4.24′) for instance probably does not represent the best possible upper bound in the general case either.

5. There is no difficulty in extending all results in the previous sections to the case where we consider instead functions in the hyperbolic plane (we are using the complex upper half-plane model) which transform in the way

$$g(Mz) = \chi(M) e^{ik \arg(cz+d)} g(z) ,$$

where

$$\chi(M) e^{ik \arg(cz+d)}$$

is a multiplier system for Γ, k being an arbitrary real number, and we consider the eigenfunctions and eigenvalues of the operator

$$y^2 \left(\frac{\partial^2}{\partial x^2} + \frac{\partial^2}{\partial y^2}\right) - iky\frac{\partial}{\partial x}$$

on this space of functions. We can again if the multiplier is singular with respect to one or more cusps of Γ, define Eisenstein series which give rise to the continuous spectrum on $\sigma = \frac{1}{2} + it$.

However, if we begin to look at higher dimensional cases, significant differences appear. We may, for example, look at the situation for the general n-dimensional hyperbolic space when $n > 2$. There is then no great difficulty in repeating our arguments until we come to Section 4. There are changes, the leading term on the righthand side in the general version (2.4) is of order T^n and our best estimate of the remainder term will be of order $\frac{T^{n-1}}{\log T}$, while (1.7) and (1.8) will be essentially unchanged in nature. However, the attempt to produce a reasonably sharp version of Theorem 4 fails. For $n = 3$ the continuous group of motions is essentially given by $SL(2, C)$, and if we consider the group Γ where the elements are matrices $\begin{pmatrix} a & b \\ c & d \end{pmatrix}$ where a, b, c, d are integers from the field $K = k(\sqrt{-1})$ with $ad - bc = 1$, and take $\chi = \chi_0$, we find our $\varphi(s, \chi_0)$ in this case is

$$\frac{\pi\Gamma(s-1)}{\Gamma(s)} \frac{\zeta_K(s-1)}{\zeta_K(s)} = \frac{\pi}{s-1} \frac{\zeta_K(s-1)}{\zeta_K(s)},$$

where ζ_K is the Dedekind zeta-function of the field K, so our $L(s, \chi_0)$ in this case is

$$L(s, \chi_0) = \frac{\zeta_K(s-1)}{\zeta_K(s)}.$$

In this case $\varphi(s, \chi_0)\varphi(2 - s, \chi_0) = 1$ and the pole is located at $s = 2$.

It is true that for this $L(s, \chi_0)$ we have, for $\frac{3}{2} \leq \sigma \leq 3$

$$\int_1^T |L(\sigma + it, \chi_0)|^2 \, dt = O\left(T \min\left(\frac{1}{(\sigma - \frac{3}{2})^2}, \log^2 T\right)\right),$$

and one would like to show that something like that would hold in general. However, for a general $L(s, \chi)$ in the case $n = 3$, we can only prove such a thing if we replace $\frac{3}{2}$ by a somewhat larger number < 2.

To some extent the failure to extend the results of Section 4 is academic in the case when $n \geq 3$, since the remainder term in (2.4) is not only of higher order than $T \log \log T$, but also than $T \log T$.

In view of several recent papers, one may well believe that unless Γ and χ are such that the subgroup of Γ where $\chi = 1$, is arithmetical (that is, in the case of $n = 2$, a congruence subgroup of the modular group), the continuous spectrum is large and probably the expression $N_\chi(\frac{n}{2}, T)$, the number of zeros of $\varphi(s, \chi)$ in $\sigma > \frac{n}{2}$; $|t| < T$, is of much higher order than $T \log T$, and might well even be of order T^n, while the discrete spectrum might be finite or even nonexistent.

References

Hejhal, D.A.: *The Selberg Trace Formula for PSL (2,R)*, vol. 2, Springer Lecture Notes 1001, Springer-Verlag 1983

Littlewood, J.E.: *On the Zeros of the Riemann Zeta-function*, Proc. Camb. Phil. Soc. 22, (1924) pp. 295–318

Selberg, A.: *Collected Papers* vol. 1, Springer-Verlag 1989

Addendum

After this manuscript was completed, I noticed that a change in the treatment of the lefthand side of (2.13) is not only simpler, but also leads to a slightly sharper result.

We see that the lefthand side of (2.13) is less than

$$\sum_{h\le|T'-\gamma|<1}\frac{4(\beta-\frac{1}{2})}{(\sigma-\beta)^2+(T'-\gamma)^2}\,. \tag{1}$$

Since

$$\int_{\frac{1}{2}}^{\infty}\frac{4(\beta-\frac{1}{2})}{(\sigma-\beta)^2+(T'-\gamma)^2}d\sigma<4\pi\frac{\beta-\frac{1}{2}}{|T'-\gamma|}\le\frac{4\pi}{h}\left(\beta-\frac{1}{2}\right)\,,$$

we get in (2.15) the term

$$\frac{4\pi}{h}\sum_{|T'-\gamma|<1}\left(\beta-\frac{1}{2}\right)=O\left(\frac{\log T}{h}\right)\,,$$

instead of $O(\frac{\log^2 T}{h})$. We therefore modify our choices (2.10) and (2.10′), so

$$h=(TR_2(T))^{-\frac{1}{3}}(\log T)^{\frac{2}{3}}\,, \tag{2}$$

and

$$l=\left[\left(\frac{R_2^2(T)}{T\log T}\right)^{\frac{1}{3}}\right]\,, \tag{3}$$

and take $h^*=2lh$ as before. The last term in (2.17) now becomes

$$O\left(\sqrt{T\log T}\left(\frac{R_2(T)}{\sqrt{T\log T}}\right)^{\frac{1}{3}}\right)\,, \tag{4}$$

and so in Theorem 2 we get

$$R_2(T) = \max\left(\sqrt{T \log T}, R_1(T)\right) , \tag{5}$$

instead of

$$R_2(T) = \max\left(\sqrt{T} \log T, R_1(T)\right) .$$

Added in proof: The observant reader will have noticed that in the inequality (4.23) a term $\kappa_1 T \log \kappa_1$ is missing on the right-hand side. As a consequence the terms

$$\frac{\kappa_1}{2\pi} T \log \kappa_1, -\frac{\kappa_1}{4\pi} T \log \kappa_1 \quad \text{and} \quad -\frac{\kappa_1}{\pi} T \log \kappa_1$$

are missing on the right-hand side of (4.24), (4.25) and (4.26) respectively. Due to the somewhat hasty preparation of the manuscript this was overlooked. However, these statements are nevertheless true in the form given here, though the proof of this is rather longer and more complicated in detail.

The inequality in footnote 5 gives

$$\sum_{j=1}^{\kappa_1} |\varphi_{i,j}(s,\chi)|^2 < \left(1 + \frac{2\sigma - 1}{|t|}\right)^2 ,$$

for $\sigma > \frac{1}{2}$, $t \neq 0$. If we for the poles σ_r on $\frac{1}{2} < \sigma \leq 1$, write

$$\phi_r(\chi) = \lim_{s \to \sigma_r} (s - \sigma_r)\phi(s, \chi) ,$$

then the $\phi_r(\chi)$ are Hermitian and we have

$$\phi_r(\chi) \geq 0 \quad \text{and} \quad \sum_r \phi_r(\chi) < E .$$

We replace Lemma 3 with a more general statement referring to a set of functions $f_j(s)$, $j = 1, ..., \kappa_1$, of the form (4.1), which satisfy the inequality

$$\sum_{j=1}^{\kappa_1} |f_j(s)|^2 < 8(1 + |t|)\left(1 + \frac{\sigma}{|t|}\right)^2$$

for $\sigma \geq 0$, $t \neq 0$; as well as the inequality

$$\sum_{j=1}^{\kappa_1} \left| \sum_r c_r^{(j)} \right|^2 < 8 ,$$

where $c_r^{(j)}$ denotes the residue of $f_j(s)$ at the simple pole $s = \sigma_r$, $\sigma_r > 0$.

We can then show that Theorem 4 holds if we replace the left-hand side of (4.19) with the expression

$$\int_1^T \sum_{j=0}^{\kappa_1} |L_{i,j}(\sigma + it, \chi)|^2 dt .$$

This implies that (4.23) is true in the form given in this paper, thus (4.24), (4.25) and (4.26) are similarly true without any change.

44.
Old and new conjectures and results about a class of Dirichlet series

Presented at the Amalfi Conference on Number Theory,
September 1989

0. Introduction

This lecture deals with four different aspects of a subject to which I have given quite a bit of thought at various times. First, from the middle 1940's to the early fifties, I was concerned with investigating the distribution of the values of $\log \zeta(s)$ and the logarithm of similar functions on or near the critical line. At first I dealt just with the imaginary part of the logarithm (and some of that was published at the time), but later also with the real part. In a later period from the early to late seventies I was interested in studying the distribution of the zeros of $\zeta(s) - a$ where $a \neq 0$, particularly the distribution in the neighborhood of the critical line. I again also looked at the same problem for more general functions. In recent years I have, stimulated by the work of Bombieri and Hejhal on the zeros of Epstein zeta-funcions of definite binary forms, studied value distribution of finite linear combinations of dirichlet series with functional equations and Eulerproduct, as well as the distribution of the zeros and a-points of such linear combinations. Finally, throughout this time I was led to formulate conjectures of which some perhaps are of more interest than any of the results I could prove.

1. Definitions and conjectures

We shall consider dirichlet series

$$F(s) = \sum_n \frac{a_n}{n^s}, \tag{1.1}$$

which are absolutely convergent for $\sigma > 1$, where $s = \sigma + it$. We assume $F(s)$ to be meromorphic, but so that $(s-1)^m F(s)$ is an integral function of finite order for some integer $m \geq 0$. We also assume that $F(s)$ has a functional equation of the form

$$\Phi(s) = \overline{\Phi(1-\bar{s})}, \tag{1.2}$$

where

$$\Phi(s) = \varepsilon Q^s \prod_{i=1}^{k} \Gamma(\lambda_i s + \mu_i) F(s) \tag{1.3}$$

and

$$|\varepsilon| = 1 \,, \quad Q > 0 \,, \quad \lambda_i > 0 \,, \quad \operatorname{Re} \mu_i \geq 0$$

are constants.

(1.2) implies that $\Phi(s)$ is real for $s = \frac{1}{2} + it$. It also follows that $(s-1)^m F(s)$ is of order 1 .

The zeros of $F(s)$ can be divided into two classes: trivial zeros, located at poles of the $\Gamma(\lambda_i z + \mu_i)$, $i = 1, 2, \ldots, k$; and the rest, which we refer to as nontrivial. The latter will all be located in some vertical strip $1 - A \leq \sigma \leq A$.

We shall write

$$\Phi\left(\frac{1}{2} + it\right) = \kappa(t) e^{i\vartheta(t)} F\left(\frac{1}{2} + it\right) , \tag{1.4}$$

where $\kappa(t)$ and $\vartheta(t)$ are real and $\kappa(t) > 0$, and

$$\Lambda = \sum_{i=1}^{k} \lambda_i \,. \tag{1.5}$$

Let us denote by $N(T)$ the number of nontrivial zeros in the region $0 \leq t \leq T$ for T positive,[1] counting zeros on the boundary with half multiplicity. Then

$$\begin{aligned} N(T) &= \frac{\vartheta(T)}{\pi} + \frac{1}{\pi} \arg F\left(\frac{1}{2} + iT\right) + c \\ &= \frac{\Lambda}{\pi} T(\log T + c') + O(\log T) \,, \end{aligned} \tag{1.6}$$

where c and c' are constants.

We now make some further assumptions

$$a_1 = 1 \,, \quad a_n = O(n^\delta) \,, \tag{1.7}$$

for any fixed positive δ, and

$$\log F(s) = \sum_n \frac{b_n}{n^s} \,, \tag{1.8}$$

with $b_n = 0$ except when n is of the form p^r where p is a prime and r a positive integer. Further, we assume

$$b_n = O(n^\theta) \tag{1.9}$$

for some $\theta < \frac{1}{2}$.

---

[1] For simplicity, I shall deal only with t and $T \geq 0$ here and later. Similar results hold for the case t and $T < 0$. These can also be derived by looking at $\overline{F(\bar{s})}$ instead of $F(s)$.

From (1.7), (1.8) and (1.9), it follows easily that the series

$$\sum_n^* \frac{|b_n|^2}{n^\sigma},$$

where $\sum^*$ designates that the series is extended only over the $n = p^r$ with $r > 1$, is convergent for

$$\sigma > \max\left(2\theta, \frac{1}{2}\right).$$

As we shall see in the next section, the value distribution of $\log F(s)$ on vertical lines or line segments very near or on the critical line $\sigma = \frac{1}{2}$ turns out to be governed by the behavior of the sum

$$\sum_{n \le x} \frac{|b_n|^2}{n} = \sum_{p \le x} \frac{|a_p|^2}{p} + O(1). \tag{1.10}$$

In all cases when the asymptotic behavior of (1.10) has been determined it turns out that

$$\sum_{p \le x} \frac{|a_p|^2}{p} = n_F \log\log x + O(1), \tag{1.11}$$

where n_F is an *integer* depending on F. *I conjecture that this is true for all F satisfying our conditions.*

To make the conjectures more precise we introduce the concept of a "primitive" function F. We say that $F(s)$ is primitive if $F(s)$ can not be written as $F(s) = F_1(s)F_2(s)$ where F_1 and F_2 again satisfy our conditions.

Conjecture 1.1. *For a primitive function F we always have $n_F = 1$, so that*

$$\sum_{p \le x} \frac{|a_p|^2}{p} = \log\log x + O(1). \tag{1.12}$$

Conjecture 1.2. *For two distinct primitive functions F and F' we have*

$$\sum_{p \le x} \frac{a_p \overline{a'_p}}{p} = O(1). \tag{1.13}$$

Thus in some sense the primitive functions are similar to an orthonormal system (though they of course do not form a Hilbert space, their number is not denumerable since for a dirichlet L function with a nonprincipal primitive character $L(s + ia)$ for any real a would be a primitive function).

We could have stated the conjectures (1.11), (1.12), (1.13) in a stronger form where each term in the sum has an additional factor $\log p$ and the $\log\log x$ on the right hand side is replaced by $\log x$.

These conjectures, which, by the way, are not unrelated to several other conjectures like the Sato-Tate conjecture, Langlands conjectures, etc.,[2] have been verified in a number of cases for dirichlet series with functional equation and Euler product that occur in number theory, by assuming that the factorizations we can give are actually in primitive factors. Only in very few cases can we actually show that a function is really primitive and can not be factorized further. If we make the additional hypothesis that the λ_i in (1.3) are half-integers, it is somewhat easier to show primitivity, but there would still not be many $F(s)$ for which it could be done.[3]

Clearly, if with $F(s)$ also

$$F^{\chi}(s) = \sum_n \frac{a_n}{n^s}\chi(n) ,$$

satisfies our conditions, where χ is a primitive dirichlet character, then F and F^{χ} are either both primitive or both not primitive since they have the same n_F. It is not unreasonable to *conjecture that if they are not primitive, they will factor in the same way into primitive factors, in the sense that we would get the factorization of* F^{χ} *by putting the character* χ *into all the series that define the factors of* F.

It seems finally natural to conjecture that for the class of functions $F(s)$ which satisfy our conditions, the Riemann hypothesis holds so that the nontrivial zeros would all lie on the critical line $\sigma = \frac{1}{2}$.

For the investigation of the value-distribution of $\log F(\sigma + it)$ for $\sigma = \frac{1}{2}$ or σ very near to $\frac{1}{2}$, we need, beside (1.11), either to assume the Riemann hypothesis (hereafter referred to as R.H.) or some weaker hypotheses concerning the nontrivial zeros.

If we denote, as is usual, by $N(\sigma, T)$ the number of zeros $\varrho = \beta + i\gamma$ with $\beta \geq \sigma$ and γ between 0 and T, and assume that for $\sigma > \frac{1}{2}$ we have

[2] One obvious consequence is that any $F(s)$ with a pole of order n at $s = 1$, contains $(\zeta(s))^n$ as a factor. When I first formulated these conjectures in the late forties, I had mostly examples of Dedekind zeta-functions and the dirichlet series derived from modular forms that were eigenfunctions of Hecke operators. Earlier in the 40's I had investigated the series that arose by convolution from two modular forms whose dimension (in the Hecke-terminology) differed by an integer or a half integer, and in particular convolution with a theta series. This enabled me to show that $\zeta(2s)\sum_n \frac{c_n^2}{n^{k-1+s}}$ where the c_n are the fourier-coefficients of an eigenfunction of the Hecke operator of dimension k is actually divisible by $\zeta(s)$. Later in the middle fifties, when I had gained much more understanding of Eisenstein series, I saw that if χ is a primitive character then also $L(2s, \chi^2)\sum_n \frac{\chi(n)c_n^2}{n^{k-1+s}}$ was divisible by $L(s, \chi)$.

[3] In all instances known to me of functions $F(s)$ satisfying our conditions it is true that $2\lambda_i$ is an integer.

$$N(\sigma, T) = O\left(|T|^{1-\alpha(\sigma-\frac{1}{2})} \frac{\log|T|}{\sqrt{\log\log|T|}}\right) \tag{1.14}$$

uniformly in σ as $|T| \to \infty$, where α is some positive constant, we will be able to prove essentially as sharp results as on the R.H.. Neither R.H. nor (1.14) has been proved to hold for any $F(s)$ in our class.

However, a slightly weaker form of (1.14), namely that

$$N(\sigma, T) = O\left(|T|^{1-\alpha(\sigma-\frac{1}{2})} \log|T|\right) , \tag{1.14'}$$

uniformly for $\sigma > \frac{1}{2}$, can actually be proved to hold in some instances. In the early 40's I verified it for the Riemann zeta-function[4] and dirichlet L-series. Today, using techniques more recently developed by A. Good, H. Iwaniec and J. Hafner, one can prove (1.14′) to hold for the L-functions of quadratic fields and for the dirichlet series formed with the fourier coefficients of a cusp form (analytic or eigenfunction of the hyperbolic Laplacian) which is an eigenfunction of the Hecke operators for the modular group or some congruence subgroup. Using (1.14′) we do not obtain quite as sharp results (*though in a few instances we do*), but for the classes for which (1.14′) can be proved and also (1.11) established, we then have results that are proved unconditionally.

2. Approximate formulas for $\log F(s)$ in the critical strip, and value distribution of $\log F(s)$ on vertical lines σ near or equal to $\frac{1}{2}$

We shall state the formulas first in the form we use if we assume the truth of the R.H. for $F(s)$. If we only assume (1.14) or (1.14′) we use analogous but somewhat more complicated formulas.

For $x \geq 2$, we define

$$\theta_x(n) = \begin{cases} 1 & \text{for} \quad 1 \leq n \leq x , \\ 2 - \frac{\log n}{\log x} & \text{for} \quad x \leq n \leq x^2 , \\ 0 & \text{for} \quad n \geq x^2 . \end{cases} \tag{2.1}$$

We also write $b_x(n) = b_n \theta_x(n)$. Further, let $2 \leq x \leq t^2$ and put $\sigma_x = \frac{1}{2} + \frac{1}{\log x}$. Then, for $\sigma_x \leq \sigma^* \leq \sigma$, we have

---

[4] Selberg [1].

(2.2)

$$\log F(\sigma+it) = \sum_{n<x^2} \frac{b_x(n)}{n^{\sigma+it}} + O\left(\frac{x^{\frac{1}{2}-\sigma}}{(\sigma^*-\frac{1}{2})\log^2 x}\left(\left|\sum_{n<x^2} \frac{b_x(n)\log n}{n^{\sigma^*+it}}\right| + \log|t|\right)\right).$$

Also, for $\frac{1}{2} \le \sigma \le \sigma_x$, we have, if we write $\eta_t = \min_\varrho |t-\gamma|$, that

(2.2′)

$$\log F(\sigma+it) = \sum_{n<x^2} \frac{b_x(n)}{n^{\sigma+it}} + O\left(\frac{1+\log^+ \frac{1}{\eta_t \log x}}{\log x}\left(\left|\sum_{n<x^2} \frac{b_x(n)\log n}{n^{\sigma_x+it}}\right| + \log|t|\right)\right).$$

The constants implied by the O symbols in (2.2) and (2.2′) depend on F only. We should also remark that if we take the imaginary part of (2.2′) we may drop the term $\log^+ \frac{1}{\eta_1 \log x}$ in the remainder term.

If we instead of R.H. assume only (1.14) or (1.14′) we have similar formulas where σ_x is replaced by a $\sigma_{x,t}$ which depends not only on x but also on t and which is defined in a way which depends beside x also on the zeros ϱ with imaginary parts close to t. Also the $\theta_x(n)$ then has a more complicated definition.[5]

These formulas enable us to prove

Theorem 1. *Let k be a positive integer, $0 < a < 1$, and $T^{\frac{a}{k}} \le x \le T^{\frac{1}{k}}$, then for $\sigma \ge \frac{1}{2}$ we have*

$$\frac{1}{T}\int_0^T \left|\log F(\sigma+it) - \sum_{p<x} \frac{a_p}{p^{\sigma+it}}\right|^{2k} dt = O(k^{4k}e^{Ak}), \tag{2.3}$$

with a constant A depending on a and F, and if we do not assume R.H., also on the constant α in (1.14′). If we in (2.3) replace the expression in the $|\,|$ symbol by its imaginary part we may replace the k^{4k} in the O-symbol by k^{2k}.

This enables us to get hold on the distribution of $\log F(\sigma+it)$ when $\sigma = \frac{1}{2}$ or when $\sigma > \frac{1}{2}$ and so close to $\frac{1}{2}$ that

[5] See Selberg [1], Section 4, the new $\theta_x(n)$ is the factor with which $\Lambda(n)$ is multiplied there to produce $\Lambda_x(n)$ in Lemma 10. Also $\sigma_{x,t}$ is defined by (4.1) and (4.2).

$$\sum_{p<T^{\frac{1}{k}}} \frac{|a_p|^2}{p^{2\sigma}} ,$$

is of the same order of magnitude as

$$\sum_{p<T^{\frac{1}{k}}} \frac{|a_p|^2}{p} ,$$

which is essentially when

$$\frac{1}{2} \leq \sigma \leq \frac{1}{2} + (\log T)^{-\delta} , \tag{2.4}$$

for some positive fixed δ.

We find in this way

Theorem 2. *For $\frac{1}{2} \leq \sigma \leq \frac{1}{2} + (\log T)^{-\delta}$, the function*

$$\kappa(\sigma, t) = \kappa_F(\sigma, t) = \frac{\log F(\sigma + it)}{\sqrt{\pi \sum_{p<t} \frac{|a_p|^2}{p^{2\sigma}}}} , \tag{2.5}$$

has a normal gaussian distribution in the complex plane. Also, the real and the imaginary parts of $\kappa(\sigma, t)$ have a normal gaussian distribution on the line.

More precisely,[6] *let $\chi_{a,b}(n)$ denote the characteristic function of an interval (a, b), then*

$$\int_0^T \chi_{a,b} \left(\operatorname{Re} \kappa(\sigma, t)\right) dt = T \int_a^b e^{-\pi u^2} du + O\left(T \frac{(\log\log\log T)^2}{\sqrt{\log\log T}}\right) , \tag{2.6}$$

and

$$\int_0^T \chi_{a,b} \left(\operatorname{Im} \kappa(\sigma, t)\right) dt = T \int_a^b e^{-\pi u^2} du + O\left(T \frac{\log\log\log T}{\sqrt{\log\log T}}\right) . \tag{2.6'}$$

Since in the region defined by (2.4) we have

$$\sum_{p<T} \frac{|a_p|^2}{p^{2\sigma}} = n_F \log\left(\min\left(\frac{1}{\sigma - \frac{1}{2}}, \log T\right)\right) + O(1) ,$$

we see that all primitive functions F have the same distribution. We also find, using (1.13), that distinct primitive functions are statistically idependent.

[6] For the larger σ in the range, say $\sigma > \frac{1}{2} + \frac{\log\log T}{\log T}$, the remainder terms in (2.6) and (2.6′) can be improved and will be the same for (2.6) and (2.6′).

We may now look at $\log|F(\sigma+it)|$ also for $\sigma<\frac{1}{2}$, using the functional equation (1.2) to relate it to $\log|F(1-\sigma+it)|$, we see that in the region

$$\frac{1}{2}-(\log T)^{-1+\varepsilon}<\sigma<\frac{1}{2},$$

where $\varepsilon\to 0$ as $T\to\infty$, and where

$$\sum_{p<T}\frac{|a_p|^2}{p^{2(1-\sigma)}}\sim n_F\log\log T\ ,$$

we have the relation

$$\frac{\log|F(\sigma+it)|}{\sqrt{\pi n_F\log\log t}}=\frac{\log|F(1-\sigma+it)|}{\sqrt{\pi n_F\log\log t}}+\Lambda\frac{(\frac{1}{2}-\sigma)\log|t|}{\sqrt{\pi n_F\log\log t}}+o(1)\ .$$

Thus for

$$\sigma=\frac{1}{2}-\mu\frac{\sqrt{\log\log T}}{\log T}, \tag{2.7}$$

where $\mu>0$ is a constant, we see that

$$\frac{\log|F(\sigma+it)|}{\sqrt{\pi n_F\log\log t}}$$

has a gaussian distribution that has been shifted to the right by the amount

$$\mu'=\frac{\Lambda\mu}{\sqrt{\pi n_F}}\ . \tag{2.8}$$

Thus results similar to (2.6) can be proved (they may be derived directly from (2.6)) with fairly precise remainder terms.

3. The distribution of the a-points for $F(s)$ or the zeros of $F(s)-a$, for $a\neq 0$

The results we have indicated about the distribution of the values of $\log|F(\sigma+it)|$ on vertical lines with σ very near to $\frac{1}{2}$, make it possible to evaluate asymptotically integrals like

$$\int_0^T\log^+|F(\sigma+it)|\,dt \tag{3.1}$$

for $\sigma=\frac{1}{2}$ or σ near $\frac{1}{2}$. Here it turns out to matter more whether we assume R.H. or (1.14), or only (1.14′). If we only assume (1.14), we can prove for instance

$$\int_0^T \log^+ \left| F\left(\frac{1}{2}+it\right)\right| dt = \frac{\sqrt{n_F}}{2\sqrt{\pi}} T\sqrt{\log\log T} + O(T)\,, \tag{3.2}$$

while if we assume (1.14) or the R.H. we get instead of $O(T)$ a remainder-term

$$O\left(T\frac{(\log\log\log T)^3}{\sqrt{\log\log T}}\right)\,. \tag{3.3}$$

Similar evaluations can be given for

$$\frac{1}{2} < \sigma < \frac{1}{2} + (\log T)^{-\delta}$$

(with somewhat better remainder terms for the larger σ), and also for $\sigma < \frac{1}{2}$ and of the form

$$\sigma = \frac{1}{2} - \mu\frac{\sqrt{\log\log T}}{\log T}\,, \tag{3.4}$$

only in this case the main term involves a more complicated constant depending on μ (or rather the μ' given by (2.8)) because of the shift of the gaussian distribution.

Our main tool for coming to grips with the distribution of the a-points of $F(s)$, or zeros of $F(s)-a$, with $a \neq 0$, is a well known formula due to Littlewood[7] which relates the sum ($\varrho = \beta + i\gamma$ will now denote the zeros of $F(s)-a$)

$$\sum_{\substack{0<\gamma<T\\ \beta>\sigma}} (\beta - \sigma)\,,$$

to the integral[8]

$$\frac{1}{2\pi}\int_0^T \log\left|\frac{F(\sigma+it)-a}{1-a}\right| dt\,, \tag{3.5}$$

with an error of $O(\log T)$.

To estimate the integral (3.5), of which the nontrivial part is

$$\frac{1}{2\pi}\int_0^T \log|F(\sigma+it)-a|\, dt\,, \tag{3.5'}$$

we divide $(0,T)$ into three sets:

$$S_1\,,\quad \text{the set of } t \text{ with } |F(\sigma+it)| > \frac{3}{2}|a|\,,$$

[7] J.E. Littlewood [1].

[8] We assume here $a \neq 1$. For $a = 1$ or near 1 a modified formula can be used. There is a family of zeros of $F(s)-a$ which move out to infinity to the right as $a \to 1$.

$$S_2\ ,\quad \text{the set of } t \text{ with } |F(\sigma+it)| < \frac{1}{2}|a|\ ,$$

and

$$S_3\ ,\quad \text{the set of } t \text{ with} \frac{1}{2}|a| \le |F(\sigma+it)| \le \frac{3}{2}|a|\ .$$

From (2.6) (or if $\sigma < \frac{1}{2}$ and of the form (2.7) from the corresponding result mentioned at the end of Section 2), we get for the measure of S_3,

$$m(S_3) = O\left(T\frac{(\log\log\log T)^2}{\sqrt{\log\log T}}\right)\ . \tag{3.6}$$

In S_1 we have

$$\log|F(\sigma+it)-a| = \log|F(\sigma+it)| + O\left(\frac{1}{|F(\sigma+it)|}\right)\ ,$$

and in S_2

$$\log|F(\sigma+it)-a| = \log|a| + O\left(|F(\sigma+it)|\right)\ .$$

It is now easy to show

$$\begin{aligned}\int_{S_1}\log|F(\sigma+it)-a|\,dt = &\int_0^T \log^+|F(\sigma+it)|\,dt\\ &+O\left(T\frac{(\log\log\log T)^2}{\sqrt{\log\log T}}\right)\end{aligned} \tag{3.7}$$

and

$$\begin{aligned}\int_{S_2}\log|F(\sigma+it)-a|\,dt =&m(S_2)\log|a|\\ &+O\left(T\frac{(\log\log\log T)^2}{\sqrt{\log\log T}}\right)\ .\end{aligned} \tag{3.7$'$}$$

The first terms on the right hand side of these equations are easily evaluated by using what we know of the distribution of $\log|F(\sigma+it)|$.

The integral over the set S_3 causes more problems, although we know that the measure of S_3 is comparatively small. Our problem is to show that $F(\sigma+it)$ does not stay long very close to a on the line. For $\sigma = \frac{1}{2}$ we have no problem and can show

$$\int_{S_3}\log\left|F\left(\frac{1}{2}+it\right)-a\right|dt = O\left(T\frac{(\log\log\log T)^3}{\sqrt{\log\log T}}\right) \tag{3.8}$$

for all $a \ne 0$. The same can be shown for $\sigma < \frac{1}{2}$ and of the form (3.4) if we assume R.H.. For $\sigma > \frac{1}{2}$ the R.H. does not seem to help, but it is easy, without any hypothesis, to show that for *almost all* a (in a very strong sense, which I shall not elaborate on here) we have, whether $\sigma > \frac{1}{2}$ or $\sigma < \frac{1}{2}$ that

$$\int_{S_3} \log|F(\sigma+it)-a|\,dt = O\left(T\frac{(\log\log\log T)^3}{\sqrt{\log\log T}}\right) . \tag{3.8'}$$

In this way we get estimates for the sum

$$\sum_{\substack{0<\gamma<T\\ \beta>\sigma}} (\beta-\sigma) . \tag{3.9}$$

For $\sigma=\frac{1}{2}$, this estimate is

$$\begin{aligned}\frac{\sqrt{n_F}}{4\pi^{\frac{3}{2}}}T\sqrt{\log\log T}+\frac{T}{4\pi}\log\frac{|a|}{|1-a|^2}\\ +O\left(T\frac{(\log\log\log T)^3}{\sqrt{\log\log T}}\right) ,\end{aligned} \tag{3.10}$$

if we assume the R.H. or (1.14) but only

$$\frac{\sqrt{n_F}}{4\pi^{\frac{3}{2}}}T\sqrt{\log\log T}+O(T) , \tag{3.10'}$$

if we assume (1.14′).

For $\sigma>\frac{1}{2}$ one finds one has to increase σ a bit before a change in the estimate is noticeable. More specifically, assuming R.H. or (1.14) and writing

$$\sigma=\frac{1}{2}+\frac{\omega(T)}{\log T} ,$$

where $0<\omega(T)<e^{\sqrt{\log\log T}}$, we find the change in the estimate for (3.9) to be of the order

$$T\frac{\log\omega(T)}{\sqrt{\log\log T}} ,$$

and so would not be noticeable compared to the remainder term in (3.10) for

$$\omega(T)<e^{(\log\log\log T)^3} .$$

For $\sigma<\frac{1}{2}$ and of the form (3.4), we get on the other hand, for $a\neq 1$,

$$\begin{aligned}\sum_{\substack{0<\gamma<T\\ \beta>\sigma}} (\beta-\sigma) = &\frac{\sqrt{n_F}}{2\sqrt{\pi}}\left(\frac{e^{-\pi\mu'^2}}{2\pi}+\mu'-\mu'\int_{\mu'}^{\infty}e^{-\pi u^2}du\right)T\sqrt{\log\log T}\\ &+\frac{\log|a|}{2\pi}\int_{\mu'}^{\infty}e^{-\pi u^2}du\cdot T-\frac{\log|1-a|}{2\pi}T\\ &+O\left(T\frac{(\log\log\log T)^3}{\sqrt{\log\log T}}\right) ,\end{aligned} \tag{3.11}$$

with the stronger hypothesis R.H. or (1.14), if we only assume (1.14′) we again have a remainder term $O(T)$ (which swallows the second and third term on the right hand side of (3.11)).

It is easily shown that in the region $0 < t < T$, $|\sigma| < A$ for large enough A, there are about as many a-points of $F(s)$ as there are zeros.[9] From (3.11) it is possible to show that *about half of the a-points lie to the left of the line* $\sigma = \frac{1}{2}$, *and statistically well distributed at distances of order*

$$\frac{\sqrt{\log\log T}}{\log T}$$

from the line $\sigma = \frac{1}{2}$. From (3.11) it follows that, *if we define* $N_a(\sigma, T)$ *and* $N_a(T)$ *for a-points analogously to* $N(\sigma, T)$ *and* $N(T)$ *for the zeros, we have for* σ *of the form (3.4)*

$$N_a(\sigma, T) \sim N_a(T) \cdot \int_{-\mu'}^{\infty} e^{-\pi u^2} du \,, \tag{3.12}$$

for $\mu > 0$, *so this family of a-points is distributed to the left of* $\sigma = \frac{1}{2}$ *with densities corresponding to one half of a gaussian distribution, and accounts as we said for about half of* $N_a(T)$. *Of the rest most lie quite close to the line* $\sigma = \frac{1}{2}$ *at distances of order not more than*

$$\frac{(\log\log\log T)^3}{\log T \sqrt{\log\log T}}$$

(assuming R.H. or (1.14), assuming only (1.14′) rather weaker conclusions hold). For this second family of zeros that lie mostly quite close to $\sigma = \frac{1}{2}$, it seems difficult to describe how they distribute to the left and the right of the line,[10] without making some strong (although rather plausible) additional conjecture about the distribution of the zeros of $F(s)$ on $\sigma = \frac{1}{2}$. However, it is possible to *prove* that *if we consider two values a and a′ such that their ratio is real and negative then*

$$(F(s) - a)\,(F(s) - a')$$

for almost all values of the argument of a has about $\frac{3}{4}$ *of its zeros on the left side of* $\sigma = \frac{1}{2}$ *and about* $\frac{1}{4}$ *of its zeros on the right side of* $\sigma = \frac{1}{2}$.

It is not unreasonable to conjecture that this is true for $F(s) - a$ *itself and for all* $a \neq 0$.

---

[9] There is a difference of $O(T)$ for $a = 1$, otherwise the difference is $O(\log T)$ at most for large T.

[10] For $a \neq 0$, $F(s) - a$ probably can only have a finite number of zeros on $\sigma = \frac{1}{2}$.

4. Linear combinations of dirichlet-series with functional equation and Euler product

We shall now turn to functions which are linear combinations of the kind of functions we have considered so far, say

$$F(s) = \sum_{i=1}^{n} c_i F_i(s) , \tag{4.1}$$

where $n > 1$ and we assume the functions $F_i(s)$ are linearly independent and satisfy the same functional equation.

We shall need some properties that are easily established. For simplicity we shall assume the F_i relatively prime,[11] then it is easily shown that the $\log F_i(s) = \log F_i(\sigma + it)$ are statistically independent.

In the region

$$\left|\sigma - \frac{1}{2}\right| < (\log T)^{-1+\varepsilon}$$

where $\varepsilon \to 0$ as $T \to \infty$, we can show that for $i \neq j$,

$$|\log |F_i(\sigma + it)| - \log |F_j(\sigma + it)|| > (\log\log\log T)^2 , \tag{4.2}$$

outside a subset of $(0, T)$ of measure not more than

$$O\left(T\frac{(\log\log\log T)^2}{\sqrt{\log\log T}}\right) . \tag{4.3}$$

This means that outside of this subset we have in $(0, T)$

$$\begin{aligned} \log |F(\sigma + it)| &= \max_i \log |c_i F_i(\sigma + it)| \\ &\quad + O\left(e^{-(\log\log\log T)^2}\right) . \end{aligned} \tag{4.4}$$

One obvious consequence of this is that *the distribution for*

$$\frac{\log |F(\sigma + it)|}{\sqrt{\pi \log\log t}}$$

can be determined from those of the

$$\frac{\log |F_i(\sigma + it)|}{\sqrt{\pi n_{F_i} \log\log t}} . \tag{4.5}$$

One conclusion that can be easily drawn is that *on the line* $\sigma = \frac{1}{2}$, $|F(\frac{1}{2}+it)|$ *is large except for a subset of* $(0, T)$ *of measure asymptotic to* $2^{-n}T$.

[11] By which we mean that no two contain the same primitive factor.

We also find that *for "almost all" sets of* c_i[12] *we have*

$$\text{(4.6)}\qquad \int_0^T \log|F(\sigma+it)|\,dt = \int_0^T \log\left(\max_i |F_i(\sigma+it)|\right)dt + CT + O\left(T\frac{(\log\log\log T)^3}{\sqrt{\log\log T}}\right),$$

and similarly, with a constant $a \neq 0$,

$$\text{(4.6')}\qquad \int_0 \log|F(\sigma+it)-a|\,dt = \int_0^T \log\left(\max_i\left(|F_i(\sigma+it)|\,,|a|\right)\right)dt + C'T + O\left(T\frac{(\log\log\log T)^3}{\sqrt{\log\log T}}\right).$$

Our knowledge of the distribution functions of the (4.5) enables us to evaluate the integrals on the right hand side of (4.6) and (4.6′). It is evident that the right hand sides of (4.6) and (4.6′) tend to increase as n increases, indicating that the zeros of $F(s)$ and of $F(s)-a$ in general shift somewhat to the right as n increases.

If we try to assess how large a proportion of the zeros of $F(s)$ lies to the left of the critical line $\sigma = \frac{1}{2}$, and what proportion lies to the right, we are led to *the conjecture that for* $a \neq 0$ *about*

$$\frac{1+2^{-n}}{2}$$

of the total number of zeros of $F(s)-a$ *in* $0<t<T$, *lie to the left of the line* $\sigma=\frac{1}{2}$, *and about*

$$\frac{1-2^{-n}}{2}$$

of the total number to the right of the line. We can not prove this, but a weaker result, similar to that given for $n=1$, can be proved in this case also.

If we only assume *the* $F_i(s)$ *to be linearly independent, but not all relatively prime, the results are similar, but the numerical factors entering in our conclusions above,* 2^{-n}, $\frac{1+2^{-n}}{2}$ *and* $\frac{1-2^{-n}}{2}$, *become rather more complicated and difficult to compute.*

If we make the additional assumption that in (4.1) the coefficients c_i are all real, we get that

[12] For instance, in the sense that if we normalize the c_i so that $\sum|c_i|^2=1$, then the equalities (4.6) and (4.6′) hold except for a set of c_i of measure zero on this unitary sphere. Actually it holds even in a stronger sense.
In (4.6) and (4.6′), C and C' are constants depending on the $|c_i|$ and the F_i, and in case of C' also on $|a|$.

$$\varepsilon Q^s \prod_{i=1}^{k} \Gamma(\lambda_i s + \mu_i) F(s) ,$$

is real on the critical line $\sigma = \frac{1}{2}$.

We may then expect that there could be quite many zeros of $F(s)$ on the line $\sigma = \frac{1}{2}$, though for $n > 1$ the fact that $|F(\frac{1}{2}+it)|$ is more often large than small indicates that the sum

$$\sum_{\substack{0<\gamma<T \\ \beta>\frac{1}{2}}} \left(\beta - \frac{1}{2}\right)$$

is actually of the order

$$T\sqrt{\log\log T} ,$$

(this can actually be proved for almost all choices of the c_i). Thus the number of zeros to the right of the line $\sigma = \frac{1}{2}$ would have to be at least of the order

$$T\sqrt{\log\log T} .$$

Now it is easily established that $\log F_i(\frac{1}{2} + it)$ and $\log F_i(\frac{1}{2} + it')$ on $(0,T)$ are not statistically independent if

$$|t - t'| < (\log T)^{-1+\varepsilon} , \tag{4.7}$$

where $\varepsilon > 0$ tends to zero as $T \to \infty$. We may call (4.7) the "region of inertia", and have then for almost all t and t' satisfying (4.7) that

$$\left| \log \left| F_i\left(\frac{1}{2} + it\right)\right| - \log \left| F_i\left(\frac{1}{2} + it\right)\right| \right| = o\left(\sqrt{\log\log T}\right) . \tag{4.8}$$

In order to utilize this fact to show that $F(s)$ has many zeros on $\sigma = \frac{1}{2}$, we have to make some additional hypothesis. We shall make the assumption that the zeros of the $F_i(s)$ are "well spaced", say that for some fixed positive $\theta \le 1$,

$$\limsup_{T\to\infty} \frac{\#(|\gamma_{n+1} - \gamma_n| < \frac{\delta}{\log T}, 0 < \gamma_n < T)}{T \log T} = O(\delta^\theta) , \tag{4.9}$$

uniformly as $\delta \to 0$.[13]

We now divide $(0,T)$ in intervals of length

$$H = \frac{\omega(T)}{\log T} ,$$

where

[13] For the Riemann zeta function, for instance, the pair correlation hypothesis of H.L. Montgomery gives that this ratio is $O(\delta^3)$.

$$\omega(t) = \exp\left(\sqrt{\log\log T}\right) .$$

Let I_r denote the interval $((r-1)H, rH)$ for $r = 1, 2, \ldots, [\frac{T}{H}]$. We find that except for $o(\frac{T}{H})$ of these intervals, they are so that one $\log|F_i(\frac{1}{2}+it)|$ dominates over the others in I_r except for a subset of I_r of measure $o(H)$. Since the zeros of $F_i(\frac{1}{2}+it)$ are well spaced, we see that the oscillations of the real function

$$e^{i\vartheta(t)} c_i F_i\left(\frac{1}{2}+it\right)$$

will also dominate over the other terms in

$$e^{i\vartheta(t)} F\left(\frac{1}{2}+it\right) ,$$

so that $F(\frac{1}{2}+it)$ will have about as many zeros in I_r as $F_i(\frac{1}{2}+it)$ has, which is

$$\frac{\Lambda}{\pi} H \log T + O\left((H \log T)^{\frac{3}{4}}\right) ,$$

except for $o(\frac{T}{H})$ of the I_r. The hypothesis of well spacing of the zeros of the $F_i(s)$ thus gives us the conclusion that *almost all zeros of $F(s)$ are on the critical line $\sigma = \frac{1}{2}$.*

For the case of the Epstein zeta-function of a rational binary form, this result was first conjectured by H.L. Montgomery and then proved on the hypothesis of well spacing by E. Bombieri and D. Hejhal.[14]

5. Concluding remarks

We may finally speculate on what is the situation for an $F(s)$ which is not of the form (4.1) but which is the limit of expressions (4.1) or of such expressions multiplied by some factor N^s where N may grow as n goes to infinity. Examples would be, for instance, the Epstein zeta-function of an irrational definite binary form.

It seems reasonable to *conjecture that such a function is almost always large on the critical line, and that its a-points for $a \neq 0$ are about equally many on each side of the critical line.*

It is much more doubtful whether, if the c_i in the approximations are real, such a function would have almost all its zeros on the critical line. This might be dependent on how well it could be approximated compared with the growth of n.

Numerical calculations are unlikely to throw much light on this since we are dealing at best with extremely slow convergence. Factors like

[14] see Bombieri and Hejhal [1].

$$\sqrt{\log \log T}$$

hardly change much in the range accessible by computation.

Let me close by mentioning another conjecture, also beyond any numerical evidence that could be expected in a foreseeable time.

If $F(s)$ is a function of the kind considered in Sections 1, 2 and 3, which is primitive or a product of distinct primitive factors, then *the number of sign changes of* $\arg F(\frac{1}{2}+it)$ *in* $(0,T)$ *is asymptotic to*

$$\frac{2\sqrt{\pi}}{\sqrt{n_F \log\log T}} N(T) \sim \frac{2\Lambda T \log T}{\sqrt{\pi n_F \log\log T}} \,.$$

We can prove an upper bound which differs from this only by a factor

$$(\log\log\log T)^2$$

in order of magnitude, from below we can not get that close.

A more complete account with proofs is under preparation and will in time appear elsewhere.

References

Bombieri, E., Hejhal, D.: *Sur les zéros des fonctions zêta d'Epstein,* Comptes Rendus Acad. Sci. Paris 304 (1987) pp. 213–217

Littlewood, E.: *On th Zeros of the Riemann Zeta-Function,* Porc. Camb. Phil. Soc. 22 (1924) pp. 295–318

Selberg, A.: *Contributions to the theory of the Riemann zeta-function,* Archiv f. Mathematik og Naturvidenskab. 48 (1946) No. 5, pp. 89–155

45.
Lectures on sieves

Dedicated to the memory of Viggo Brun

Contents

1. Introduction

When I wrote up my lectures at Stony Brook in 1969 [Selberg 4], I stated in the introduction that "this material always seemed awkward to handle, since it did not seem easy to develop a good system of notation and terminology." Because of this, while I at the time had left out, for instance, the full theory of the Buchstab–Rosser sieve (on which I lectured in the following years at the Pennsylvania State University in the spring of 1974 and also a couple of times at the Institute for Advanced Study) as well as the Λ^2 sieve and the lower bound sieves that can be associated with it, I was reluctant to return to give a fuller account of my work on the subject.

In recent years I have recognized that the main difficulty was a failure to distinguish between the properties that pertained to the problem, the sifting problem, as I now call it, and those that pertained to the tool, the sieve. This disposes of the rather unsatisfactory notions of "k–residue sieve" or "k dimensional sieve" in favor of the general concept of "sifting density".

As a consequence much of the general theory will here be developed for variable rather than just constant density. Also in the proof arrangements there are some changes compared to the Stony Brook lectures, in that I have gone back to an earlier form of proof which avoids the introduction of the $\tilde{\Lambda}$ concept used there. Besides the general theory, I give a full account of my theory of the Buchstab–Rosser sieve, the results are of course in agreement with those independently found by Iwaniec [Iwaniec 1], as well as more details on the Λ^2 sieve and the $\Lambda^2\Lambda^-$ sieve. While the latter is also carried out for variable density, for the Buchstab–Rosser sieve I consider only the case of constant density (while it is possible to modify the Buchstab–Rosser sieve to adapt to variable density, the resulting analytic problems do not in general permit explicit solutions, though it is possible that for some other special density–distribution explicit solutions might be obtained).

I have included some material concerning sifting of an interval instead of on a general weighted set, as well as on limitations of the sieve method for certain problems. In connection with the Fourier analysis which is used when sifting an interval, I have also included some other applications on which I have lectured from time to time. The outline of a theory for sifting with weights is sketched out, but without the detail given for the standard or original type of sifting. Finally, I have included two applications to number theory on which I have also lectured in the past.

2. The sifting problem

The original problem as considered by Viggo Brun [1], can be formulated as follows:

We have a given interval $\mathcal{I}_x$ of length x, and a finite set of primes $\mathcal{P}$ which we call the "sifting range", its elements we denote by p. With each p in $\mathcal{P}$ is associated a set of $u(p) < p$ distinct residue classes modulo p, and we refer to $u(p)$ as the sifting density with respect to p.

Our problem now is to estimate the number of integers remaining in $\mathcal{I}_x$ after we have excluded every integer which for some p in $\mathcal{P}$ lies in one of the $u(p)$ residue classes associated with p. The information given is clearly not enough to estimate this number exactly, so upper and lower bounds are what we aim for.

Since it seems difficult to develop a quite satisfactory theory for the problem in this formulation, we replace it with a more general formulation. The methods developed to handle this more general problem, will also apply to the original problem (but may not necessarily lead to best possible results there).

The general problem we formulate as follows:

We assume we have given a set of non negative numbers, "weights", w_n, associated with each integer n and such that $\sum_n w_n = x$. We call this a "weighted set" and denote it by W_x. We also have a finite set of primes $\mathcal{P}$[1] and with each p in this sifting range $\mathcal{P}$ we assume there is associated a sifting property or relation $p \times n$ (read "p excludes n" or "p crosses out n") which for each individual n may be valid or not. If for some n the relation $p \times n$ is false for all p in $\mathcal{P}$, we write $\mathcal{P} \bar{\times} n$ (read "$\mathcal{P}$ does not exclude n" or "$\mathcal{P}$ does not cross out n").

In this setting the sifting problem consists of estimating the quantity

$$M(W_x, \mathcal{P}) = \sum_{\mathcal{P} \bar{\times} n} w_n \, . \tag{2.1}$$

We have to assume that some information, beyond what has so far been said, is available. The information given is at first assumed to be of a rather general nature. Later we shall make more special assumptions, modeled on what would have been immediately available for the original sifting problem for the interval $\mathcal{I}_x$.

We introduce the notation $(\mathcal{P})$ to denote the set of squarefree positive integers all of whose prime factors belong to $\mathcal{P}$. The elements of $(\mathcal{P})$ we shall denote by the letters d or δ.

If the relation $p \times n$ holds for all p that divide d, we shall write $d \times n$ (observe that $1 \times n$ holds for all n). We define

$$N_d = N_d(W_x, \mathcal{P}) = \sum_{d \times n} w_n \, , \tag{2.2}$$

and assume that some information is available about the N_d; at first only that we for the N_d have certain upper and lower bounds

$$N_d^- \le N_d \le N_d^+ \, . \tag{2.3}$$

[1] It will be obvious to the reader that in this more abstract setting we could have dispensed with the prime numbers altogether; however, some valuation on the set of sifting relations would be necessary.

We do not exclude the possibility that for some d the information contained in (2.3) may be trivial or redundant. We have of course always $N_d \geq 0$ and also when $d_1 \mid d_2$ that $N_{d_1} \geq N_{d_2}$.

If $\mu(d)$ denotes the Möbius function we have

$$M(W_x, \mathcal{P}) = \sum_n w_n \sum_{\substack{d \times n \\ d \in (\mathcal{P})}} \mu(d) \tag{2.4}$$

$$= \sum_{d \in (\mathcal{P})} \mu(d) \sum_{d \times n} w_n = \sum_{d \in (\mathcal{P})} \mu(d)\, N_d\,.$$

If we here replace N_d by $N_d^{\operatorname{sgn}\mu(d)}$ we would get an upper bound, but in general one that would be useless because of the many terms involved, and the same is true for the lower bound we would obtain by replacing N_d by $N_d^{\operatorname{sgn}-\mu(d)}$.

In order to bring the number of terms down, we try to replace the $\mu(d)$ in (2.4) by another set of numbers λ_d such that we get an inequality instead of an identity, but such that the numbers λ_d become zero or at least small for most of the d in $(\mathcal{P})$.

By a Λ–system $\Lambda(\mathcal{P})$ we shall understand a set of real numbers λ_d associated with each d in $(\mathcal{P})$. If $\Lambda(\mathcal{P})$ for all d in $(\mathcal{P})$ satisifies the condition

$$\theta_d = \sum_{\delta \mid d} \lambda_\delta \geq \sum_{\delta \mid d} \mu(\delta)\,, \tag{2.5}$$

we refer to $\Lambda(\mathcal{P})$ as an upper bound sieve and denote it by $\Lambda^+(\mathcal{P})$. If on the other hand $\Lambda(\mathcal{P})$ for all d in $(\mathcal{P})$ satisfies the condition

$$\theta_d = \sum_{\delta \mid d} \lambda_\delta \leq \sum_{\delta \mid d} \mu(\delta)\,, \tag{2.5$'$}$$

we refer to it as a lower bound sieve and denote it by $\Lambda^-(\mathcal{P})$.

Clearly for a $\Lambda^+(\mathcal{P})$ we have

$$M(W_x, \mathcal{P}) \leq \sum_n w_n \sum_{\substack{d \times n \\ d \in (\mathcal{P})}} \lambda_d \tag{2.6}$$

$$= \sum_{d \in (\mathcal{P})} \lambda_d\, N_d \leq \sum_{d \in (\mathcal{P})} \lambda_d\, N_d^{\operatorname{sgn}\lambda_d}\,,$$

and in the same way for a $\Lambda^-(\mathcal{P})$

$$M(W_x, \mathcal{P}) \geq \sum_{d \in (\mathcal{P})} \lambda_d\, N_d^{\operatorname{sgn}-\lambda_d}\,. \tag{2.6$'$}$$

We shall show that (2.6) and (2.6′) by suitable choice of $\Lambda^+(\mathcal{P})$ and $\Lambda^-(\mathcal{P})$ actually give the best possible bounds for $M(W_x, \mathcal{P})$ obtainable from the information (2.3) available.

This can be seen as follows. For all sets of $\{N_d^+, N_d^-\}$ which are consistent, and all W_x such that $N_d(W_x)$ lies between these bounds, we consider the points with coordinates $M(W_x, \mathcal{P})$, $\{N_d^+, N_d^-\}$ (in a $2^{r+1}+1$ dimensional space, if r is the number of elements of $\mathcal{P}$). These clearly define a convex region in the space, since if we have two weighted sets W_x and W_x' we may form $W_x'' = \alpha W_x + \beta W_x'$ for α, β non negative and $\alpha + \beta = 1$. A convex region is always completely defined by a (in general infinite) set of linear inequalities corresponding to the equations of the supporting hyperplanes, and for any point of the boundary at least one of these inequalities is sharp; the boundary point lies in the supporting plane so that the corresponding inequality becomes an equality.

For any consistent set $\{N_d^+, N_d^-\} = \mathcal{N}$ let us denote by $M^+(W_x, \mathcal{P}, \mathcal{N})$ the maximum value of $M(W_x, \mathcal{P})$, we have therefore that there exists a general linear inequality

$$M(W_x, \mathcal{P}) \le \sum_{d \in (\mathcal{P})} (\rho_d N_d^+ + \sigma_d N_d^-) \tag{2.7}$$

which holds for all W_x, $\{N_d^+, N_d^-\}$ and which for the given $\mathcal{N} = \{N_d^+, N_d^-\}$ gives the precise upper bound for $M(W_x, \mathcal{P})$ or

$$M^+(W_x, \mathcal{P}, \mathcal{N}) = \sum_{d \in (\mathcal{P})} (\rho_d N_d^+ + \sigma_d N_d^-)\,. \tag{2.8}$$

Clearly the ρ_d are all non negative and the σ_d all non positive (since we otherwise could improve our bound by increasing the N_d^+ with negative ρ_d or decreasing the N_d^- with positive σ_d, clearly absurd). Now (2.7) must in particular hold if we let $p \times n$ mean $p \mid n$ and for some particular d define $w_d = 1$ and all other $w_n = 0$, let this be our W, and put $N_\delta^+ = N_\delta^- = 1$, for $\delta \mid d$, and $N_\delta^+ = N_\delta^- = 0$, for $\delta \nmid d$. (2.7) then gives

$$\sum_{\delta \mid d} \mu(\delta) \le \sum_{\delta \mid d} (\rho_\delta + \sigma_\delta)\,,$$

which shows that if we write $\lambda_d = \rho_d + \sigma_d$ then this forms a $\Lambda^+(\mathcal{P})$ system. We see that unless one of the two numbers ρ_d, σ_d is equal to zero, so that λ_d equals the other, (2.6) would give a better upper bound than (2.7) which is impossible. This proves that the precise upper bounds are obtainable from (2.6) by suitable choice of the $\Lambda^+(\mathcal{P})$. A similar argument works for the best lower bound of $M(W_x, \mathcal{P})$, which for a given set of $\{N_d^+, N_d^-\} - \mathcal{N}$, we refer to as $M^-(W_x, \mathcal{P}, \mathcal{N})$.

It is quite clear that if the information $N_d^- \le N_d \le N_d^+$ is redundant for some d, the optimal (not necessarily unique) $\Lambda^+(\mathcal{P})$ or $\Lambda^-(\mathcal{P})$ in (2.6) or (2.6′) will have the corresponding $\lambda_d = 0$. For an upper bound sieve, we see from (2.5) that $\lambda_1 \ge 1$. If $\lambda_1 > 1$ we clearly get a better $\Lambda^+(\mathcal{P})$ by dividing through by λ_1, so we may in the future consider only $\Lambda^+(\mathcal{P})$ with $\lambda_1 = 1$. Similarly (2.5′) implies that $\lambda_1 \le 1$ for any lower bound sieve $\Lambda^-(\mathcal{P})$. If it is possible to obtain a positive value for $M^-(W_x, \mathcal{P}, \mathcal{N})$ this clearly means that

λ_1 is positive. We may then divide through by λ_1 and obtain a larger lower bound unless $\lambda_1 = 1$. It may be, however, that the information provided only implies the trivial lower bound zero, which is obtained by choosing the λ_d in $\Lambda^-(\mathcal{P})$ all identically zero. We may thus in the future limit our attention to $\Lambda^-(\mathcal{P})$ with $\lambda_1 = 1$, since any non trivial bound is obtained by such a sieve.

We may summarize these results as

$$M^+(W_x, \mathcal{P}, \mathcal{N}) = \min_{\Lambda^+(\mathcal{P})} \sum_{d \in (\mathcal{P})} \lambda_d N_d^{\operatorname{sgn} \lambda_d}, \tag{2.9}$$

and

$$M^-(W_x, \mathcal{P}, \mathcal{N}) = \max_{\Lambda^-(\mathcal{P})} \left(0, \sum_{d \in (\mathcal{P})} \lambda_d N_d^{\operatorname{sgn} -\lambda_d}\right). \tag{2.9'}$$

Here we assume always $\lambda_1 = 1$, it is convenient to introduce the notation

$$M^=(W_x, \mathcal{P}, \mathcal{N}) = \max_{\Lambda^-(\mathcal{P})} \sum_{d \in (\mathcal{P})} \lambda_d N_d^{\operatorname{sgn} -\lambda_d}, \tag{2.9''}$$

(where the $=$ is understood as a reinforced minus sign), $M^=$ is actually the solution of a more general problem. If we in W_x require non negativity of w_n only for those n for which the relation $\mathcal{P} \bar{\times} n$ does not hold, then $M^=$ gives the best lower bound for

$$\sum_{\mathcal{P} \bar{\times} n} w_n$$

in this more general situation.

3. More specific assumptions

We shall in the following consider a situation where W_x depends on a parameter x which we may let tend to infinity; the sifting range $\mathcal{P}$ will be generally of the form $\mathcal{P}(\xi)$ (the set of primes $< \xi$) or of the form $\mathcal{P}(\xi_1, \xi_2)$ (the set of primes p with $\xi_1 \leq p < \xi_2$) here the ξ are functions of x, usually of the form $\xi = x^\alpha$ with some fixed α. We also make the assumption that

$$N_d = N_d(W_x, \mathcal{P}) = \sum_{d \times n} w_n = \frac{u(d)}{d} x + R_d, \tag{3.1}$$

where $u(d)$ is a normal multiplicative function on $(\mathcal{P})$ (in the sense defined in [Selberg 6]) with $0 \leq u(p) < p$,[2] and $|R_d| \leq u(d)$.[3] We refer, as in the case

[2] If we permitted $u(p) = p$ for some p in $\mathcal{P}$ the problem is clearly trivial, we could also exclude $u(p) = 0$, by simply omitting p from the sifting range.

[3] Assumption (3.1) and the bound on $|R_d|$ are modeled on what holds for the original sifting problem for the interval $\mathcal{I}_x$.

of the problem mentioned at the beginning of Section 2, to $u(p)$ as the *sifting density* with respect to p.

(3.1) and the bound on $|R_d|$ implies a specific set $\{N_d^+, N_d^-\} = \mathcal{N}$ that is derived from u. We shall therefore in this situation write $M_u^+(W_x, \mathcal{P})$ for the left hand side of (2.9), and modify the notation for the left hand side of (2.9′) and (2.9″) in the same way.

We get from (2.9), (2.9′) and (2.9″)

$$M_u^+(W_x,\mathcal{P}) = \min_{\Lambda^+(\mathcal{P})} \left(x \sum_{d\in(\mathcal{P})} \lambda_d \frac{u(d)}{d} + \sum_{d\in(\mathcal{P})} |\lambda_d|\, u(d) \right), \tag{3.2}$$

$$M_u^-(W_x,\mathcal{P}) = \max(0, M_u^=(W_x,\mathcal{P})), \tag{3.2′}$$

$$M_u^=(W_x,\mathcal{P}) = \max_{\Lambda^-(\mathcal{P})} \left(x \sum_{d\in(\mathcal{P})} \lambda_d \frac{u(d)}{d} - \sum_{d\in(\mathcal{P})} |\lambda_d|\, u(d) \right). \tag{3.2″}$$

We shall for brevity often write $1/f(d)$ for $u(d)/d$; $f(d)$ is again a multiplicative function; if some $u(d) = 0$ this will lead to $f(d) = \infty$. We could avoid this by dropping the p's for which $u(p) = 0$ from the sifting range, but properly interpreted, all formulas in which the $f(d)$ enter later will be meaningful even if we retain these p in the sifting range.

We introduced in the previous section the expression

$$\theta_d = \sum_{\delta\,|\,d} \lambda_\delta\,,$$

from which we get

$$\lambda_d = \sum_{\delta\,|\,d} \mu(d/\delta)\,\theta_\delta$$

this gives

$$\begin{aligned}
\sum_{d\in(\mathcal{P})} \frac{u(d)}{d}\lambda_d &= \sum_{d\in(\mathcal{P})} \frac{\lambda_d}{f(d)} = \sum_{d_1 d_2\in(\mathcal{P})} \frac{\theta_{d_1}\,\mu(d_2)}{f(d_1)\,f(d_2)} \\
&= \sum_{d_1\in(\mathcal{P})} \frac{\theta_{d_1}}{f(d_1)} \sum_{\substack{d_2\in(\mathcal{P})\\ (d_2,d_1)=1}} \frac{\mu(d_2)}{f(d_2)} \\
&= \sum_{d_1\in(\mathcal{P})} \frac{\theta_{d_1}}{f(d_1)} \prod_{p\in\mathcal{P}} \left(1-\frac{1}{f(p)}\right) \prod_{p\,|\,d_1} \left(1-\frac{1}{f(p)}\right)^{-1} \\
&= \prod_{p\in\mathcal{P}} \left(1-\frac{1}{f(p)}\right) \sum_{d\in(\mathcal{P})} \frac{\theta_d}{f'(d)}\,.
\end{aligned} \tag{3.3}$$

Here $f'(d)$ is the multiplicative function defined by $f'(p) = f(p) - 1$.

If we write

$$e_u(\mathcal{P}) = \prod_{p\in\mathcal{P}} \left(1-\frac{1}{f(p)}\right), \tag{3.4}$$

and

$$T_u(\Lambda(\mathcal{P})) = \sum_{d\in(\mathcal{P})} \frac{\theta_d}{f'(d)}, \tag{3.5}$$

and finally

$$R_u(\Lambda(\mathcal{P})) = \sum_{d\in(\mathcal{P})} |\lambda_d|\, u(d), \tag{3.6}$$

we can rephrase (3.2) and (3.2″) as

$$M_u^+(W_x, \mathcal{P}) = \min_{\Lambda^+(\mathcal{P})} (x\, e_u(\mathcal{P})\, T_u(\Lambda^+(\mathcal{P})) + R_u(\Lambda^+(\mathcal{P}))) \tag{3.7}$$

and

$$M_u^-(W_x, \mathcal{P}) = \max_{\Lambda^-(\mathcal{P})} (x\, e_u(\mathcal{P})\, T_u(\Lambda^-(\mathcal{P})) - R_u(\Lambda^-(\mathcal{P}))). \tag{3.7′}$$

We call $E_u(W_x, \mathcal{P}) = x\, e_u(\mathcal{P})$ the "expected" value of $M_u(W_x, \mathcal{P})$ and $e_u(\mathcal{P})$ the "expectation". Since $T_u(\Lambda^+(\mathcal{P})) \geq 1$, and $T_u(\Lambda^-(\mathcal{P})) \leq 1$ (with equality only if $\Lambda^+(\mathcal{P})$ or $\Lambda^-(\mathcal{P})$ equals the sieve with $\lambda_d = \mu(d)$) we have

$$M_u^-(W_x, \mathcal{P}) \leq x\, e_u(\mathcal{P}) \leq M_u^+(W_x, \mathcal{P}),$$

and the problem of minimizing the right hand side of (3.7) or maximizing the right hand side of (3.7′), can be stated as the problem of making $T_u(\Lambda)$ as close to one as possible, while at the same time keeping $R_u(\Lambda)$ essentially smaller than $x\, e_u(\mathcal{P})$. We refer to $x\, e_u(\mathcal{P})\, T_u(\Lambda)$ as the main term and $R_u(\Lambda)$ as the remainder term. $T_u(\Lambda)$ is important since it measures the ratio of the main term and the expected value.

For a $\Lambda^+(\mathcal{P})$ it is clear that the expression $T_u(\Lambda^+(\mathcal{P}))$ increases if we increase the $u(p)$, it is also clear that the ratio

$$\frac{R_u(\Lambda^+(\mathcal{P}))}{E_u(W_x, \mathcal{P})},$$

increases if the $u(p)$ increase (since R_u increases while E_u decreases). From this we see that the ratio

$$\frac{M_u^+(W_x, \mathcal{P})}{E_u(W_x, \mathcal{P})} \tag{3.8}$$

is strictly increasing if we increase any $u(p)$, since if we compare u_1 and u_2 with $u_1(p) \leq u_2(p)$ for all p we may choose an optimal $\Lambda^+(\mathcal{P})$ for u_2 and apply this sieve to both problems.

In the same way we can see that the ratio

$$\frac{M_u^-(W_x, \mathcal{P})}{E_u(W_x, \mathcal{P})} \tag{3.8′}$$

is strictly decreasing if we increase any of the $u(p)$, and so the ratio

$$\frac{M_u^-(W_x,\mathcal{P})}{E_u(W_x,\mathcal{P})} \tag{3.8''}$$

is strictly decreasing if any $u(p)$ are increased, as long as the ratio is positive. We refer to this as *monotonicity* of the ratios with respect to u.

Finally, we mention a more general sifting problem, that of sifting with weights. Assume that we wish to estimate not the expression (2.1), but, introducing the notation d_n for the largest d in $(\mathcal{P})$ for which $d \times n$ holds, the sum

$$M(W_x,\mathcal{P},\sigma) = \sum_n w_n\, \sigma(d_n)\,, \tag{3.9}$$

where $\sigma(d) \geq 0$ is a weight function defined on $(\mathcal{P})$ and such that $\sigma(d) = 0$ if d contains more than r_0 prime factors, and for $d = p_1 \dots p_r$ with $r \leq r_0$, $\sigma(d)$ is some function depending on r and the prime factors of d.

If we define a $\Lambda^+(\mathcal{P},\sigma)$ by the inequalities

$$\sum_{\delta \mid d} \lambda_\delta \geq \sigma(d) \tag{3.10}$$

for all d in $(\mathcal{P})$, and similarly a $\Lambda^-(\mathcal{P},\sigma)$ by the inequalities

$$\sum_{\delta \mid d} \lambda_\delta \leq \sigma(d) \tag{3.10'}$$

for all d in $(\mathcal{P})$, and further define

$$M_u^+(W_x,\mathcal{P},\sigma),\ M_u^-(W_x,\mathcal{P},\sigma)\,,$$

in an analogous way as before, we can repeat the development in Section 2 and in this section: The only difference is that we cannot stick to the convention $\lambda_1 = 1$, in general it might seem that we could replace this with the convention $\lambda_1 = \sigma(1)$, but this in some cases would not lead to the optimal bounds, so we stick just to (3.10) and (3.10′) with no further restrictions.[4]

It is easily seen that the expected value on the assumption (3.1) is

$$E_u(W_x,\mathcal{P},\sigma) = x\, e_u(\mathcal{P}) \sum_{d\in(\mathcal{P})} \frac{\sigma(d)}{f'(d)}\,. \tag{3.11}$$

The results about monotonicity also hold in this situation.[5]

The proofs of these more general results are quite similar to those given in the previous section and in this one.

---

[4] We could show, however, that for the d with $\sigma(d) > 0$ for the optimal Λ^+ the smallest of the ratios $\theta(d)/\sigma(d)$ equals 1, a similar result for the largest ratio holds for the optimal Λ^- if a positive lower bound is obtainable.

[5] But for the ratios $(M_u^\pm(W_x,\mathcal{P},\sigma) - E_u(W_x,\mathcal{P},\sigma))/E_u(W_x,\mathcal{P})$.

4. Further assumptions, definitions and main objects

Our main object is to study the behavior of the ratios (3.8), (3.8′), and (3.8″) as $x \to \infty$, in a situation where $\mathcal{P}$ is either of the form $\mathcal{P}(x^\alpha)$ with constant α or $\mathcal{P}(x^\alpha, x^{\alpha'})$ with $0 < \alpha < \alpha'$.

Clearly we can say nothing without further assumptions about the $u(p)$. First of all, we will allow the $u(p)$ to depend on x, rather than require them to be a fixed sequence which remains the same as $x \to \infty$.

We shall use the notation

$$S_u(\mathcal{P}) = \sum_{p \in \mathcal{P}} \frac{u(p)}{p} \log p \tag{4.1}$$

and if $\mathcal{P} = \mathcal{P}(x^\alpha)$, we shall write $S_u(x^\alpha)$ for $S(\mathcal{P})$ and for $\mathcal{P} = \mathcal{P}(x^\alpha, x^{\alpha'})$ we write $S_u(x^\alpha, x^{\alpha'})$. We now assume that for $\beta > 0$,

$$S_u(x^\beta) = g(\beta) \log x + o(\log x)\,, \tag{4.2}$$

where $g(\beta)$ is an increasing continuous function with $g(0) = 0$. We assume for simplicity that $g(\beta)$ has a continuous derivative $g'(\beta)$ for $\beta > 0$ except possibly at isolated points where we may assume right and left derivatives exist.[6] Where $g'(\beta)$ exists we call it the sifting density around x^β. The case most considered before is the case of constant sifting density when $g(\beta) = k\,\beta$ with some positive constant k. One sees at once that except when $g(\beta) = k\,\beta$, the requirement (4.2) necessitates that the sequence $u(p)$ depends on x in some way.

When $\mathcal{P} = \mathcal{P}(x^\alpha)$ or $\mathcal{P}(x^\alpha, x^{\alpha'})$ we shall for brevity write $e_u(\mathcal{P}) = e_u(x^\alpha)$ or $e_u(\mathcal{P}) = e_u(x^\alpha, x^{\alpha'})$.

From (4.2) it is clear that $u(p)/p = o(1)$. We also have using (4.2), that

$$\begin{aligned}
\sum_{x^\alpha \le p \le x^{\alpha'}} \frac{u(p)}{p} &= \int_\alpha^{\alpha'} \frac{dS_u(x^\beta)}{\beta \log x} \\
&= \left[\frac{S_u(x^\beta)}{\beta \log x}\right]_\alpha^{\alpha'} + \int_\alpha^{\alpha'} \frac{S_u(x^\beta)}{\beta^2 \log x}\, d\beta \\
&= \left[\frac{g(\beta)}{\beta}\right]_\alpha^{\alpha'} + \int_\alpha^{\alpha'} \frac{g(\beta)}{\beta^2}\, d\beta + o(1) \\
&= \int_\alpha^{\alpha'} \frac{d\,g(\beta)}{\beta} + o(1)\,.
\end{aligned}$$

Since $u(p)/p = o(1)$, we have for $p \ge x^\eta$ with $\eta > 0$

$$\log\left(1 - \frac{u(p)}{p}\right) = -\frac{u(p)}{p} + o\left(\frac{u(p)}{p}\right).$$

[6] Considerably less would do.

Combining this with the preceding, we get

$$\sum_{x^\alpha \le p \le x^{\alpha'}} \log\left(1 - \frac{u(p)}{p}\right) = -\int_\alpha^{\alpha'} \frac{d\,g(\beta)}{\beta} + o(1)\,,$$

or

$$e_u(x^\alpha, x^{\alpha'}) = (1 + o(1)) \exp\left(-\int_\alpha^{\alpha'} \frac{d\,g(\beta)}{\beta}\right). \tag{4.3}$$

Thus all expressions $e_u(x^\alpha)$ with fixed $\alpha > 0$ are of the same order of magnitude, which will depend on the very early factors $(1 - u(p)/p)$ with $p < x^\varepsilon$ for any fixed positive ε. (4.2) does not tell us enough about $u(p)/p$ to prevent $e_u(x^\alpha)$ from being extremely small, *we therefore make the assumption that*

$$e_u(x) > x^{-\delta}\,, \tag{4.4}$$

for any positive δ and $x > x(\delta)$.

The expected value thus satisfies the inequality

$$E_u(W_x, \mathcal{P}(x^\alpha)) = x\, e_u(x^\alpha) > x^{1-\delta} \tag{4.4$'$}$$

for any positive δ for $x > x(\delta)$.

Thus in particular if $R_u(\Lambda^\pm(\mathcal{P}(x^\alpha))) = O(x^{1-\varepsilon})$ for some $\varepsilon > 0$, the remainder term R_u in (3.7) and (3.7$'$) will ultimately be negligible compared to the main term, except in the case that $T_u(\Lambda^-(\mathcal{P}(x^\alpha)))$ is zero or tends to zero as $x \longrightarrow \infty$.

5. Remarks concerning remainder terms

Our ultimate object is to study the behavior of the ratios (3.8) and (3.8$'$) when $\mathcal{P} = \mathcal{P}(x^\alpha)$ as $x \longrightarrow \infty$, and to prove that these ratios actually under fairly general assumptions tend to limit functions that depend only on the function g and the parameter α.

In preparation for this study, we have to discuss certain principles for constructing sieves.

The usefulness of a sieve for a specific sifting problem be it a Λ^+ or a Λ^- depends on two things:

(a) The $T_u(\Lambda)$ should deviate from 1 as little as possible, while at the same time (b): the term $R_u(\Lambda)$ is small enough not to interfere with the efficiency ratio $T_u(\Lambda)$.

For large x we therefore want $R_u(\Lambda)$ to be of less order than $x\, e_u(p)$, which with the assumption (4.4) we have made before, certainly would be true if

$$R_u(\Lambda) < x^{1-\varepsilon} \tag{5.1}$$

for $x > x_0$ and some $\varepsilon > 0$. On the other hand $R_u(\Lambda)$ certainly would ruin any estimation if $R_u(\Lambda) > x$, so any sieve that does not satisfy

$$(5.1') \qquad R_u(\Lambda) \leq x,$$

would certainly be disqualified for the sifting problem.

We shall denote a $\Lambda^{\pm}$ which satisfies

$$R_u(\Lambda) = \sum_{d \in (\mathcal{P})} u(d)\,|\lambda_d| < z$$

by the symbol $\Lambda_u^{\pm}(\mathcal{P}, z)$.

One way of trying to keep $R_u(\Lambda)$ down is to require the λ_d to be zero for $d > z$, and we shall denote such a $\Lambda^{\pm}$ by $\Lambda^{\pm}[\mathcal{P}, z]$. That this implies a bound on $R_u(\Lambda)$ if $T_u(\Lambda)$ is not too large can be seen as follows.

We have

$$|\lambda_d| = \left|\sum_{\delta \mid d} \mu\,(d/\delta)\,\theta_\delta\right| \leq \sum_{\delta \mid d} |\theta_\delta|\,.$$

Thus

$$(5.2) \qquad \begin{aligned} \sum_d \frac{u(d)}{d}\,|\lambda_d| &\leq \sum_d \frac{1}{f(d)} \sum_{\delta \mid d} |\theta_\delta| \\ &\leq \sum_\delta \frac{|\theta_\delta|}{f(\delta)} \sum_d \frac{1}{f(d)} = \prod_{p \in \mathcal{P}} \left(1 + \frac{1}{f(p)}\right) \sum_\delta \frac{|\theta_\delta|}{f(\delta)} \\ &< e_u^{-1}(\mathcal{P}) \sum_d \frac{|\theta_d|}{f(d)}\,. \end{aligned}$$

If we are dealing with a Λ^+ we get (since $|\theta_d| = \theta_d$) that

$$(5.3) \qquad \sum_d \frac{u(d)}{d}\,|\lambda_d| < e_u^{-1}(\mathcal{P})\,T_u(\Lambda^+)\,,$$

from which we get for a $\Lambda^+[\mathcal{P}, z]$

$$(5.4) \qquad R_u(\Lambda^+) < z \sum_d \frac{u(d)}{d}\,|\lambda_d| < z\,e_u^{-1}(\mathcal{P})\,T_u(\Lambda^+)\,.$$

For a Λ^- we get (since $\theta_1 = 1$ and $|\theta_d| = -\theta_d$ for $d > 1$) that

$$(5.5) \qquad \sum_d \frac{u(d)}{d}\,|\lambda_d| < e_u^{-1}(\mathcal{P})\,(2 - T_u(\Lambda^-))\,,$$

from which it follows in the same way for a $\Lambda^-[\mathcal{P}, z]$

$$(5.6) \qquad R_u(\Lambda^-) < z\,e_u^{-1}(\mathcal{P})\,(2 - T_u(\Lambda^-))\,.$$

In particular, if $T_u(\Lambda^-) > 0$, that is, if we get a positive lower bound at all, then

$$R_u(\Lambda^-) < 2\,z\,e_u^{-1}(\mathcal{P})\,. \tag{5.6'}$$

We shall show in Section 7 that under the assumptions made about $u(p)$, $e_u(\mathcal{P})$, and $g(\beta)$, we can construct sieves for which $T_u(\Lambda^{\pm})$ remains bounded as $x \to \infty$, and which are of the form $\Lambda^+[\mathcal{P}(x^\alpha), x^\beta]$ for any $\beta > 0$, or of the form $\Lambda^-[\mathcal{P}(x^\alpha), x^\beta]$ for any $\beta \geq \alpha$. Thus if we take $\beta < 1$ (which for the lower bound problem requires $\alpha < 1$) we will have remainder terms $R_u(\Lambda)$ which because of (5.3) or (5.3′) will be of smaller order than $x\,e_u(x^\alpha)$.

However, we do not know yet whether there exist sieves which for our sifting problem are "optimal" as $x \to \infty$ (in the sense of leading asymptotically to the best bounds for the ratios (3.8), (3.8′), or (3.8″)), and at the same time are of the type $\Lambda^{\pm}[\mathcal{P}(x^\alpha), z]$ for any $z \leq x$. For the purpose of developing the general theory we must therefore still consider the class $\Lambda_u^{\pm}(\mathcal{P}(x^\alpha), z)$ where some limitation is put on z. Clearly we have to demand at least $z \leq x$, since otherwise the remainder term would be too large. $z = x^{1-\varepsilon}$ for some fixed $\varepsilon > 0$, we have seen, would be enough to ensure that the remainder term $R_u(\Lambda)$ is of smaller order than $x\,e_u(x^\alpha)$ as $x \to \infty$, and we might then let $\varepsilon \to 0$ in the end. It is a priori not clear whether this leads to the same results as staying with the class $\Lambda_u^{\pm}(\mathcal{P}(x^\alpha), x)$ or not.

6. Some general principles for constructing, combining and modifying sieves

Let $\mathcal{P}_1$ and $\mathcal{P}_2$ be two sets of primes and let $\mathcal{P}_3$ denote their intersection $\mathcal{P}_1 \cap \mathcal{P}_2$ (which may be empty) and write $\mathcal{P}_1' = \mathcal{P}_1 - \mathcal{P}_3$, $\mathcal{P}_2' = \mathcal{P}_2 - \mathcal{P}_3$. Suppose we have two Λ's, $\Lambda^{(1)}$ and $\Lambda^{(2)}$, defined for $(\mathcal{P}_1)$ and $(\mathcal{P}_2)$ respectively.

We can now form a new Λ defined for $\mathcal{P} = \mathcal{P}_1 \cup \mathcal{P}_2$ in the following way. Each d in $(\mathcal{P})$ can be written as $d = d_1\,d_2\,d_3$ where $d_1 \in (\mathcal{P}_1')$, $d_2 \in (\mathcal{P}_2')$, and $d_3 \in (\mathcal{P}_3)$, we now define

$$\lambda_d = \sum_{[\delta_1', \delta_2'] = d_3} \lambda^{(1)}_{d_1\delta_1'}\,\lambda^{(2)}_{d_2\delta_2'}\,, \tag{6.1}$$

where $[\delta_1, \delta_2]$ denotes the smallest common multiple of δ_1 and δ_2. We now have

$$\begin{aligned}\theta_d = \sum_{\delta \mid d} \lambda_d &= \sum_{\substack{\delta_i \mid d_i \\ i=1,2,3}} \; \sum_{[\delta_1', \delta_2'] = \delta_3} \lambda^{(1)}_{\delta_1\delta_1'}\,\lambda^{(2)}_{\delta_2\delta_2'} \\ &= \sum_{\substack{\delta_1 \mid d_1 \\ \delta_1' \mid d_3}} \lambda^{(1)}_{\delta_1\delta_1'} \sum_{\substack{\delta_2 \mid d_2 \\ \delta_2' \mid d_3}} \lambda^{(2)}_{\delta_2\delta_2'} = \theta^{(1)}_{d_1 d_3}\,\theta^{(2)}_{d_2 d_3}\,.\end{aligned} \tag{6.2}$$

I. We first consider the case that $\mathcal{P}_1$ and $\mathcal{P}_2$ are such that the intersection $\mathcal{P}_3$ is empty, and that we have given for each of the sets $\mathcal{P}_1$ and $\mathcal{P}_2$ an upper bound sieve $\Lambda_1^+(\mathcal{P}_1)$ and $\Lambda_2^+(\mathcal{P}_2)$. From (6.2) it is then clear that the $\Lambda(\mathcal{P})$ defined according to (6.1) in this case is again an upper bound sieve $\Lambda^+(\mathcal{P})$, we have

$$\lambda_d = \lambda_{d_1}^{(1)}\,\lambda_{d_2}^{(2)}$$

if $d = d_1 d_2$ where $d_1 \in (\mathcal{P}_1)$ and $d_2 \in (\mathcal{P}_2)$, and from (6.2)

$$\theta_d = \theta_{d_1}^{(1)}\,\theta_{d_2}^{(2)}\,.$$

We also see that

$$T_u(\Lambda^+) = T_u(\Lambda_1^+)\,T_u(\Lambda_2^+) \tag{6.3}$$

and

$$R_u(\Lambda^+) = R_u(\Lambda_1^+)\,R_u(\Lambda_2^+)\,. \tag{6.4}$$

Thus if $\Lambda_1^+ = \Lambda_u^+(\mathcal{P}_1, z_1)$, $\Lambda_2^+ = \Lambda_u^+(\mathcal{P}_2, z_2)$ we get

$$\Lambda^+ = \Lambda_u^+(\mathcal{P}, z_1 z_2)\,.$$

Also if $\Lambda_1^+ = \Lambda_1^+[\mathcal{P}_1, z_1]$, $\Lambda_2^+ = \Lambda_2^+[\mathcal{P}_2, z]$, we get

$$\Lambda^+ = \Lambda^+[\mathcal{P}, z_1\, z_2]\,.$$

If we for $\mathcal{P}_1$ and $\mathcal{P}_2$ are given $\Lambda_1^+(\mathcal{P}_1)$, $\Lambda_1^-(\mathcal{P}_1)$, $\Lambda_2^+(\mathcal{P}_2)$, $\Lambda_2^-(\mathcal{P}_2)$, denoting the λ in $\Lambda_i^\pm(\mathcal{P})$ by $\lambda_d^{(i)\,\pm}$, and writing for $d = d_1\, d_2$ with $d_1 \in (\mathcal{P}_1)$, $d_2 \in (\mathcal{P}_2)$

$$\lambda_d = \lambda_{d_1}^{(1)\,+}\,\lambda_{d_2}^{(2)\,-} + \lambda_{d_1}^{(1)\,-}\,\lambda_{d_2}^{(2)\,+} - \lambda_{d_1}^{(1)\,+}\,\lambda_{d_2}^{(2)\,+}\,, \tag{6.5}$$

we get (still assuming the intersection $\mathcal{P}_1 \cap \mathcal{P}_2$ to be empty) that for the new $\Lambda(\mathcal{P})$

$$\begin{aligned} \theta_d = \sum_{\delta\,|\,d} \lambda_d &= \theta_{d_1}^{(1)\,+}\,\theta_{d_2}^{(2)\,-} + \theta_{d_1}^{(1)\,-}\,\theta_{d_2}^{(2)\,+} - \theta_{d_1}^{(1)\,+}\,\theta_{d_2}^{(2)\,+} \\ &= \theta_{d_1}^{(1)\,-}\,\theta_{d_2}^{(2)\,-} - (\theta_{d_1}^{(1)\,+} - \theta_{d_1}^{(1)\,-})\,(\theta_{d_2}^{(2)\,+} - \theta_{d_2}^{(2)\,-})\,. \end{aligned} \tag{6.6}$$

Thus we see that the $\Lambda(\mathcal{P})$ defined by (6.5) is a $\Lambda^-(\mathcal{P})$. Furthermore, we see that

(6.7)

$$R_u(\Lambda) = \sum_{d \in (\mathcal{P})} u(d)\,|\lambda_d| \le R_u(\Lambda_1^+)\,R_u(\Lambda_2^-) + R_u(\Lambda_1^-)\,R_u(\Lambda_2^+) + R_u(\Lambda_1^+)\,R_u(\Lambda_2^+)\,,$$

and

$$T_u(\Lambda) = T_u(\Lambda_1^-)\,T_u(\Lambda_2^-) - \{T_u(\Lambda_1^+) - T_u(\Lambda_1^-)\}\,\{T_u(\Lambda_2^+) - T_u(\Lambda_2^-)\}\,. \tag{6.8}$$

If the $\Lambda_i^{\pm}(\mathcal{P}_i)$, $i = 1, 2$, are of the form $\Lambda_u(\mathcal{P}_i, z_i)$, we see from (6.7) that

$$R_u(\Lambda^-(\mathcal{P})) \leq 3\, z_1 z_2\,,$$

so the new Λ^- is of the form

$$\Lambda_u(\mathcal{P},\, 3\, z_1 z_2)\,.$$

If the $\Lambda_i^{\pm}(\mathcal{P}_i)$, $i = 1, 2$, are of the type $\Lambda_i^{\pm}[\mathcal{P}_i, z_i]$, we get from (6.5) that the new $\Lambda^-(\mathcal{P})$ is of the type $\Lambda^-[\mathcal{P}, z_1 z_2]$.

We shall later mostly use this principle for constructing upper and lower bound sieves for $\mathcal{P}$ in situations where $\mathcal{P} = \mathcal{P}(x^\alpha)$, $\mathcal{P}_1 = \mathcal{P}(x^\eta)$, and $\mathcal{P}_2 = \mathcal{P}(x^\eta, x^\alpha)$ and where η is very small, and with sieves $\Lambda_{(1)}^{\pm}[\mathcal{P}_1, x^\delta]$ and $\Lambda_{(2)\,u}^{\pm}(\mathcal{P}_2, x^{1-\delta-\eta})$, where δ also is small but the ratios δ/η and $\delta/g(\eta)$ are large. For $\mathcal{P}_1$ we can produce such sieves where $T_u(\Lambda^{\pm}(\mathcal{P}_1))$ is very close to 1 as will be evident from the results (7.20) and (7.20′) in the next section. Thus we will be able to reduce the problem of sieves for the range $\mathcal{P}(x^\alpha)$ to that of the range $\mathcal{P}(x^\eta, x^\alpha)$ which is a somewhat simpler problem.

II. We now consider the situation where $\mathcal{P}_1 = \mathcal{P}_2 = \mathcal{P}$ in the construction (6.1). It then takes the form

$$\lambda_d = \sum_{[\delta_1, \delta_2] = d} \lambda_{\delta_1}^{(1)}\, \lambda_{\delta_2}^{(2)}\,, \tag{6.9}$$

and we have $\theta_d = \theta_d^{(1)}\, \theta_d^{(2)}$. We express (6.9) symbolically by $\Lambda = \Lambda^{(1)}\, \Lambda^{(2)}$.

Clearly if $\Lambda^{(1)}$ and $\Lambda^{(2)}$ are both upper bound sieves or both lower bound sieves the resulting Λ is an upper bound sieve, while if one is an upper bound sieve and the other a lower bound sieve Λ will be a lower bound sieve. We see that if $\Lambda^{(1)}$ is of the type $\Lambda^{(1)}[\mathcal{P}, z_1]$, $\Lambda^{(2)}$ of the type $\Lambda^{(2)}[\mathcal{P}, z_2]$, then Λ is of the type $\Lambda[\mathcal{P}, z_1\, z_2]$.

Finally, we consider the case $\Lambda^{(1)} = \Lambda^{(2)}$, in this case, since $\theta_d = (\theta_d^{(1)})^2 \geq 0$, we see that $\Lambda = {\Lambda^{(1)}}^2$ is an upper bound sieve for any $\Lambda^{(1)}$ regardless of whether $\Lambda^{(1)}$ is a sieve or not. We shall in the future refer to such upper bound sieves as Λ^2 sieves. If $\Lambda^{(1)} = \Lambda^{(1)}[\mathcal{P}, z]$ then $\Lambda = {\Lambda^{(1)}}^2$ is a $\Lambda[\mathcal{P}, z^2]$.

We may of course combine more than two Λ's in this way, we may for instance combine a, by itself useless, lower bound sieve with a Λ^2 upper bound sieve, and produce an effective lower bound sieve which we denote as a $\Lambda^2\,\Lambda^-$ sieve. If $\Lambda_1 = \Lambda_1[\mathcal{P}, z_1]$, $\Lambda_2 = \Lambda_2^-[\mathcal{P}, z_2]$ then

$$\Lambda = \Lambda_1^2\, \Lambda_2^- = \Lambda^-[\mathcal{P}, z_1^2\, z_2]\,.$$

We now turn to two other principles which we shall make use of.

Suppose $\Lambda_1 = \Lambda_u(\mathcal{P}, z)$ and that $d_0 \in (\mathcal{P})$, $d_0 > 1$. Let $\mathcal{P}_{d_0}$ denote the set of primes in $\mathcal{P}$ which do not divide d_0. Let $\Lambda_2 = \Lambda_u(\mathcal{P}_{d_0}, z/d_0)$ and define a new $\Lambda_u(\mathcal{P}, z')$ as follows

$$\lambda_d = \lambda_d^{(1)} - \theta_{d_0}^{(1)}\, \lambda_{d/d_0}^{(2)}\,, \tag{6.10}$$

for $d \in (\mathcal{P})$, here $\lambda^{(2)}_{d/d_0}$ is interpreted as zero if d_0 does not divide d, and

$$z' = z\left(1 + \frac{u(d_0)}{d_0}\,|\theta^{(1)}_{d_0}|\right). \tag{6.11}$$

From (6.10) we see that

$$\theta_d = \sum_{\delta\,|\,d} \lambda_\delta = \theta^{(1)}_d - \theta^{(1)}_{d_0}\,\theta^{(2)}_{d/d_0}\,, \tag{6.12}$$

where again $\theta^{(2)}_{d/d_0}$ is interpreted as zero if d_0 does not divide d. Furthermore, we have

$$T_u(\Lambda) = T_u(\Lambda_1) - \frac{\theta^{(1)}_{d_0}}{f'(d_0)}\,T_u(\Lambda_2)\,. \tag{6.13}$$

If we instead have given a $\Lambda_1 = \Lambda_1[\mathcal{P}, z]$, and a $\Lambda_2 = \Lambda_2[\mathcal{P}_{d_0}, z/d_0]$, then the Λ resulting from (6.10) will be a $\Lambda[\mathcal{P}, z]$. (6.13) will still hold.

III. Suppose now that $\Lambda_1 = \Lambda^+_u(\mathcal{P}, z)$ and that for some $d_0 \in (\mathcal{P})$, $d_0 > 1$, but so that z/d_0 is large, we have that $\theta^{(1)}_{d_0} > 0$. If we have another sieve $\Lambda_2 = \Lambda^-_u(\mathcal{P}_{d_0}, z/d_0)$, then our construction (6.10) leads to a $\Lambda^+_u(\mathcal{P}, z')$ where z' is given by (6.11), for this new upper bound sieve $\theta_{d_0} = 0$ as can be seen from (6.12), and from (6.13) we see that the new sieve is an improvement over the original one if $T_u(\Lambda^-_2) > 0$, since we then get

$$T_u(\Lambda^+) < T_u(\Lambda^+_1)\,.$$

Similarly, if we had a $\Lambda_1 = \Lambda^-_u(\mathcal{P}, z)$ and a $d_0 > 1$ with $\theta^{(1)}_{d_0} < 0$, the construction (6.10) gives a $\Lambda = \Lambda^-_u(\mathcal{P}, z')$ such that

$$T_u(\Lambda^-) > T_u(\Lambda^-_1)\,,$$

if $T_u(\Lambda^-_2) > 0$. This principle applies in the same way to $\Lambda[\mathcal{P}, z]$'s.

IV. If we have a $\Lambda_1(\mathcal{P})$ which is "almost" a sieve except that some of the $\theta^{(1)}_d$ with $d \in (\mathcal{P})$ and not too large, have the wrong sign, we can make use of the construction (6.10) to correct the signs.

To illustrate this assume $\Lambda_1 = \Lambda_u(\mathcal{P}, z)$ is such that for $d \in (\mathcal{P})$ and $d > 1$, $\theta^{(1)}_d \le 0$ except for $d = d_0 > 1$, while $\theta^{(1)}_{d_0} > 0$. If $\Lambda_2 = \Lambda^+_u(\mathcal{P}_{d_0}, z/d_0)$ then (6.10) defines a $\Lambda^-_u(\mathcal{P}, z')$ such that $\theta_{d_0} = 0$. The same procedure can be used to produce an upper bound sieve from a Λ which has $\theta_d \ge 0$ except for $d = d_0$, while $\theta_{d_0} < 0$.

The procedure is equally applicable for Λ's of the type $\Lambda[\mathcal{P}, z]$. By repeated application of this principle one can correct the signs of several θ_d. The most frequent use of this principle is to extend the range of a lower bound sieve. If we have a sieve $\Lambda^-(\mathcal{P})$, and we apply it to the range $\mathcal{P}'$ that we get by adding one prime q which is not in $\mathcal{P}$, we see that over the new range all the θ_d for $d \in (\mathcal{P}')$ have the correct sign, except θ_q which equals 1 instead of being ≤ 0. Thus Principle IV applies with $d_0 = q$.

To extend the range of an upper bound sieve if we again add a prime q which is not in $\mathcal{P}$ to the range, is simpler. We may just use our $\Lambda^+(\mathcal{P})$ as a $\Lambda^+(\mathcal{P}')$, while it is true that this leads to $\theta_q = 1$ which is somewhat large, it does have the right sign. One sees that

$$T_u(\Lambda^+(\mathcal{P}')) = \frac{f(q)}{f'(q)}\, T_u(\Lambda^+(\mathcal{P}))\,.$$

If q is not too large, Principle III may be used to improve on the $\Lambda^+(\mathcal{P}')$, and the procedure is then quite similar to that used to extend the range of a lower bound sieve.

It is possible to do this extension of the range using the constructions given in I, but I does not cover the general situations considered in III and IV.

V. A general combinatorial principle for construction of sieves is found (though very differently expressed) already in Brun's fundamental paper [Brun 1]. If we define Λ recursively according to the following rules:

If $d = p_1 \dots p_r;\ p_1 < p_2 < \dots < p_r$ we define

$$\lambda_d = -\lambda_{d/p_1}\,, \qquad \text{for } r \text{ odd}, \tag{6.14}$$

and for r even we have the choice

$$\lambda_d = \begin{cases} -\lambda_{d/p_1} \\ 0 \end{cases} \tag{6.14$'$}$$

according to some convention.

We see that the λ_d that are not zero will have the value $\mu(d)$. For any $d > 1$ we can write $d = p_1\, d'$, where p_1 is the smallest prime dividing d, and we have

$$\theta_d = \sum_{\delta\,|\,d} \lambda_\delta = \sum_{\delta\,|\,d'} (\lambda_\delta + \lambda_{p_1\,\delta})\,. \tag{6.15}$$

From (6.14) and (6.14$'$) we easily see that each bracket $(\lambda_\delta + \lambda_{p_1\,\delta})$ is either zero or -1, thus $\theta_d \le 0$, so our Λ is a Λ^-.

Similarly, we may define a Λ recursively by the rules:

For $d = p_1 \dots p_r;\ p_1 < p_2 < \dots < p_r$ we define

$$\lambda_d = -\lambda_{d/p_1}\,, \qquad \text{for } r \text{ even}, \tag{6.16}$$

and for r odd, we have the choice

$$\lambda_d = \begin{cases} -\lambda_{d/p_1} \\ 0 \end{cases} \tag{6.16$'$}$$

according to some convention.

If we again for $d > 1$ write θ_d as in (6.15) we see now that the brackets $(\lambda_\delta + \lambda_{p_1\delta})$ are either zero or 1, so that $\theta_d \geq 0$, and our Λ is a Λ^+.

For the Λ^+ construction, we can clearly make the Λ^+ a $\Lambda^+[\mathcal{P}, z]$ by simply requiring that all λ_d with $d > z$ vanish in addition to (6.16) and (6.16′); while for the Λ^- construction we can make a $\Lambda^-[\mathcal{P}, z]$ by imposing this requirement if and only if $\mathcal{P}$ contains no prime larger than z.

This construction is of an extremely general nature, it becomes useful only when definite criteria are given for the choices to be made in (6.14′) or (6.16′). We shall consider that question in Section 11.

7. Preliminary results using Λ^2 and $\Lambda^2\Lambda^-$ sieves

For use in the later sections we shall need to know that it is possible to construct $\Lambda^+[\mathcal{P}(x^\eta), x^\delta]$ and $\Lambda^-[\mathcal{P}(x^\eta), x^\delta]$ for which $T_u(\Lambda^+)$ and $T_u(\Lambda^-)$ differ very little from 1 if both η and δ are small, but such that δ/η is large. We shall first consider the Λ^+ case. For brevity we shall write $\xi = x^\eta$ and $z = x^{\delta/2}$. We now consider a $\Lambda = \Lambda(\mathcal{P}(\xi))$ with $\lambda_1 = 1$ and $\lambda_d = 0$ for $d > z$, while the other λ_d for $d \in (\mathcal{P})$ are left to our free disposal. We form $\Lambda' = \Lambda^2$ and find that the λ'_d are of the form

$$\lambda'_d = \sum_{[d_1, d_2] = d} \lambda_{d_1} \lambda_{d_2} ,$$

we therefore get

$$\sum_d \frac{u(d)}{d} \lambda'_d = \sum_{d_1, d_2} \frac{\lambda_{d_1} \lambda_{d_2}}{f[d_1, d_2]} , \tag{7.1}$$

where we for convenience here and later write $f[d_1, d_2]$ for $f([d_1, d_2])$, where $[d_1, d_2]$ as before denotes the smallest common multiple. We shall similarly write $f(d_1, d_2)$ for $f((d_1, d_2))$, where (d_1, d_2) denotes the greatest common divisor of d_1 and d_2. We transform (7.1) further

$$\begin{aligned} \sum_d \frac{u(d)}{d} \lambda'_d &= \sum_{d_1, d_2} \frac{\lambda_{d_1} \lambda_{d_2}}{f(d_1)\, f(d_2)}\, f(d_1, d_2) \\ &= \sum_{d_1, d_2} \frac{\lambda_{d_1}}{f(d_1)} \frac{\lambda_{d_2}}{f(d_2)} \sum_{\substack{\rho \mid d_1 \\ \rho \mid d_2}} f'(\rho) \\ &= \sum_\rho f'(\rho) \left\{ \sum_{\rho \mid d} \frac{\lambda_d}{f(d)} \right\}^2 . \end{aligned} \tag{7.2}$$

We now introduce the notation

$$\sum_{\rho \mid d} \frac{\lambda_d}{f(d)} = \mu(\rho) \frac{y_\rho}{f'(\rho)} , \tag{7.3}$$

which implies

$$\sum_{d \mid \rho} \frac{y_\rho}{f'(\rho)} = \mu(d)\, \frac{\lambda_d}{f(d)}\,, \tag{7.3$'$}$$

and in particular

$$\sum_{\rho} \frac{y_\rho}{f'(\rho)} = \lambda_1 = 1\,. \tag{7.3$''$}$$

Clearly our convention that $\lambda_d = 0$ for $d > z$, implies $y_\rho = 0$ for $\rho > z$ and vice versa, and we may regard the y_ρ as our variables to be chosen, subject only to the restriction (7.3$''$), instead of the λ_d's. (7.2) now takes the form

$$\sum_{d} \frac{u(d)}{d}\, \lambda_d' = \sum_{\rho} \frac{y_\rho^2}{f'(\rho)}\,, \tag{7.4}$$

and we now determine the y_ρ so as to make $\sum_\rho y_\rho^2/f'(\rho)$ a minimum subject to the condition (7.3$''$). We find that the minimizing y_ρ for $\rho \in (\mathcal{P})$ and $\rho \le z$ are given by

$$y_\rho = \frac{1}{\sum_z}\,, \tag{7.5}$$

where we have written

$$\sum\nolimits_z = \sum_{\substack{\rho \le z \\ \rho \in (\mathcal{P})}} \frac{1}{f'(\rho)}\,. \tag{7.6}$$

This gives

$$\sum_{d} \lambda_d' \frac{u(d)}{d} = \frac{1}{\sum_z}\,. \tag{7.7}$$

From this we see that

$$T_u(\Lambda^2) = \frac{1}{e_u(\xi) \sum_z}\,, \tag{7.8}$$

so to find an upper bound for $T_u(\Lambda^2)$ we need to find a lower bound for

$$e_u(\xi) \sum\nolimits_z\,.$$

We have, if $\Delta > 0$, that

$$\begin{aligned} e_u(\xi) \sum\nolimits_z &> e_u(\xi) \left\{ \sum_{d} \frac{1}{f'(d)} - z^{-\Delta} \sum_{d} \frac{d^\Delta}{f'(d)} \right\} \\ &= 1 - z^{-\Delta}\, e_u(\xi) \prod_{p \le \xi} \left(1 + \frac{p^\Delta}{f'(p)}\right) \\ &= 1 - z^{-\Delta} \prod_{p \le \xi} \left(1 + \frac{p^\Delta}{f(p) - 1}\right) \left(1 - \frac{1}{f(p)}\right) \\ &= 1 - z^{-\Delta} \prod_{p \le \xi} \left(1 + \frac{p^\Delta - 1}{f(p)}\right)\,. \end{aligned} \tag{7.9}$$

Here

$$z^{-\Delta}\prod_{p\le\xi}\left(1+\frac{p^{\Delta}-1}{f(p)}\right)<\exp\left(-\Delta\,\log z+\sum_{p\le\xi}\frac{p^{\Delta}-1}{f(p)}\right)$$
$$<\exp\left(-\Delta\,\log z+\frac{\xi^{\Delta}-1}{\log\xi}\sum_{p\le\xi}\frac{\log p}{f(p)}\right)$$
$$=\exp\left(-\Delta\,\log z+\frac{\xi^{\Delta}-1}{\log\xi}\,S_u(\xi)\right).$$

If $\log z > S_u(\xi)$, we may now choose $\Delta>0$ so that $\xi^{\Delta}=\log z/S_u(\xi)$, and get

$$z^{-\Delta}\prod_{p\le\xi}\left(1+\frac{p^{\Delta}-1}{f(p)}\right)<\exp\left(-\frac{\log z}{\log\xi}\log\frac{\log z}{e\,S_u(\xi)}-\frac{S_u(\xi)}{\log\xi}\right). \tag{7.10}$$

Combining this with (7.9) and (7.8) we get

$$T_u(\Lambda^2)<\left(1-\exp\left(-\frac{\log z}{\log\xi}\log\frac{\log z}{e\,S_u(\xi)}-\frac{S_u(\xi)}{\log\xi}\right)\right)^{-1} \tag{7.11}$$

for $S_u(\xi)<\log z$. From (5.4) we get a bound for $R_u(\Lambda^2)$ as

$$R_u(\Lambda^2)<z^2\,e_u^{-1}(\xi)\,T_u(\Lambda^2)\,. \tag{7.11$'$}$$

We next turn to the lower bound and the $\Lambda^2\,\Lambda^-$ sieve. Let $\xi=x^{\eta}$ as before, $\zeta=x^{\eta'}$ with $\eta'\ge\eta$; and $z=x^{(\delta-\eta')/2}$, where δ now is supposed to be $>\eta'$.

We define Λ as before and assume that we have a lower bound sieve $\Lambda'^-=\Lambda'^-[\mathcal{P}(\xi),x^{\eta'}]$ and form the sieve

$$\Lambda''^-=\Lambda^2\,\Lambda'^-\,,$$

this will be a $\Lambda''^-[\mathcal{P}(\xi),x^{\delta}]$ sieve and the λ''_d are given by

$$\lambda''_d=\sum_{[d_1,d_2,d_3]=d}\lambda'_{d_1}\,\lambda_{d_2}\,\lambda_{d_3}\,,$$

where the square bracket again denotes the least common multiple of the numbers inside. We get from this

$$\sum_d\frac{u(d)}{d}\,\lambda''_d=\sum_{d_1,d_2,d_3}\frac{\lambda'_{d_1}\,\lambda_{d_2}\,\lambda_{d_3}}{f[d_1,d_2,d_3]} \tag{7.12}$$
$$=\sum_{d_1}\frac{\lambda'_{d_1}}{f(d_1)}\sum_{d_2,d_3}\frac{\lambda_{d_2}\,\lambda_{d_3}}{f_{d_1}[d_2,d_3]}\,.$$

Here and in the following we use the notation $f_a(d)=f[a,d]/f(a)$, it is seen that $f_a(d)$ is multiplicative, and that $f_a(p)=f(p)$ for $(p,a)=1$ while

$f_a(p) = 1$ for $p \mid a$. We use $f'_a(d)$ to denote the multiplicative function defined by $f'_a(p) = f_a(p) - 1$ so it is seen that $f'_a(d) = f'(d)$ if $(d,a) = 1$ while $f'_a(d) = 0$ for $(d,a) > 1$.

The inner double sum on the right hand side of (7.12) is built quite analogously to the right hand side of (7.1) so we can transform it in the same way and get

$$\sum_{d_2,d_3} \frac{\lambda_{d_2}\lambda_{d_3}}{f_{d_1}[d_2,d_3]} = \sum_{\rho} f'_{d_1}(\rho) \left\{ \sum_{\rho \mid d} \frac{\lambda_d}{f_{d_1}(d)} \right\}^2$$
$$= \sum_{(\rho,d_1)=1} f'(\rho) \left\{ \sum_{\rho \mid d} \frac{\lambda_d}{f_{d_1}(d)} \right\}^2,$$

noting here that $f_{d_1}(d) = f(d)/f(d_1,d)$, and using

$$f(d_1,d) = \sum_{\substack{\delta \mid d_1 \\ \delta \mid d}} f'(\delta),$$

we see that for $(\rho, d_1) = 1$,

$$\sum_{\rho \mid d} \frac{\lambda_d}{f_{d_1}(d)} = \sum_{\rho \mid d} \frac{f(d_1,d)}{f(d)} \lambda_d$$
$$= \sum_{\delta \mid d_1} f'(\delta) \sum_{\delta\rho \mid d} \frac{\lambda_d}{f(d)}$$
$$= \frac{\mu(\rho)}{f'(\rho)} \sum_{\delta \mid d_1} \mu(\delta)\, y_{\rho\delta},$$

using the notation of (7.3). From this we now get

$$\sum_{d_2,d_3} \frac{\lambda_{d_2}\lambda_{d_3}}{f_{d_1}[d_2,d_3]} = \sum_{(\rho,d_1)=1} \frac{1}{f'(\rho)} \left\{ \sum_{\delta \mid d_1} \mu(\delta)\, y_{\rho\delta} \right\}^2,$$

and so finally from (7.12)

$$\sum_{d} \frac{u(d)}{d} \lambda''_d = \sum_{d_1} \frac{\lambda'_{d_1}}{f(d_1)} \sum_{(\rho,d_1)=1} \frac{1}{f'(\rho)} \left\{ \sum_{\delta \mid d_1} \mu(\delta)\, y_{\rho\delta} \right\}^2. \tag{7.13}$$

For our present purposes our $\Lambda'^{-}[\mathcal{P}(\xi), x^{\eta'}]$ is to be given by $\lambda'_1 = 1$, $\lambda'_p = -1$ for $p < \xi$ and $\lambda_d = 0$ for all other d, thus $\eta' = \eta$; and (7.13) takes with this choice the form

$$\sum_{d} \frac{u(d)}{d} \lambda''_d = \sum_{\rho \le z} \frac{y_\rho^2}{f'(\rho)} - \sum_{\substack{p<\xi \\ (p,\rho)=1}} \frac{1}{f(p)\, f'(\rho)} \{y_\rho - y_{\rho p}\}^2. \tag{7.13'}$$

As before the fact that $\lambda_d = 0$ for $d > z$ implies that $y_\rho = 0$ for $\rho > z$ and vice versa. If we on the right hand side of (7.13′) write $\frac{y_\rho}{\sum_d y_d/f'(d)}$ instead of y_ρ, we see that we can drop the condition $\sum_\rho y_\rho/f'(\rho) = 1$, and the right hand side of (7.13′) becomes

$$\frac{\displaystyle\sum_\rho \frac{y_\rho^2}{f'(\rho)} - \sum_{\substack{p<\xi \\ (\rho,p)=1}} \frac{1}{f(p)\,f'(\rho)} \{y_\rho - y_{\rho p}\}^2}{\left(\displaystyle\sum_\rho \frac{y_\rho}{f'(\rho)}\right)^2}. \tag{7.14}$$

Here we choose y_ρ as follows:

$$y_\rho = \begin{cases} 1 & \text{for } \rho \le z/\xi, \\ \frac{\log(z/\rho)}{\log \xi} & \text{for } z/\xi \le \rho \le z, \\ 0 & \text{for } \rho \ge z. \end{cases}$$

We see then that $y_\rho - y_{\rho p} = 0$ for $\rho \le z/\xi^2$, and that for $z/\xi^2 \le \rho \le z$, $0 \le y_\rho - y_{\rho p} \le \frac{\log p}{\log \xi}$. Also we have

$$\sum_\rho \frac{y_\rho}{f'(\rho)} < \sum_\rho \frac{1}{f'(\rho)} = e_u^{-1}(\xi)\,, \tag{7.15}$$

and for the first term in the numerator of (7.14) we have that

$$\sum_\rho \frac{y_\rho^2}{f'(\rho)} > \sum_{\rho<z/\xi} \frac{1}{f'(\rho)} > e_u^{-1}(\xi) - \sum_{\rho>z/\xi^2} \frac{1}{f'(\rho)}\,.$$

For the second term in the numerator we get the upper bound

$$\begin{aligned} \frac{1}{\log^2 \xi} \sum_{p<\xi} \frac{\log^2 p}{f'(p)} \sum_{z/\xi^2<\rho<z} \frac{1}{f'(\rho)} &< \frac{1}{\log \xi} \sum_{p<\xi} \frac{\log p}{f(p)} \sum_{\rho>z/\xi^2} \frac{1}{f'(\rho)} \\ &= \frac{S_u(\xi)}{\log \xi} \sum_{\rho>z/\xi^2} \frac{1}{f'(\rho)}\,. \end{aligned}$$

For the numerator in (7.14) we thus get the lower bound

$$e_u^{-1}(\xi) - \left(1 + \frac{S_u(\xi)}{\log \xi}\right) \sum_{\rho>z/\xi^2} \frac{1}{f'(\rho)}\,. \tag{7.16}$$

Here we have, for $\Delta > 0$ that

$$\begin{aligned} \sum_{\rho>z/\xi^2} \frac{1}{f'(\rho)} &< (z/\xi^2)^{-\Delta} \sum_\rho \frac{\rho^\Delta}{f'(\rho)} \\ &= (z/\xi^2)^{-\Delta} \prod_{p<\xi} \left(1 + \frac{p^\Delta - 1}{f(p)}\right) \cdot e_u^{-1}(\xi)\,, \end{aligned}$$

apart from the term $e_u^{-1}(\xi)$ this expression is identical to the one we estimated in the case of the Λ^2 sieve except that we have z/ξ^2 instead of z, so we may use (7.10) with z/ξ^2 instead of z, provided $\log z/\xi^2 > S_u(\xi)$. Inserting the resulting estimation in (7.16) we get the lower bound for the numerator of (7.14)

$$(7.17)\quad e_u^{-1}(\xi)\left(1-\left(1+\frac{S_u(\xi)}{\log\xi}\right)\exp\left(-\frac{\log z/\xi^2}{\log\xi}\log\frac{\log z/\xi^2}{e\,S_u(\xi)}-\frac{S_u(\xi)}{\log\xi}\right)\right)$$

and, provided this is not negative, we get using (7.15), the lower bound for (7.14) and also for (7.13′):

(7.18)

$$\sum_d \frac{u(d)}{d}\lambda_d'' > e_u(\xi)\left(1-\left(1+\frac{S_u(\xi)}{\log\xi}\right)\exp\left(-\frac{\log z/\xi^2}{\log\xi}\log\frac{\log z/\xi^2}{e\,S_u(\xi)}-\frac{S_u(\xi)}{\log\xi}\right)\right)$$

and for our $\Lambda''^- = \Lambda^2\,\Lambda'^-$

$$(7.19)\quad T_u(\Lambda''^-) > 1-\left(1+\frac{S_u(\xi)}{\log\xi}\right)\exp\left(-\frac{\log z/\xi^2}{\log\xi}\log\frac{\log z/\xi^2}{e\,S_u(\xi)}-\frac{S_u(\xi)}{\log\xi}\right)$$

valid for $\log z/\xi^2 > S_u(\xi)$, provided the right hand side is not negative. From (5.6) we get since our sieve Λ'' is a $\Lambda[\mathcal{P},\, z^2\,\xi]$ that

$$(7.19')\qquad R_u(\Lambda'') < 2\,z^2\,\xi\,e_u^{-1}(\xi)\,.$$

We note that the estimations (7.11), (7.11′), and (7.19), (7.19′), are obtained using only the value of $S_u(\xi)$ and do not require the assumption of (4.2) or anything else about the distribution of the $u(p)$. In a later section we shall make more precise estimations of the type of sums involved by making use of (4.2), but the above results suffice for the applications we now have in mind.

Using in case of the Λ^2 sieve the assumption $\xi = x^\eta$ and $z = x^{\delta/2}$ as well as (4.2) we obtain from (7.11) as $x \to \infty$

$$(7.20)\qquad \limsup_{x\to\infty} T_u(\Lambda^2) \le \left(1-\exp\left(-\frac{\delta}{2\eta}\log\frac{\delta}{2\,e\,g(\eta)}-\frac{g(\eta)}{\eta}\right)\right)^{-1}$$

and from (7.19) for the $\Lambda^2\,\Lambda^-$ sieve, using (4.2) and the assumptions $\xi = x^\eta$, $z = x^{(\delta-\eta)/2}$, that

(7.20′)

$$\liminf_{x\to\infty} T_u(\Lambda^2\,\Lambda^-) \ge 1-\left(1+\frac{g(\eta)}{\eta}\right)\exp\left(-\frac{\delta-5\,\eta}{2\eta}\log\frac{\delta-5\,\eta}{2\,e\,g(\eta)}-\frac{g(\eta)}{\eta}\right).$$

Both sieves are of the type $\Lambda[\mathcal{P}(x^\eta),\, x^\delta]$. (7.20) is valid for $\delta > 2\,g(\eta)$, and (7.20′) for $\delta > 2\,g(\eta)+5\,\eta$, always assuming that the second term on the right hand side of (7.20′) is ≤ 1.

8. General theory and existence theorems

We now introduce the notation for $\alpha \geq 0$,

$$\liminf_{x\to\infty} T_u(\Lambda^+) = T_g^+(\alpha)\,, \tag{8.1}$$

where $\Lambda^+ = \Lambda_u^+(\mathcal{P}(x^\alpha),\, x)$, and for $0 \leq \alpha \leq 1$,

$$\limsup_{x\to\infty} T_u(\Lambda^-) = T_g^-(\alpha)\,, \tag{8.1$'$}$$

where $\Lambda^- = \Lambda_u^-(\mathcal{P}(x^\alpha),\, x)$.

In the same way, we define for $0 < \eta < \alpha$,

$$\liminf_{x\to\infty} T_u(\Lambda^+) = T_g^+(\eta,\alpha)\,, \tag{8.1$''$}$$

where $\Lambda^+ = \Lambda_u^+(\mathcal{P}(x^\eta,\, x^\alpha),\, x)$, and for $0 < \eta < \alpha \leq 1$

$$\limsup_{x\to\infty} T_u(\Lambda^-) = T_g^-(\eta,\alpha)\,, \tag{8.1$'''$}$$

where $\Lambda^- = \Lambda_u^-(\mathcal{P}(x^\eta,\, x^\alpha),\, x)$.

We here consider as $x \to \infty$, all possible sequences of $u(p)$ that satisfy (4.2) and the requirements $0 \leq u(p) < p$, as well as (4.4), and that $g(\beta)$ and $g'(\beta)$ satisfy the requirements specified in Section 4. These requirements imply that $g'(\beta)$ is bounded in any interval $0 < \eta \leq \beta \leq \alpha$, while on the other hand $g'(\beta)$ may well tend to infinity as $\beta \to 0$.

That (8.1$''$) and (8.1$'''$) define finite quantities is seen by using trivial sieves: $\Lambda^+(\mathcal{P})$ with $\lambda_1 = 1$ and all other $\lambda_d = 0$; and the $\Lambda^-(\mathcal{P})$ with $\lambda_1 = 1$, $\lambda_p = -1$ for $p \in \mathcal{P}$ and $\lambda_d = 0$ for all other p. That the last sieve is of the type $\Lambda_u^-(\mathcal{P}(x^\eta,\, x^\alpha),\, x)$ follows from (4.2), since we have for $\beta < 1$ that

$$\begin{aligned}\sum_{p<x} u(p) &\leq \frac{x^\beta}{\beta \log x}\, S_u(x^\beta) + \frac{x}{\log x}\,(S_u(x) - S_u(x^\beta)) \\ &= (g(1) - g(\beta))\,x + o(x)\,.\end{aligned}$$

Letting here $\beta \to 1$ we get since g is continuous

$$\sum_{p\leq x} u(p) = o(x)\,.$$

That (8.1) defines a finite quantity can be concluded from (7.20), since we may use the Λ^+ sieve constructed there for the range $\mathcal{P}(x^\eta)$ and some $\delta < 1$ over the wider range $\mathcal{P}(x^\alpha)$ as a $\Lambda^+[\mathcal{P}(x^\alpha), x^\delta]$, this only brings in a factor $e_u(x^\eta)/e_u(x^\alpha)$ in the $T_u(\Lambda^+)$. In the case of (8.1$'$), for α sufficiently small we can use the sieve constructed to get (7.20$'$), for α larger but < 1, we may use Principle IV from Section 6 to extend the range of our $\Lambda^-[\mathcal{P}(x^\eta),\, x^\delta]$ successively over the remaining range $\mathcal{P}(x^\eta,\, x^\alpha)$, for each prime $p < x^\alpha$ that we add to the range we use as $\Lambda_2 = \Lambda^+[\mathcal{P}(p), (x/p)^{2/3}]$, a sieve constructed like that used in (7.20) $\Lambda^+[\mathcal{P}((x/p)^\eta), (x/p)^{2/3}]$ over the range $\mathcal{P}(p)$. This again shows that for $\alpha < 1$, (8.1′) defines a finite quantity.

Lemma 1. *We have*

(i) $T_g^+(\alpha)$ *is continuous in* α *for* $0 \le \alpha$,

(ii) $T_g^-(\alpha)$ *is continuous in* α *for* $0 \le \alpha < 1$,

(iii) $T_g^+(\eta, \alpha)$ *is continuous in both* η *and* α *for* $0 < \eta \le \alpha$,

(iv) $T_g^-(\eta, \alpha)$ *is continuous in both* η *and* α *for* $0 < \eta \le \alpha \le 1$,

(v) if $\limsup_{\eta\to 0} g(\eta))/\eta < 1$, $T_g^-(\alpha)$ *exists also for* $\alpha = 1$ *and is continuous for* $0 \le \alpha \le 1$.

Let $0 < \alpha_1 < \alpha_2$, from the definition of $T_g^+(\alpha_2)$ it follows that there exists a sequence of $x \to \infty$ such that for each x in this sequence there is a set of $u(p)$ satisfying our conditions and a $\Lambda_u^+(\mathcal{P}(x^{\alpha_2}), x)$ such that when x tends to ∞ through this sequence

$$\lim T_u(\Lambda_u^+(\mathcal{P}(x^{\alpha_2}), x)) = T_g^+(\alpha_2)\,.$$

If we for each such x take the same set of $u(p)$ and restrict the sieve $\Lambda_u^+(\mathcal{P}(x^{\alpha_2}), x)$ to the range $\mathcal{P}(x^{\alpha_1})$ by omitting the λ_d where d does not belong to $(\mathcal{P}(x^{\alpha_1}))$ we have

$$T_u(\Lambda_u^+(\mathcal{P}(x^{\alpha_1}), x)) \le T_u(\Lambda_u^+(\mathcal{P}(x^{\alpha_2}), x))\,.$$

Letting now $x \to \infty$ through our sequence we see that

$$T_g^+(\alpha_1) \le T_g^+(\alpha_2)\,. \tag{8.2}$$

Again from the definition of $T_g^+(\alpha_1)$ we have that there exists a sequence of x tending to ∞ such that with each x in the sequence is associated a set of $u(p)$ and a sieve $\Lambda_u^+(\mathcal{P}(x^{\alpha_1}), x)$ and such that when $x \to \infty$ through this sequence

$$\lim T_u(\Lambda_u^+(\mathcal{P}(x^{\alpha_1}), x)) = T_g^+(\alpha_1)\,.$$

We may use this sieve unchanged for the larger range as a $\Lambda_u^+(\mathcal{P}(x^{\alpha_2}), x)$, and we see that

$$T_u(\Lambda_u^+(\mathcal{P}(x^{\alpha_2}), x)) = \frac{e_u(x^{\alpha_1})}{e_u(x^{\alpha_2})} T_u(\Lambda_u^+(\mathcal{P}(x^{\alpha_1}), x))\,.$$

As we let $x \to \infty$ through this sequence we get using (4.3), that

$$T_g^+(\alpha_2) \le \exp\left(\int_{\alpha_1}^{\alpha_2} \frac{dg(t)}{t}\right) T_g^+(\alpha_1)\,. \tag{8.3}$$

From (8.2) and (8.3) the continuity of $T_g^+(\alpha)$ is clear for $\alpha > 0$. From our result (7.20) it is evident that $T_g^+(\alpha) < 1 + \varepsilon$ for any fixed $\varepsilon > 0$ for $\alpha \le$

$\eta(\varepsilon) > 0$. Since $T_g^+(0) = 1$ by the definition and $T_g^+(\alpha)$ always is ≥ 1, the continuity at $\alpha = 0$ follows. This proves (i).

For $T_g^+(\eta, \alpha)$ arguments similar to those above give if $0 < \eta_2 \leq \eta_1 \leq \alpha_1 \leq \alpha_2$, that

$$T_g^+(\eta_1, \alpha_1) \leq T_g^+(\eta_2, \alpha_2)\,, \tag{8.4}$$

and

$$\exp\left(-\int_{\eta_1}^{\alpha_1} \frac{dg(t)}{t}\right) T_g^+(\eta_1, \alpha_1) \geq \exp\left(-\int_{\eta_2}^{\alpha_2} \frac{dg(t)}{t}\right) T_g^+(\eta_2, \alpha_2)\,. \tag{8.5}$$

From these statements the continuity in η and α claimed in (iii) follows.

Turning to $T_g^-(\alpha)$, we can show in much the same way as we showed (8.2) that if $0 < \alpha_1 \leq \alpha_2 < 1$, we have

$$T_g^-(\alpha_1) \geq T_g^-(\alpha_2)\,. \tag{8.6}$$

To obtain an inequality which in combination with (8.6) will prove (ii), we have to proceed in a rather different way and use Principle IV of Section 6.

By definition, there exists a sequence of $x \to \infty$ such that with each such x is associated a set of $u(p)$ satisfying our requirements and a $\Lambda_1^- = \Lambda_u^-(\mathcal{P}(x^{\alpha_1}), x)$ such that as $x \to \infty$ through this sequence, $\lim T_u(\Lambda_1^-) = T_g^-(\alpha_1)$. We now use Principle IV to increase the range of this sieve by adding the range $\mathcal{P}(x^{\alpha_1}, (2\,x)^{\alpha_2})$ and using this time a sequence of $x' = 2\,x$ tending to ∞, and keep the set of $u(p)$ associated with x as associated with x'. We extend the range of our Λ_1^- from the range $\mathcal{P}(x^{\alpha_1})$ to the range $\mathcal{P}(x'^{\alpha_2})$ by for each prime p with $x^{\alpha_1} \leq p \leq (2\,x)^{\alpha_2}$ taking as $\Lambda_2^+ = \Lambda^+[\mathcal{P}(p), (x'/p)^{2/3}]$ a sieve constructed like that used in (7.20) $\Lambda^+[\mathcal{P}((x'/p)^{\eta}), (x'/p)^{2/3}]$ using it over the larger range $\mathcal{P}(p)$ (since η will be chosen small, we can assume $p > (x'/p)^{\eta}$ for $p \geq x^{\alpha_1}$).

The λ_d in our $\Lambda^-(\mathcal{P}(x'^{\alpha_2}))$ are thus given by

$$\lambda_d = \lambda_d^{(1)} - \sum_{x^{\alpha_1} \leq p < (2\,x)^{\alpha_2}} \lambda_{d/p}^{(2)}\,,$$

where the $\lambda_d^{(1)}$ are taken from our Λ_1^- and for each p the $\lambda_{d/p}^{(2)}$ are taken from the Λ_2^+ associated with p. Here terms are taken as zero when they are not defined. We now have

(8.7)

$$\begin{aligned}
&T_u(\Lambda^-(\mathcal{P}(x'^{\alpha_2})))\\
&= T_u(\Lambda^-(\mathcal{P}(x^{\alpha_1})))\, e_u^{-1}(x^{\alpha_1}, x'^{\alpha_2})\\
&\qquad - \sum_{x^{\alpha_1} \leq p < x'^{\alpha_2}} \frac{1}{f(p)}\, e_u^{-1}\left((x'/p)^{\eta}, x'^{\alpha_2}\right) T_u(\Lambda^+[\mathcal{P}((x'/p)^{\eta}),\, (x'/p)^{2/3}])\,.
\end{aligned}$$

It is easily seen that the $\Lambda^-(\mathcal{P}(x'^{\alpha_2}))$ so constructed has $R_u(\Lambda^-) \le x+o(x) \le x'$ for x large enough if α_2 is bounded away from 1, say $\alpha_2 \ge 1-\Delta$ with some $\Delta > 0$.

Letting now x' tend to infinity through the sequence $2x$, and using (4.3) and (7.20) with η chosen so small that $g(\eta\,\Delta) \le \Delta/30$, we get

$$(8.8) \qquad T_g^-(\alpha_2) \ge T_g^-(\alpha_1)\exp\left(\int_{\alpha_1}^{\alpha_2} \frac{dg(t)}{t}\right) - \exp\left(\int_{\eta\,\Delta}^{\alpha_2} \frac{dg(t)}{t}\right)\int_{\alpha_1}^{\alpha_2} \frac{dg(t)}{t}\,(1-e^{-1/(3\,\eta)})^{-1}\,.$$

(8.8) combined with (8.6) proves the continuity of $T_g^-(\alpha)$ for $0 < \alpha < 1$.

By definition we have $T_g^-(0) = 1$ and $T_g^-(\alpha) \le 1$ and (7.20′) shows that as $\alpha \to 0$, $T_g^-(\alpha) \to 1$. Thus (ii) is proved.

For $T_g^-(\eta,\alpha)$ we use similar arguments as for $T_g^-(\alpha)$, but the situation is simpler since $T_g(\eta,\alpha)$ always exists for $\alpha = 1$. We get for $0 < \eta_2 \le \eta_1 \le \alpha_1 \le \alpha_2 \le 1$, that

$$T_g^-(\eta_1,\alpha_1) \ge T_g^-(\eta_2,\alpha_2)\,.$$

In a construction similar to that which led to (8.8) we this time simply use the Λ_2^+ which has $\lambda_1^{(2)} = 1$ and all other $\lambda_d^{(2)} = 0$. In this way we obtain, again for $0 < \eta_2 \le \eta_1 \le \alpha_1 \le \alpha_2 \le 1$, the inequality

$$T_g^-(\eta_2,\alpha_2) \ge \exp\left\{\int_{\eta_2}^{\eta_1} \frac{dg(t)}{t} + \int_{\alpha_1}^{\alpha_2} \frac{dg(t)}{t}\right\}\left(T_g^-(\eta_1,\alpha_1) - \int_{\eta_2}^{\eta_1} \frac{dg(t)}{t} - \int_{\alpha_1}^{\alpha_2} \frac{dg(t)}{t}\right)$$

This inequality combined with the one given above, proves (iv).

It remains to consider (v). If we for $\alpha \le 1$ have a sequence $x \to \infty$ and associated with each x in the sequence a set of $u(p)$ and a $\Lambda_u^-(\mathcal{P}(x^\alpha), x)$ such that when x goes to ∞ through this sequence

$$\lim T_u(\Lambda_u^-(\mathcal{P}(x^\alpha), x)) = T_g^-(\alpha)\,,$$

we clearly can modify the sequence of $u(p)$ by decreasing or omitting some of them, as long as we do not interfere with the validity of (4.2), that is we should not lower the expression $S_u(x)$ by more than an amount which is $o(\log x)$ as $x \to \infty$. Such a modification can only raise the $T_u(\Lambda_u^-(x^\alpha), x)$, so by the definition of $T_g^-(\alpha)$, they would still have to converge to the same limit.

The assumption

$$\limsup_{\beta\to 0} \frac{g(\beta)}{\beta} < 1\,,$$

is seen to imply that there exists a positive β_0 and a $\theta < 1$, such that for a $u(p)$ sequence so modified,

$$(8.9) \qquad S_u(\xi) < \theta \log \xi\,,$$

for $1 < \xi \le x'^{\beta_0}$, and an a such that

$$S_u(\xi) < a \log \xi \,, \tag{8.9$'$}$$

for $1 < \xi \le x'$.

If we now go back to (8.7) and under the new assumption make a more careful estimation of the right hand side, using (7.11) to give an upper bound for the $T_u(\Lambda^+)$ we get first

$$T_u(\Lambda^+) < (1 - e^{-\frac{1}{3\eta}\log\frac{1}{3ea\eta}})^{-1} < 2\,, \tag{8.10}$$

for $\eta \le \eta_0$. Next we consider for $1 < \xi \le (x'/2)^\eta$ with $\eta \le \beta_0$ the expression

$$\log(e_u(\xi,\, x'^{\alpha_2}))^{-1} \le \sum_{\xi \le p < x'} \log\left(1 - \frac{u(p)}{p}\right)^{-1} .$$

In accordance with the remarks made above about modifying the set of $u(p)$'s by omitting some of them, we see easily that if we for some fixed positive δ omit the $u(p)$ for which $u(p)/p > \delta$, we do not interfere with the validity of (4.2). Thus

$$\begin{aligned} \log e_u(\xi,\, x'^{\alpha_2})^{-1} &< \frac{1}{1-\delta} \sum_{\xi \le p < x'} \frac{u(p)}{p} = \frac{1}{1-\delta}\int_\xi^{x'} \frac{d\,S_u(t)}{\log t} \\ &= \frac{1}{1-\delta}\left(\frac{S_u(x')}{\log x'} - \frac{S_u(\xi)}{\log \xi} + \int_\xi^{x'} \frac{S_u(t)}{t\log^2 t}\,dt\right) . \end{aligned}$$

Using here (8.9) and (8.9$'$), we get

$$\log e_u(\xi,\, x'^{\alpha_2})^{-1} < \frac{1}{1-\delta}\left(a' + \theta \int_\xi^{x'} \frac{dt}{t\,\log t}\right) < a'' + \theta' \log \frac{\log x'}{\log \xi}\,,$$

where if we have chosen δ small enough, $\theta' < 1$. Thus

$$e_u(\xi,\, x'^{\alpha_2})^{-1} < A\left(\frac{\log x'}{\log \xi}\right)^{\theta'} ,$$

for $1 < \xi < (x'/2)^\eta$ and $\eta \le \min(\eta_0, \beta_0)$. Using the above and (8.10), (8.7) now takes the form

$$\begin{aligned} T_u(\Lambda_u^-(\mathcal{P}(x'^{\alpha_2}))) &> T_u(\Lambda_u^-(\mathcal{P}(x^{\alpha_1})))\,\frac{1}{e_u(x^{\alpha_1}, x'^{\alpha_2})} \\ &\quad - 2\,\frac{A}{\eta^{\theta'}} \sum_{x^{\alpha_1} \le p < x'^{\alpha_2}} \frac{u(p)}{p}\left(1 - \frac{\log p}{\log x'}\right)^{-\theta'} . \end{aligned} \tag{8.11}$$

We have now
(8.12)

$$\sum_{x^{\alpha_1}\le p<x'^{\alpha_2}} \frac{u(p)}{p}\left(1-\frac{\log p}{\log x'}\right)^{-\theta'} = \int_{x^{\alpha_1}}^{x'^{\alpha_2}} \frac{d(S_u(t)-S_u(x'^{\alpha_2}))}{\log t\left(1-\dfrac{\log t}{\log x'}\right)^{\theta'}}$$

$$< \frac{S_u(x'^{\alpha_2})-S_u(x^{\alpha_1})}{\alpha_1\,\log x\left(1-\alpha_1\dfrac{\log x}{\log x'}\right)^{\theta'}} + \frac{\theta'}{\log x'}\int_{x^{\alpha_1}}^{x'^{\alpha_2}} \frac{S_u(x'^{\alpha_2})-S_u(t)}{t\left(1-\dfrac{\log t}{\log x'}\right)^{\theta'+1}\log t}\,dt$$

$$= \frac{S_u(x'^{\alpha_2})-S_u(x^{\alpha_1})}{\alpha_1\,\log x\left(1-\alpha_1\dfrac{\log x}{\log x'}\right)^{\theta'}} + \frac{\theta'}{\log x'}\int_{\alpha_1(1-\log 2/\log x')}^{\alpha_2} \frac{S_u(x'^{\alpha_2})-S_u(x'^{v})}{v\,(1-v)^{\theta'+1}}\,dv\,.$$

For $\alpha_2 < 1$, we get if we let x tend to ∞ and use (4.2) that the right hand side of (8.12) tends to

$$\frac{g(\alpha_2)-g(\alpha_1)}{\alpha_1\,(1-\alpha_1)^{\theta'}}+\theta'\int_{\alpha_1}^{\alpha_2}\frac{g(\alpha_2)-g(v)}{v\,(1-v)^{\theta'+1}}\,dv < \frac{g(1)-g(\alpha_1)}{\alpha_1\,(1-\alpha_1)^{\theta'}}+\theta'\int_{\alpha_1}^{\alpha_2}\frac{g(1)-g(v)}{v\,(1-v)^{\theta'+1}}\,dv$$

Now since the left derivative of g at 1 exists, the ratio

$$\frac{g(1)-g(v)}{1-v}$$

has an upper bound B_g for $0 \le v < 1$, thus the above expression is less than

$$\frac{B_g}{\alpha_1}\,(1-\alpha_1)^{1-\theta'} + \theta'\,B_g\int_{\alpha_1}^{1}\frac{dv}{v\,(1-v)^{\theta'}} < B_g'\,(1-\alpha_1)^{1-\theta'}$$

for $1/2 \le \alpha_1 < 1$. (8.11) now gives, combining the above bound with (8.12) for $\alpha_2 < 1$ as $x \to \infty$ that

$$T_g^-(\alpha_2) \ge T_g^-(\alpha_1)\,\exp\left(\int_{\alpha_1}^{\alpha_2}\frac{dg(t)}{t}\right) - B_g''\,(1-\alpha_1)^{1-\theta'}\,. \tag{8.13}$$

This proves that $T_g^-(\alpha)$ remains bounded from below as $\alpha \to 1$, and with (8.6) proves that it tends to a limit. That this limit really is $T_g^-(1)$ can be seen as follows. Consider the expression

$$S_u(x') - S_u(t) - g(1)\,\log x' + g\left(\frac{\log t}{\log x}\right)\log x$$

for $x^{1/2} \le t \le x'$ and denote its maximum by $\varepsilon(x)\,\log x$, it follows from (4.2) that $\varepsilon(x)$ tends to zero as $x \to \infty$ through the sequence we are considering. We now modify our set of $u(p)$ by taking $u(p) = 0$ for $x^{1-\varepsilon'(x)} \le p < x'$, and for $p < x^{\varepsilon'(x)}$, where $\varepsilon'(x) = \varepsilon(x) + 1/\log\log\log x$. If m is an upper bound for $g'(\beta)$ for $1/2 \le \beta \le 1$, we see that for the modified sequence

$$S_u(x') - S_u(t) < (m+1)\,\log(x'/t)$$

for $\sqrt{x} \le t < x'$, and x large enough. Thus we have for $1 < t < x'$

$$S_u(x') - S_u(t) < B_g'''\,\log(x'/t)\,. \tag{8.14}$$

It is now easily seen that the $\Lambda_u^-(\mathcal{P}(x'^{\alpha_2}))$ constructed to get (8.7) now will be of type $\Lambda_u^-(\mathcal{P}(x'),\, x')$ for $\alpha_2 = 1$. Using (8.14), we get for the right hand side of (8.12) the upper bound

$$\frac{B_g'''\log\dfrac{x'}{x^{\alpha_1}}}{\alpha_1\log x\left(1-\alpha_1\dfrac{\log x}{\log x'}\right)^{\theta'}} + B_g'''\,\theta'\int\limits_{\alpha_1\left(1-\frac{\log 2}{\log x'}\right)}^{1}\frac{dv}{v\,(1-v)^{\theta'}} < B_g^{IV}\,(1-\alpha_1)^{1-\theta'}$$

valid for $1/2 \le \alpha_1 < 1$. If we insert this bound for (8.12) in (8.11) for $\alpha_2 = 1$, we get

$$T_u(\Lambda_u^-(\mathcal{P}(x'),\, x')) > T_u(\Lambda_u^-(\mathcal{P}(x^{\alpha_1}),\, x))\,\frac{1}{e_u(x^{\alpha_1},\, x')} - B_g^{V}\,(1-\alpha_1)^{1-\theta'}\,.$$

Letting now $x \to \infty$ through our sequence of x such that $T_u(\Lambda_u^-(\mathcal{P}(x^{\alpha_1}),\, x))$ tends to $T_g^-(\alpha_1)$, we see that the right hand side of the above expression remains bounded and tends to a limit, this shows that $T_g^-(1)$ exists and we get finally

$$T_g^-(1) \ge T_g^-(\alpha_1)\,\exp\left(\int_{\alpha_1}^{1}\frac{dg(t)}{t}\right) - B_g^{V}\,(1-\alpha_1)^{1-\theta'}\,.$$

This combined with $T_g^-(1) \le T_g^-(\alpha_1)$, shows that $T_g^-(1)$ is the limit of $T_g^-(\alpha_1)$ as $\alpha_1 \to 1$. Lemma 1 is thus completely proved.

It is clear from the previous arguments that if we have enough information about g; more precisely, if we have a majorant $g^*(t)$ of $g'(t)$ which is such that $\int_0^1 g^*(t)dt < \infty$, then we can find effective measures of continuity for $T_g^+(\alpha)$ and $T_g^-(\alpha)$ (in case of the latter for $0 \le \alpha \le a < 1$). For $T_g^+(\eta,\alpha)$ and $T_g^-(\eta,\alpha)$ the same is true as long as η is bounded away from zero. Clearly $T_g^+(\eta,\alpha) \le T_g^+(\alpha)$ and $T_g^-(\eta,\alpha) \ge T_g^-(\alpha)$, so since $T_g^+(\eta,\alpha)$ is nondecreasing as η decreases and $T_g^-(\eta,\alpha)$ nonincreasing as η decreases, these two functions both tend to certain limit functions as $\eta \to 0$, we shall write $T_g^+(0,\alpha)$ and $T_g^-(0,\alpha)$ for these limit functions. Clearly

$$T_g^+(0,\alpha) \le T_g^+(\alpha)\,,$$

and

$$T_g^-(0,\alpha) \ge T_g^-(\alpha)\,.$$

The natural guess is that equality holds in both cases.

In the general situation considered here it is not possible to prove this along the lines Lemma 2 in my Stony Brook lectures was proved. That proof could be carried over quite simply if we assume $g'(t)$ nondecreasing with increasing t, and with a little more effort if we assume $g(kt) \geq k\,g(t)$ for $k > 1$; but these are strong limitations on $g(t)$.

Instead we shall use arguments which incorporate also ideas from Lemma 4 of that paper, and our first objective is not the equality of $T_g^{\pm}(\alpha)$ and $T_g^{\pm}(0, \alpha)$.

We shall need some new definitions and notations. If we have given a sequence of functions F_r, for $r = 0, 1, 2, \ldots$, such that $F_0 = 1$, while F_r for $r \geq 1$ is a function of r variables $v_1, v_2, \ldots, v_r$, which range over an interval $(0, \alpha)$ or (η, α), we may associate with such a sequence a Λ system $\Lambda_F(\mathcal{P}(x^\alpha))$ or $\Lambda_F(\mathcal{P}(x^\eta,\, x^\alpha))$ as follows:

For $d = p_1 \ldots p_r$ we define

$$\lambda_d = F_r\left(\frac{\log p_1}{\log x}, \frac{\log p_2}{\log x}, \ldots, \frac{\log p_r}{\log x}\right), \tag{8.15}$$

and for convenience we shall use the notation $\Lambda_F(\alpha)$ or $\Lambda_F(\eta, \alpha)$ for the Λ system. If the F_r are such that they vanish whenever the sum of the arguments exceeds a we shall denote the Λ–system, which is then a $\Lambda[\mathcal{P}(x^\alpha),\, x^a]$ or a $\Lambda[\mathcal{P}(x^\eta,\, x^\alpha),\, x^a]$ by $\Lambda_F[\alpha;\, a]$ or $\Lambda_F\,[\eta, \alpha;\, a]$.

We shall also introduce the notation $\bar{g}'(\eta, \alpha)$ for the least upper bound of $g'(t)$ for $\eta \leq t \leq \alpha$; with the assumptions we have made earlier about $g'(t)$ this value is actually attained and is the maximum of $g'(t)$ in the interval. We shall also write for $\eta < \alpha$

$$L(\eta, \alpha) = \int_\eta^\alpha \frac{dg(t)}{t}\,. \tag{8.16}$$

Lemma 2. *Let $0 < \eta < \alpha$ and let η' and α' be defined by $L(\eta, \eta') = L(\alpha', \alpha) = \frac{2}{N+2}\,L(\eta, \alpha)$, where N is a positive integer, and*

$$\delta = \frac{L(\eta, \alpha)}{(N+2)\,\bar{g}'(\eta, \alpha)}\,, \tag{8.17}$$

then there exists a $\Lambda_F^+[\,\eta, \alpha';\, e^{-\delta}]$ such that

$$\lim_{x\to\infty} T_u(\Lambda_F^+[\,\eta, \alpha';\, e^{-\delta}]) = T_g^+(\eta', \alpha)\,. \tag{8.18}$$

Furthermore, for $\alpha < 1$, there exists a $\Lambda_F^-[\,\eta, \alpha';\, e^{-\delta}]$ such that

$$\lim_{x\to\infty} T_u(\Lambda_F^-[\,\eta, \alpha';\, e^{-\delta}]) = T_g^-(\eta', \alpha)\,. \tag{8.19}$$

In (8.18) and (8.19) x may go to infinity through any sequence.

We begin by dividing the interval (η, α) in $N+2$ parts (t_{i-1}, t_i), $i = 1, 2, \ldots, N+2$, and such that $t_0 = \eta$, $t_2 = \eta'$, $t_N = \alpha'$, $t_{N+2} = \alpha$, and $L(t_{i-1}, t_i) = \frac{1}{N+2} L(\eta, \alpha)$.

From the relation

$$\int_{t_{i-1}}^{t_i} \frac{g'(t)}{t}\, dt = \frac{L(\eta, \alpha)}{N+2},$$

we see, since the left hand side is less than or equal to

$$\bar{g}'(\eta, \alpha) \int_{t_{i-1}}^{t_i} \frac{dt}{t} = \bar{g}'(\eta, \alpha) \log \frac{t_i}{t_{i-1}},$$

that

$$\log \frac{t_i}{t_{i-1}} \geq \frac{L(\eta, \alpha)}{(N+2)\, \bar{g}'(\eta, \alpha)} = \delta\,. \tag{8.20}$$

We introduce the notation Δ_i for (t_{i-1}, t_i) for $i = 1, 2, \ldots, N$, and similarly Δ_i' for (t_{i+1}, t_{i+2}) for $i = 1, 2, \ldots, N$.

From the definition of $T_g^+(\eta', \alpha)$ we have that there exists a sequence of x_j tending to infinity such that with each x_j there is associated a set of $u(p)$ satisfying (4.2) and a $\Lambda_u^+(\mathcal{P}(x_j^{\eta'}, x_j^{\alpha}),\, x_j)$ such that

$$\lim_{j \to \infty} T_u(\Lambda_u^+(\mathcal{P}(x_j^{\eta'}, x_j^{\alpha}),\, x_j)) = T_g^+(\eta', \alpha)$$

as $x_j \to \infty$ through this sequence. We shall use the notation $\lambda_d^{(j)}$ for the λ in the Λ_u^+ corresponding to x_j. In the following we also use the notation $\sum_{(i_1,\ldots,i_r)}^{(j)}$ to indicate that a sum is taken over all p_ν such that $\log p_\nu / \log x_j \in \Delta_{i_\nu}'$ for $\nu = 1, 2, \ldots, r$. We now consider the ratio for $r \geq 1$,

$$\ell_j(i_1, i_2, \ldots, i_r) = \frac{\displaystyle\sum_{(i_1,\ldots,i_r)}^{(j)} \frac{\lambda_{p_1 \cdots p_r}^{(j)}}{f(p_1) \ldots f(p_r)}}{\displaystyle\sum_{(i_1,\ldots,i_r)}^{(j)} \frac{1}{f(p_1) \ldots f(p_r)}}, \tag{8.21}$$

as usual terms for which a symbol is not defined are omitted, so if some of the i_ν are equal we omit the terms in the numerator where the same p occurs more than once.

The denominator in (8.21) is simply

$$\prod_{\nu=1}^{r} \left(\sum_{x_j^{t_{i_\nu+1}} \leq p < x_j^{t_{i_\nu+2}}} \frac{1}{f(p)} \right)$$

which tends to

$$\prod_{\nu=1}^{r} L(t_{i_\nu+1}, t_{i_\nu+2}) = \left(\frac{L(\eta, \alpha)}{N+2} \right)^r$$

as $j \to \infty$.

The numerator we can bound in two ways, first we have since our sieve $\Lambda^{(j)}$ has $R_u(\Lambda) \le x_j$, that the numerator in (8.21) is less than

$$x_j^{-\Sigma_{\nu=1}^r t_{i_\nu+1}} \sum_{(i_1,\ldots,i_r)}^{(j)} u(p_1)\ldots u(p_r)\,|\lambda^{(j)}_{p_1\ldots p_r}| \le r!\,x_j^{1-\Sigma_{\nu=1}^r t_{i_\nu+1}},$$

where the factor $r!$ was put in in case some of the i_ν are equal (if all are equal the same term $u(d)\,|\lambda_d^{(j)}|$ could occur $r!$ times). From this we see that

$$\ell_j(i_1,\ldots,i_r) \to 0\,, \tag{8.22}$$

as $j \to \infty$ if $\sum_{\nu=1}^r t_{i_\nu+1} > 1$. Since $\sum_{\nu=1}^r t_{i_\nu+1} > r\,\eta$, we see we have only a finite number of the expressions (8.21) left to consider, less than

$$\sum_{r=1}^{[1/\eta]} N^r < (N+1)^{1/\eta}$$

in fact. We now wish to show that the remaining expressions remain bounded. From (5.3) we have

$$\begin{aligned}\sum_d \frac{|\lambda_d^{(j)}|}{f(d)} &< e_u^{-1}(x_j^{\eta'}, x_j^{\alpha})\;T_u(\Lambda_u^+(\mathcal{P}(x_j^{\eta'}, x_j^{\alpha}),\, x_j))\\ &< 2\,\exp L(\eta',\alpha)\,T_g^+(\eta',\alpha) = B\,,\end{aligned}$$

for j large enough. This means that for j large enough the numerator in (8.21) in absolute value is less than $r!\,B$. Combining this with what we have shown about the denominator we see that the ratios (8.21) with $r < 1/\eta$ remain bounded as x_j tends to infinity. Since there are only a finite number of them it follows that we can pick out a subsequence of the x_j, say $x_{j'}$ such that each of the ratios (8.21) with $r < 1/\eta$ tends to a limit as $x_{j'}$ tends to infinity, we call this limit $\ell(i_1, i_2, \ldots, i_r)$, and we have

$$\ell(i_1, i_2, \ldots, i_r) = 0\,, \qquad \text{for } t_{i_1+1} + t_{i_2+1} + \ldots + t_{i_r+1} > 1\,.$$

We now define a set of functions $F_r(v_1, \ldots, v_r)$ for $\eta \le v_\nu \le \alpha'$, $\nu = 1, 2, \ldots, r$, in the following way: if $v_\nu \in \Delta_{i_\nu}$ for $\nu = 1, 2, \ldots, r$ then

$$F_r(v_1, \ldots, v_r) = \ell(i_1, i_2, \ldots, i_r)\,, \tag{8.23}$$

as $\Delta_{i_\nu} = (t_{i_\nu-1}, t_{i_\nu})$ and from (8.20) $t_{i_\nu} \le e^{-\delta}\,t_{i_\nu+1}$, we see that

$$F_r(v_1, \ldots, v_r) = 0\,, \tag{8.24}$$

for $v_1 + v_2 + \ldots + v_r > e^{-\delta}$. Thus if we associate with this set of functions F_r a Λ_F, it will be a $\Lambda_F[\eta, \alpha';\, e^{-\delta}]$. That it is also a Λ_F^+ can be seen as follows, let

$d = p_1 \dots p_r$ with $x^\eta \le p_\nu \le x^{\alpha'}$ and $v_\nu = \log p_\nu / \log x \in \Delta_{i_\nu}$. For a divisor of d, say $d^* = p_1 \dots p_s$, with $s \le r$, we have then that

$$\lambda_{d^*} = \lim_{x_{j'} \to \infty} \frac{\displaystyle\sum_{(i_1,\dots,i_s)}^{(j')} \frac{\lambda^{(j')}_{q_1 \dots q_s}}{f(q_1)\dots f(q_s)}}{\displaystyle\sum_{(i_1,\dots,i_s)}^{(j')} \frac{1}{f(q_1)\dots f(q_s)}},$$

where we, to avoid confusion, have used q to denote primes on the right hand side. But the right hand side is clearly also equal to

$$\lim_{x_{j'} \to \infty} \frac{\displaystyle\sum_{(i_1,\dots,i_r)}^{(j')} \frac{\lambda^{(j')}_{q_1 \dots q_s}}{f(q_1)\dots f(q_r)}}{\displaystyle\sum_{(i_1,\dots,i_r)}^{(j')} \frac{1}{f(q_1)\dots f(q_r)}}.$$

We get from this

$$\theta_d = \lim_{x_{j'} \to \infty} \frac{\displaystyle\sum_{(i_1,\dots,i_r)}^{(j')} \frac{\theta^{(j')}_{q_1 \dots q_r}}{f(q_1)\dots f(q_r)}}{\displaystyle\sum_{(i_1,\dots,i_r)}^{(j')} \frac{1}{f(q_1)\dots f(q_r)}},$$

from which $\theta_d \ge 0$ follows immediately, thus our Λ_F is a $\Lambda^+[\eta, \alpha'_j; e^{-\delta}]$. It remains to consider the expression $T_u(\Lambda_F^+)$ as $x \to \infty$; (3.3) and (3.4) give in this case

$$T_u(\Lambda_F^+) = e_u^{-1}(\mathcal{P}(x^\eta, x^{\alpha'})) \sum_{d \in (\mathcal{P})} \frac{\lambda_d}{f(d)}.$$

By introducing the expression (8.23) for λ_d if d is the product of $r \ge 1$ primes we get

$$\sum_{d \in (\mathcal{P})} \frac{\lambda_d}{f(d)} = 1 + \sum_{1 \le r < 1/\eta} \frac{1}{r!} \sum_{1 \le i_\nu \le N} \ell(i_1, i_2, \dots, i_r) \sum_{(i_1,\dots,i_r)}{}' \frac{1}{f(p_1)\dots f(p_r)},$$

where the notation $\sum'_{(i_1,\dots,i_r)}$ means that the summation is taken over $p_\nu \in \Delta_{i_\nu}$, $\nu = 1, \dots, r$, and we omit terms where two p_ν are equal. When $x \to \infty$, the expression

$$\sum_{(i_1,\dots,i_r)}{}' \frac{1}{f(p_1)\dots f(p_r)} \to \left(\frac{L(\eta, \alpha)}{N+2}\right)^r = \int_{\substack{v_\nu \in \Delta_{i_\nu} \\ \nu = 1,\dots,r}} \cdots \int \frac{g'(v_1)\dots g'(v_r)}{v_1 \dots v_r} \, dv_1 \dots dv_r \,.$$

Inserting this latter expression we see that

(8.25)

$$\lim_{x\to\infty} T_u(\Lambda_F^+) = \exp\left(\int_\eta^{\alpha'} \frac{dg(t)}{t}\right)$$
$$\times \left\{1 + \sum_{1\le r<1/\eta} \frac{1}{r!}\int_\eta^{\alpha'} \cdots \int_\eta^{\alpha'} \frac{F_r(v_1,\ldots,v_r)\, g'(v_1)\ldots g'(v_r)}{v_1\, v_2 \ldots v_r}\, dv_1 \ldots dv_r\right\}$$

On the other hand, if we consider for the sequence of $x_{j'}$ used in defining $\ell(i_1,\ldots,i_r)$, the sieve $\Lambda_u^{+\,(j')}\,(\mathcal{P}(x_{j'}^{\eta'}, x_{j'}^{\alpha}),\, x_{j'})$ and look at $T_u(\Lambda_u^{+\,(j')})$, we find as we let $x_{j'} \to \infty$ that $\lim_{x_{j'}\to\infty} T_u(\Lambda_u^{+\,(j')})$ again equals the expression on the right hand side of (8.25), on the other hand its value is by definition $T_g^+(\eta',\alpha)$. Thus (8.18) is proved.

For the second part of Lemma 2, we can proceed in a quite similar way to construct our $\Lambda_F^-[\eta,\alpha';\, e^{-\delta}]$ and establish (8.19).

For our next lemma we need to strengthen our hypothesis about $g(t)$. We shall assume[7]

(8.26) $$\lim_{\eta\to 0} \sqrt{\eta}\,\bar g'(\eta,\alpha) = 0\,.$$

Lemma 3. *If (8.26) holds then we have*

$$\lim_{\eta\to 0} T_g^+(\eta,\alpha) = T_g^+(\alpha)\,,$$

and for $\alpha < 1$,

$$\lim_{\eta\to 0} T_g^-(\eta,\alpha) = T_g^-(\alpha)\,.$$

Furthermore, if the function $g'(t)$ is known, the differences $T_g^+(\alpha) - T_g^+(\eta,\alpha)$ and $T_g^-(\eta,\alpha) - T_g^-(\alpha)$ can be effectively estimated.

Observe first that (8.26) obviously implies $\lim_{\eta\to 0} g(\eta)/\sqrt{\eta} = 0$. We now choose η so small that $g(\eta) < \varepsilon\sqrt{\eta}$ and $\bar g'(\eta,\alpha) < \varepsilon/\sqrt{\eta}$, where $0 < \varepsilon < e^{-2}$. Next we choose N so large that

$$4\,\varepsilon < \frac{L(\eta,\alpha)}{N+2} < 5\,\varepsilon\,.$$

This implies for the δ in Lemma 2 that

$$\delta = \frac{L(\eta,\alpha)}{(N+2)\,\bar g'(\eta,\alpha)} > 4\sqrt{\eta}\,,$$

and so

$$e^{-\delta} < 1 - 3\sqrt{\eta}\,,$$

if $\eta < 1/100$.

[7] We could actually manage with less, for instance $\liminf_{\eta\to 0} \sqrt{\eta}\,\bar g'(\eta,\alpha) = 0$ and $\limsup_{\eta\to 0} g(\eta)/\sqrt{\eta} < \infty$.

If we now consider the Λ^2 sieve constructed in connection with (7.11) and (7.20), with $\xi = x^\eta$ and $z = x^{\sqrt{\eta}}$, we get from (7.11)

$$T_u(\Lambda^2\,[\mathcal{P}(x^\eta), x^{\sqrt{\eta}}]) < (1 - e^{-1/\sqrt{\eta}})^{-1}\,.$$

Combining this sieve with our $\Lambda_F^+[\eta, \alpha';\, e^{-\delta}]$ from Lemma 2 according to the rules given in Section 6, Principle I, we get a sieve for the range $\mathcal{P}(x^{\alpha'})$ which is a $\Lambda^+[\mathcal{P}(x^{\alpha'}), x^{1-\sqrt{\eta}}]$, and from (6.3) we have

$$T_u(\Lambda^+[\mathcal{P}(x^{\alpha'}), x^{1-\sqrt{\eta}}]) < (1 - e^{-1/\sqrt{\eta}})^{-1}\, T_u(\Lambda_F^+[\eta, \alpha'; e^{-\delta}])\,.$$

If we use this sieve instead over the larger range $\mathcal{P}(x^\alpha)$ this adds a factor $e_u^{-1}(x^{\alpha'}, x^\alpha)$. Letting now $x \to \infty$ we obtain since $\liminf_{x\to\infty}$ of the left hand side is not less than $T_g^+(\alpha)$, that

$$\begin{aligned} T_g^+(\alpha) &\le (1 - e^{-1/\sqrt{\eta}})^{-1} \exp(L(\alpha', \alpha))\,, T_g^+(\eta', \alpha) \\ &< (1 - e^{-1/\sqrt{\eta}})^{-1}\, e^{10\,\varepsilon}\, T_g^+(\eta, \alpha)\,, \end{aligned}$$

since

$$L(\alpha', \alpha) = \frac{2\,L(\eta, \alpha)}{N+2} < 10\,\varepsilon\,.$$

Thus

$$T_g^+(\eta, \alpha) \le T_g^+(\alpha) < (1 - e^{-1/\sqrt{\eta}})^{-1}\, e^{10\,\varepsilon}\, T_g^+(\eta, \alpha)\,.$$

Since η and ε clearly may be chosen arbitrarily small, the first part of Lemma 3 is proved.

For the second part we proceed in the same way in our choice of η and N. Besides the Λ^2 sieve used above, we also need a $\Lambda^2\,\Lambda^-$ sieve of the type constructed in connection with (7.19) and (7.20′), again with $\xi = x^\eta$ and $z = x^{\sqrt{\eta}}$. This will be a

$$\Lambda^-[\mathcal{P}(x^\eta),\, x^{2\sqrt{\eta}+\eta}]$$

with

$$T_u(\Lambda^-) > 1 - e^{-1/\sqrt{\eta}}\,,$$

as we get from (7.19). Again we appeal to the rules given under Principle I in Section 6 concerning the combination of two distinct sifting ranges, this time for lower bound sieves. We see that we need both the Λ_F sieves constructed in Lemma 2. We get using (6.8) in the here more convenient form

$$T_u(\Lambda) = T_u(\Lambda_1^+)\, T_u(\Lambda_2^-) + (T_u(\Lambda_1^-) - T_u(\Lambda_1^+))\, T_u(\Lambda_2^+)\,,$$

that

$$T_g^-(\alpha') > (1 - e^{-1/\sqrt{\eta}})^{-1}\, T_g^-(\eta', \alpha) - 3\, e^{-1/\sqrt{\eta}}\, T_g^+(\eta', \alpha)\,,$$

and this lower bound of $T_g^-(\alpha')$ is obtained using as $x \to \infty$, a

$$\Lambda^-[\mathcal{P}(x^{\alpha'}),\, x^{1-\sqrt{\eta}+\eta}] = \Lambda^-[\mathcal{P}(x^{\alpha'}),\, x^{1-\sqrt{\eta}/2}]$$

sieve. Since $T_g^-(\eta',\alpha)$ may possibly be negative, we prefer to rewrite this inequality as

$$\begin{aligned}(8.27)\qquad T_g^-(\alpha') &> T_g^-(\eta',\alpha) - 3\,e^{-1/\sqrt{\eta}}(T_g^+(\eta',\alpha) + |T_g^-(\eta',\alpha)|)\\ &> T_g^-(\eta,\alpha) - 3\,e^{-1/\sqrt{\eta}}\,(T_g^+(\alpha) + |T_g^-(\alpha)| + 2)\,,\end{aligned}$$

since $1 - T_g^-(\eta',\alpha)$ is nonnegative and nondecreasing if we increase the range so that

$$|T_g^-(\eta',\alpha)| < 1 + |1 - T_g^-(\eta',\alpha)| \le 1 + |1 - T_g^-(\alpha)| \le 2 + |T_g^-(\alpha)|\,.$$

If $g'(t)$ is positive in the neighborhood to the left of α, one sees that as η is chosen smaller and N larger α' converges to α, so we might appeal to Lemma 1 (ii) to get the second part of Lemma 3. However, if $g'(t) = 0$ for $\alpha'' < t < \alpha$, one sees that the α', which is determined by the equation

$$L(\alpha',\alpha) = \frac{2}{N+2}\,L(\eta,\alpha)\,,$$

could not approach nearer than α''.[8] We therefore instead go back to the arguments used to prove (ii), in particular to (8.7) and (8.8), where we would replace the x' by x and α_1 by α', α_2 by α (and remember that the η ocurring there is some fixed positive quantity which depends on g and Δ, and has nothing to do with the η in (8.27)). (8.8) then takes the form

$$(8.28)\qquad T_g^-(\alpha) \ge T_g^-(\alpha')\exp L(\alpha',\alpha) - \exp L(\eta_0\,\Delta,\alpha)\,(1 - e^{-1/(3\,\eta_0)})^{-1}\,L(\alpha',\alpha)\,,$$

where we have written η_0 for the η ocurring in (8.8). (8.28) holds for $\alpha \le 1-\Delta$. Combining it with (8.27), and remembering that

$$L(\alpha',\alpha) = \frac{2}{N+2}\,L(\eta,\alpha)\,,$$

we see that as we choose η smaller and N larger, $T_g^-(\alpha')$ converges to $T_g^-(\alpha)$ in any case, even when α' does not converge to α, and so $T_g^-(\eta,\alpha)$ converges to $T_g^-(\alpha)$ as $\eta \to 0$. This proves the second part of Lemma 3.

It is also clear from the above arguments that if we have given an $\varepsilon > 0$, and have sufficient information about $g'(t)$, we can effectively determine an $\eta(\varepsilon)$ such that for $\eta \le \eta(\varepsilon)$ we have

$$T_g^+(\alpha) - T_g^+(\eta,\alpha) < \varepsilon\,.$$

Similarly, for $\alpha < 1 - \Delta$ with $\Delta > 0$, we can effectively determine an $\eta(\varepsilon,\Delta)$ such that for $\eta \le \eta(\varepsilon,\Delta)$, we have

$$T_g^-(\eta,\alpha) - T_g^-(\alpha) < \varepsilon\,.$$

[8] We are of course in this case not doing any sifting to speak of for the primes in the range $\mathcal{P}(x^{\alpha''},x^{\alpha})$.

It is now also clear that we may rephrase (8.1) as

$$\lim_{x\to\infty} \min_{\Lambda_u^+(\mathcal{P}(x^\alpha),x)} T_u(\Lambda^+) = T_g^+(\alpha)\,. \tag{8.29}$$

Here x may go to infinity through any sequence and the $u(p)$ may be any sequence depending on x such that (4.2) and (4.4) hold and $g'(t)$ and $g(t)$ satisfy our previous requirements in particular (8.26). Similarly, for $\alpha < 1$ we may rephrase (8.1′) as

$$\lim_{x\to\infty} \max_{\Lambda_u^-(\mathcal{P}(x^\alpha),x)} T_u(\Lambda^-) = T_g^-(\alpha)\,. \tag{8.29′}$$

If we now go back to the question of what happens to the ratios

$$\frac{M_u^+(W_x,\mathcal{P}(x^\alpha))}{E_u(W_x,\mathcal{P}(x^\alpha))}\,, \tag{8.30}$$

$$\frac{M_u^=(W_x,\mathcal{P}(x^\alpha))}{E_u(W_x,\mathcal{P}(x^\alpha))}\,, \tag{8.30′}$$

and

$$\frac{M_u^-(W_x,\mathcal{P}(x^\alpha))}{E_u(W_x,\mathcal{P}(x^\alpha))}\,, \tag{8.30″}$$

as $x \to \infty$, we see from (3.2) and (3.2″) that the optimal Λ^+ and Λ^- must have a $R_u(\Lambda) < x$, so it is clear that

$$\liminf_{x\to\infty} \frac{M_u^+(W_x,\mathcal{P}(x^\alpha))}{E_u(W_x,\mathcal{P}(x^\alpha))} \geq T_g^+(\alpha)\,, \tag{8.31}$$

and similarly,

$$\limsup_{x\to\infty} \frac{M_u^=(W_x,\mathcal{P}(x^\alpha))}{E_u(W_x,\mathcal{P}(x^\alpha))} \leq T_g^-(\alpha)\,. \tag{8.31′}$$

On the other hand, in proving Lemma 3 we constructed for any $\varepsilon > 0$ a sieve $\Lambda^+[\mathcal{P}(x^\alpha),\, x^{1-\delta}]$ with $\delta = \delta(\varepsilon) > 0$ such that

$$\lim_{x\to\infty} T_u(\Lambda^+[\mathcal{P}(x^\alpha),\, x^{1-\delta}]) < T_g^+(\alpha) + \varepsilon\,.$$

Using this sieve we find

$$\limsup_{x\to\infty} \frac{M_u^+(W_x,\mathcal{P}(x^\alpha))}{E_u(W_x,\mathcal{P}(x^\alpha))} < T_g^+(\alpha) + \varepsilon\,,$$

similarly we get using the Λ^- constructed in the proof of Lemma 3, that

$$\liminf_{x\to\infty} \frac{M_u^=(W_x,\mathcal{P}(x^\alpha))}{E_u(W_x,\mathcal{P}(x^\alpha))} > T_g^-(\alpha) - \varepsilon\,.$$

Theorem 1. *We have, under our assumptions about g and g'*

$$\lim_{x\to\infty} \frac{M_u^+(W_x, \mathcal{P}(x^\alpha))}{E_u(W_x, \mathcal{P}(x^\alpha))} = T_g^+(\alpha)\,, \tag{8.32}$$

and for $\alpha < 1$,

$$\lim_{x\to\infty} \frac{M_u^=(W_x, \mathcal{P}(x^\alpha))}{E_u(W_x, \mathcal{P}(x^\alpha))} = T_g^-(\alpha)\,. \tag{8.32'}$$

Thus also

$$\lim_{x\to\infty} \frac{M_u^-(W_x, \mathcal{P}(x^\alpha))}{E_u(W_x, \mathcal{P}(x^\alpha))} = \max(0,\, T_g^-(\alpha))\,. \tag{8.32''}$$

Similar results hold if we replace the range $\mathcal{P}(x^\alpha)$ with $\mathcal{P}(x^\eta,\, x^\alpha)$ and $T_g^\pm(\alpha)$ with $T_g^\pm(\eta,\alpha)$. These limits are effectively computable.

The first two parts of the Theorem are an immediate consequence of the definition of $T_g^+(\alpha)$ and $T_g^-(\alpha)$ combined with the two inequalities immediately preceding. (8.32″) follows from the fact that $M_u^- = \max(0,\, M_u^=)$. The part dealing with $\mathcal{P}(x^\eta, x^\alpha)$ is obvious in view of the preceding.

We wish to show that the functions $T_g^\pm(\alpha)$ and $T_g^\pm(\eta,\alpha)$ are effectively computable (assuming in the case of $T_g^-(\alpha)$ that $\alpha < 1$ and for $T_g^-(\eta,\alpha)$ that $\alpha \le 1$) if we have enough information about $g(t)$ and $g'(t)$ and they satisfy the conditions for Lemma 3 in the case of $T_g^\pm(\alpha)$, and the conditions imposed prior to (8.26) in case of the $T_g^\pm(\eta,\alpha)$.

Let us consider $T_g^+(\alpha)$. By Lemma 3 we can for a given $\varepsilon > 0$ find an $\eta > 0$ such that $T_g^+(\alpha) < T_g^+(\eta,\alpha) + \varepsilon$. Since also $T_g^+(\alpha) \ge T_g^+(\eta,\alpha)$, this shows that it is enough to show that $T_g^+(\eta,\alpha)$ can be effectively computed. We now go back to the $\ell(i_1, i_2, \ldots, i_r)$ which defined the $\Lambda_F^+(\eta,\alpha',e^{-\delta})$ constructed in Lemma 2. These are bounded by

$$|\ell(i_1,\ldots,i_r)| < 2\,r!\,\exp(L(\eta,\alpha))\,T_g^+(\eta;\alpha)\,\frac{(N+2)^r}{L(\eta,\alpha)^r} = B_r \tag{8.33}$$

for $t_{i_1} + \ldots + t_{i_r} \le e^{-\delta}$, and vanish for $t_{i_1} + \ldots + t_{i_r} > e^{-\delta}$.

By choosing N large enough we see since

$$T_g^+(\eta,\alpha') \le T_g^+(\eta',\alpha) \le T_g^+(\eta,\alpha) \le \exp\left(\frac{2\,L(\eta,\alpha)}{N+2}\right) T_g^+(\eta,\,\alpha')\,,$$

that the three numbers $T_g^+(\eta,\alpha')$, $T_g^+(\eta',\alpha)$, $T_g^+(\eta,\alpha)$, can be made to differ by less than ε for a given positive ε. From the formulas immediately preceding (8.25) we see that

(8.34)

$$\begin{aligned} T_g^+(\eta',\alpha) &= \lim_{x\to\infty} T_u^+(\Lambda_F^+) \\ &= \exp(L(\eta',\alpha))\left\{1 + \sum_{1\le r<1/\eta} \frac{1}{r!} \sum_{1\le i_\nu\le N} \ell(i_1,\ldots,i_r)\left(\frac{L(\eta,\alpha)}{N+2}\right)^r\right\}. \end{aligned}$$

We now consider sets of quantities $\ell'_r(i_1,\dots,i_r)$ with $\ell'_0 = 1$, and $\ell'_r(i_1,\dots,i_r) = 0$ for $t_{i_1} + \dots + t_{i_r} > e^{-\delta}$, while for the others the inequalities

$$|\ell'_r(i_1,\dots,i_r)| < B_r \tag{8.35}$$

hold. We define the polynomial in N variables $y_1,\dots,y_N$

$$\theta(y_1,\dots,y_N) = \sum_{\substack{0\le\nu \\ i_1,\dots,i_r}} \ell'_r(i_1,\dots,i_r)\binom{y_1}{\nu_1}\cdots\binom{y_N}{\nu_N}, \tag{8.36}$$

here we sum over all $\ell'_r(i_1,\dots,i_r) \neq 0$, and ν_i denotes the number of times i occurs among $(i_1,i_2,\dots,i_r)$. We require that

$$\theta(y_1,\dots,y_N) \ge 0, \tag{8.37}$$

for all nonnegative integral values of the variables $y_1,\dots,y_N$, and we shall refer to a set of ℓ'_r as "good" if (8.37) holds. We have then using the sieve Λ_F of the first part of Lemma 2,

$$\begin{aligned} T_g^+(\eta,\alpha') &\le \min_{\ell'_r \text{ good}} \exp(L(\eta,\alpha')) \sum_{1\le r<1/\eta} \frac{1}{r!}\,\ell'_r(i_1,\dots,i_r)\left(\frac{L(\eta,\alpha)}{N+2}\right)^r \\ &\le T_g^+(\eta',\alpha)\,. \end{aligned} \tag{8.38}$$

Now we have altogether less than $(N+1)^{1/\eta}$ different ℓ'_r which may be different from zero, and for each such ℓ'_r that it is required to lie within the range given by (8.35). We can now divide each such range $-B_r, B_r$ into $\mathcal{L}$ equal parts and test all combinations within this grid to see whether our ℓ'_r set is good or not. There would be less than $\mathcal{L}^{(N+1)^{1/\eta}}$ combinations to test. However, it is not possible to decide effectively whether (8.37) holds for all nonnegative integral y_ν, $\nu = 1,\dots,N$, but we can effectively decide that one of the following statements is true: (a) (8.37) does not hold so our set of ℓ'_r is not good, or (b) (8.37) holds if we for $r > 0$ increase each ℓ'_r by $2\,B_r/\mathcal{L}$, so the set $\ell'_r + 2\,B_r/\mathcal{L}$ is good. Of course for some sets ℓ'_r both (a) and (b) may be true.

For the cases where we find (b) to be true, we compute the expression

$$\exp(L(\eta',\alpha))\left\{1 + \sum_{1\le r<1/\eta} \frac{1}{r!} \sum_{1\le i_\nu\le N} \left(\ell'_r(i_1,\dots,i_r) + \frac{2\,B_r}{\mathcal{L}}\right)\left(\frac{L(\eta,\alpha)}{N+2}\right)^r\right\}.$$

The smallest of these will differ from (8.34) by less than

$$\frac{4}{\mathcal{L}}\exp(L(\eta',\alpha)) \sum_{1\le r\le 1/\eta} \frac{B_r}{r!} N^r \left(\frac{L(\eta,\alpha}{N+2}\right)^r$$

which if we choose $\mathcal{L}$ large enough can be made $< \varepsilon$, and will therefore differ from $T_g^+(\eta,\alpha)$ by at most $2\,\varepsilon$.

A similar argument works for the case of $T_g^-(\alpha)$ with $\alpha < 1$ and $T_g^-(\eta,\alpha)$. This establishes the effective computability of these functions and thus gives the last part of Theorem 1.

In the course of our labors to get to Theorem 1, we have seen that to approach these optimal limit functions which the T_g^+ and T_g^- represent, it is enough to consider sieves of the type $\Lambda_F^{\pm}[\eta,\alpha;\ 1-\delta]$ which are associated with a sequence of functions $F_0 = 1$, $F_1(v_1)$,..., $F_r(v_1,\dots,v_r)$,..., where the F_r are symmetric functions in the r variables and such that $F_r(v_1,\dots,v_r) = 0$ for $v_1 + v_2 + \dots + v_r > 1-\delta$, in the way that if $d \in (\mathcal{P}(x^\eta,\ x^\alpha))$, $d = p_1 \dots p_r$, then

$$\lambda_d = F_r\left(\frac{\log p_1}{\log x},\dots,\frac{\log p_r}{\log x}\right).$$

We shall refer to such a sequence with the symbol F_α if the v range over $0 < v \le \alpha$ and $F_{\eta,\alpha}$ if the v range over $\eta \le v \le \alpha$.

If the associated Λ_F is a Λ_F^+ we use the symbol F_α^+ or $F_{\eta,\alpha}^+$ for the F sequence, if it is a Λ_F^- we write F_α^- or $F_{\eta,\alpha}^-$.

Instead of $F_r(v_1,\dots,v_r)$ vanishing for $v_1+\dots+v_r > 1-\delta$, we shall require only that $F_r(v_1,\dots,v_r) = 0$ for $v_1 + \dots + v_r > 1$.

If we introduce the notation

$$\begin{aligned}(8.39)\quad \theta_F(v_1,\dots,v_r) &= 1 + \sum_{1\le i\le r} F_1(v_i) + \sum_{1\le i,j\le r} F_2(v_i,v_j)\\ &\quad + \sum_{1\le i<j<k\le r} F_3(v_i,v_j,v_k) + \dots + F_r(v_1,\dots,v_r),\end{aligned}$$

we have for a F^+ that

$$\theta_F(v_1,\dots,v_r) \ge 0$$

for all $r \ge 0$, and for a F^- that

$$\theta_F(v_1,\dots,v_r) \le 0$$

for all $r \ge 1$.

Theorem 2. *We have, with our previous assumption about $g(t)$ and $g'(t)$ that*

$$\begin{aligned}(8.40)\quad T_g^+(\alpha) = \text{l.u.b.}\Bigg\{1 &+ \int_0^\alpha \theta_F(v)\,\frac{dg(v)}{v} + \frac{1}{2!}\int_0^\alpha\!\!\int_0^\alpha \theta_F(v_1,v_2)\,\frac{dg(v_1)\,dg(v_2)}{v_1\,v_2}\\ &+ \dots + \frac{1}{r!}\int_0^\alpha\!\cdots\!\int_0^\alpha \theta_F(v_1,\dots,v_r)\,\frac{dg(v_1)\dots dg(v_r)}{v_1\dots v_r} + \dots\Bigg\},\end{aligned}$$

where F varies over the class F_α^+. For $\alpha < 1$ we have

(8.40′)

$$T_g^-(\alpha) = \text{g.l.b.} \left\{ 1 + \int_0^\alpha \theta_F(v)\,\frac{dg(v)}{v} + \frac{1}{2!} \int_0^\alpha\!\!\int_0^\alpha \theta_F(v_1, v_2)\,\frac{dg(v_1)\,dg(v_2)}{v_1\,v_2} + \ldots + \frac{1}{r!} \int_0^\alpha \cdots \int_0^\alpha \theta_F(v_1, \ldots, v_r)\,\frac{dg(v_1)\ldots dg(v_r)}{v_1 \ldots v_r} + \ldots \right\},$$

where F varies over the class F_α^-

(8.40″)

$$T_g^+(\eta, \alpha) = \text{l.u.b.} \left\{ 1 + \int_\eta^\alpha \theta_F(v)\,\frac{dg(v)}{v} + \frac{1}{2!} \int_\eta^\alpha\!\!\int_\eta^\alpha \theta_F(v_1, v_2)\,\frac{dg(v_1)\,dg(v_2)}{v_1\,v_2} + \ldots + \frac{1}{r!} \int_\eta^\alpha \cdots \int_\eta^\alpha \theta_F(v_1, \ldots, v_r)\,\frac{dg(v_1)\ldots dg(v_r)}{v_1 \ldots v_r} + \ldots \right\},$$

$$= \text{l.u.b.} \exp(L(\eta, \alpha)) \left\{ 1 + \int_\eta^\alpha F^+(v)\,\frac{dg(v)}{v} + \frac{1}{2!} \int_\eta^\alpha\!\!\int_\eta^\alpha F^+(v_1, v_2)\,\frac{dg(v_1)\,dg(v_2)}{v_1\,v_2} + \ldots + \frac{1}{r!} \int_\eta^\alpha \cdots \int_\eta^\alpha F^+(v_1, \ldots, v_r)\,\frac{dg(v_1)\ldots dg(v_r)}{v_1 \ldots v_r} + \ldots \right\},$$

where F^+ varies over $F_{\eta,\alpha}^+$. Finally, for $\alpha \le 1$,

(8.40‴)

$$T_g^-(\eta, \alpha) = \text{g.l.b.} \left\{ 1 + \int_\eta^\alpha \theta_F(v)\,\frac{dg(v)}{v} + \frac{1}{2!} \int_\eta^\alpha\!\!\int_\eta^\alpha \theta_F(v_1, v_2)\,\frac{dg(v_1)\,dg(v_2)}{v_1\,v_2} + \ldots + \frac{1}{r!} \int_\eta^\alpha \cdots \int_\eta^\alpha \theta_F(v_1, \ldots, v_r)\,\frac{dg(v_1)\ldots dg(v_r)}{v_1 \ldots v_r} + \ldots \right\},$$

$$= \text{g.l.b.} \exp(L(\eta, \alpha)) \left\{ 1 + \int_\eta^\alpha F^-(v)\,\frac{dg(v)}{v} + \frac{1}{2!} \int_\eta^\alpha\!\!\int_\eta^\alpha F^-(v_1, v_2)\,\frac{dg(v_1)\,dg(v_2)}{v_1\,v_2} + \ldots + \frac{1}{r!} \int_\eta^\alpha \cdots \int_\eta^\alpha F^-(v_1, \ldots, v_r)\,\frac{dg(v_1)\ldots dg(v_r)}{v_1 \ldots v_r} + \ldots \right\},$$

where F^- varies over the class $F_{\eta,\alpha}^-$. In the last curly brackets in (8.40″) and (8.40‴) the expressions are finite as the series break off when $r\,\eta \ge 1$, the terms being from then on identically zero.

Here (8.40″) and (8.40‴) are fairly obvious. The expression in the first bracket in each case is just the limit of $T_u(\Lambda)$ as $x \to \infty$. The sieves constructed in Lemma 2 qualify as Λ_F with $F^{\pm}_{\eta,\alpha'}$, that we can extend them to cover the whole range $\mathcal{P}(x^\eta, x^\alpha)$ and still remain of the type Λ_F with the $F_r(v_1+\ldots+v_r)$ being zero for $v_1+\ldots+v_r < e^{-\delta}$ is also easily seen, so we can find admissible F which make the right hand side as close to the left hand side as we wish. That we cannot make the right hand side of (8.40″) less than the left hand side, or the right hand side of (8.40‴) greater than the left hand side can be seen as follows.

While it is true that the Λ_F associated with our F is a $\Lambda_F[\eta, \alpha; 1]$ or a $\Lambda[\mathcal{P}(x^\eta, x^\alpha), x]$ it may not necessarily have $R_u(\Lambda) \le x$, it is however easily seen that $R_u(\Lambda) \le x^{1+\varepsilon(x)}$ where $\varepsilon(x) \to 0$ as $x \to \infty$. For the definition of $T^{\pm}_g$ we considered the class $\Lambda_u(\mathcal{P}, x)$ with $R_u(\Lambda) \le x$. The only place where we have really made use of $R_u(\Lambda) \le x$ in our study of the $T^{\pm}_g$ was in the proof of Lemma 2 where we used it to prove (8.22). However, (8.22) follows just as well if we add a factor $x^{\varepsilon(x)}$ to the right hand side of the inequality immediately preceding (8.22). From this it is clear that the T_g are not changed by allowing the wider class of F or Λ implied by $F_r(v_1+\ldots+v_r) = 0$ for $v_1+\ldots+v_r > 1$ instead of $v_1+\ldots+v_r > 1-\delta$, this establishes (8.40″) and (8.40‴).

Now, consider (8.40) and (8.40′), using the argument above, it follows that for an admissible F^+_α the right hand side of (8.40) can not be less than the left hand side, and similarly for (8.40′) we see that for an F^-_α the right hand side cannot be larger than the left hand side.

It remains to show that in both cases the right hand side can be brought arbitrarily close to the the left hand side by suitable choice of F within the proper class.

We go back to the sieves used in the proof of Lemma 3. Apart from the step where a Λ^2 sieve is brought in it is clear that everything can be arranged within the Λ_F scheme since combination of two Λ_F sieves according to Principle I, again leads to a Λ_F sieve. The Λ'^- sieve used in the combination $\Lambda^2\,\Lambda'^-$ is seen to be a Λ_F, that we can also choose our Λ in Λ^2 to be a Λ_F is a bit harder to see, but it can be done as we shall see in a later section where we will treat the Λ^2 and $\Lambda^2\,\Lambda'^-$ sieves more thoroughly. At any rate it follows that we can find admissible F^+_α and F^-_α so that the right hand sides of (8.40) and (8.40′) are within a given $\varepsilon > 0$ of the left hand side.

We might however also simply choose to interpret the right hand sides of (8.40) and (8.40′) as the limits of the right hand sides (8.40″) and (8.40‴) respectively as $\eta \to 0$. This is actually just as useful to us.

We conclude this section by giving another form for the limit as $x \to \infty$ of a $T_u(\Lambda_F)$ where $F = F^{\pm}_{\eta,\alpha}$. This form corresponds to the relation

$$(8.41) \qquad \sum_{d \in \mathcal{P}} \frac{\theta_d}{f'(d)} = 1 + \sum_{p \prec d} \frac{\lambda_{pd} + \lambda_d}{f(pd)} \, \frac{e_u(\mathcal{P}(p))}{e_u(\mathcal{P})},$$

where the notation $p \prec d$ means that p is smaller than any prime factor of d, and $\mathcal{P}(p)$ means the set of primes in $\mathcal{P}$ which are $< p$.

We have

$$\sum_{d\in\mathcal{P}} \frac{\theta_d}{f'(d)} = 1 + \sum_{p\prec d} \frac{\sum_{\delta\,|\,d}(\lambda_{p\,\delta}+\lambda_\delta)}{f'(p\,d)} = 1 + \sum_{p\prec\delta} \frac{(\lambda_{p\,\delta}+\lambda_\delta)}{f'(p\,\delta)} \sum_{\substack{(\delta,d')=1\\ p\prec d'}} \frac{1}{f'(d')}$$

$$= 1 + \sum_{p\prec\delta} \frac{(\lambda_{p\,\delta}+\lambda_\delta)}{f'(p\,\delta)} \prod_{\substack{p'\in\mathcal{P}\\ (p',\delta)=1\\ p<p'}} \left(1+\frac{1}{f'(p')}\right)$$

$$= 1 + \sum_{p\prec d} \frac{(\lambda_{p\,d}+\lambda_d)}{f(p\,d)} \frac{e_u(\mathcal{P}(p))}{e_u(\mathcal{P})} .$$

Corresponding to this formula we have

(8.42)

$$1 + \int_\eta^\alpha \theta_F(v)\,\frac{dg(v)}{v} + \ldots + \frac{1}{r!}\int_\eta^\alpha \cdots \int_\eta^\alpha \theta_F(v_1,\ldots,v_r)\,\frac{dg(v_1)\ldots dg(v_r)}{v_1\ldots v_r} + \ldots$$

$$= 1 + \int_\eta^\alpha (F_1(v)+1)\exp(L(v,\alpha))\,\frac{dg(v)}{v}$$

$$+ \int\cdots\int_{\eta\le v_1\le\ldots\le v_r\le\alpha} (F_r(v_1,\ldots,v_r)+F_{r-1}(v_2,\ldots,v_r))$$

$$\times \exp(L(v_1,\alpha))\,\frac{dg(v_1)\ldots dg(v_r)}{v_1\ldots v_r} + \ldots ,$$

which may be proved in a quite similar manner by writing, for $r \ge 1$, the $\theta_F(v_1,\ldots,v_r)$ so that we group together terms in pairs where the first contains the smallest v, v_1, while the other has exactly the same arguments except for v_1, and then carrying out the integrations with respect to the variables that do not occur in the first F of the grouping.[9]

(8.42) will be particularly useful when we later deal with the combinatorial sieves, for these it will also hold if we put $\eta = 0$. For other sieves the terms on the right hand side of (8.42) may not necessarily exist for $\eta = 0$.

9. Remarks on the preceding section and generalizations

What is unsatisfactory about the results of the previous section is the need to impose a rather strong restriction, namely (8.26),[10] in order to get our main theorems about $T_g^\pm(\alpha)$.

[9] The complete argument will be given in a later section dealing with combinatorial sieves.

[10] Which we can relax a little, but not much, and only at the cost of making the condition complicated.

There are good reasons to think that these theorems may be true if we drop (8.26) entirely.

If we in our original definition of $T_g^{\pm}(\alpha)$ replace the class $\Lambda_u^{\pm}(\mathcal{P}(x^\alpha), x)$ with $\Lambda_u^{\pm}(\mathcal{P}(x^\alpha), x^a)$ and call the resulting object $T_g^{\pm}(\alpha;\, a)$ (so that $T_g^{\pm}(\alpha;\, 1) = T_g^{\pm}(\alpha)$), this is clearly a monotonic function of a, nonincreasing as a increases in the case of $T_g^{+}(\alpha;\, a)$, nondecreasing as a increases for $T_g^{-}(\alpha;\, a)$. Thus $T_g^{\pm}(\alpha;\, a)$ is continuous almost everywhere in a; if it is continuous at $a = 1$, more precisely only continuity to the left is needed, we can easily show that $\lim_{\eta\to 0} T_g^{\pm}(\eta, \alpha) = T_g^{\pm}(\alpha)$, which is really the crucial point, and the content of Theorems 1 and 2 would be true if we omit the part about effective computability. Similarly for a $g(t)$ that satisfies our conditions except (8.26), we can show that if we replace $g(t)$ with $g(t, a) = g_a$, where $g'(t, a)$ is such that it varies monotonically with the parameter a (say, is increasing if a increases, for instance), then for almost all a we will again have

$$\lim_{\eta\to 0} T_{g_a}^{\pm}(\eta, \alpha) = T_{g_a}^{\pm}(\alpha)\,,$$

and the same conclusions would again follow as above.

If we could approach say $T_g^{+}(\alpha)$ arbitrarily close by a series of $F_r(v_1, \ldots, v_r)$ on the right hand side of (8.40) which are such that the series ocurring on the right hand side remains convergent if the rth term is multiplied by a_0^r for some $a_0 > 1$, the continuity of $T_{g_a}^{+}(\alpha)$ with $g_a = a\, g(t)$ as a function of a near $a = 1$ would be obvious, and the same conclusions would hold as above. Similarly for $T_g^{-}(\alpha)$ and the series ocurring in (8.40′). For the few cases, treated in a later section, where we can explicitly obtain the best possible results (both are cases of constant density, and an optimal set of F_r actually exists) such an $a_0 > 1$ does exist. It seems, however, quite likely that when $g'(t)$ tends strongly to infinity for $t \to 0$, the convergence of the series on the right hand side of (8.40) and (8.40′) is slower than that of any geometric series, so that no such a_0 then exists.

We now shall consider the question of generalizing the results in Section 8 to cover the situation of sifting with weights which we referred to towards the end of Section 3.

We shall make certain assumptions about the nature of the weights $\sigma(d)$ beyond those mentioned in Section 3. We assume $\mathcal{P}$ is either of the form $\mathcal{P}(x^\alpha)$ or of the form $\mathcal{P}(x^\eta, x^\alpha)$, with $\alpha < 1$ and $\eta > 0$. We assume that for $d = p_1 p_2 \ldots p_r$ with the p_i, $i = 1, \ldots, r$ in $\mathcal{P}$ we have $\sigma(p_1 \ldots p_r) = 0$ for $r > r_0$, while for $r \le r_0$,

$$\sigma(p_1 \ldots p_r) = \sigma_r \left(\frac{\log p_1}{\log x}, \ldots, \frac{\log p_r}{\log x} \right), \tag{9.1}$$

where $\sigma_r(v_1, \ldots, v_r) \ge 0$ is a continuous or piecewise continuous function of the r arguments v_i for $0 < v_i \le \alpha$ or $\eta \le v_i \le \alpha$ according to the range being $\mathcal{P}(x^\alpha)$ or $\mathcal{P}(x^\eta, x^\alpha)$. We shall first consider the case when the $\sigma_r(v_1, \ldots, v_r)$

for $r \leq r_0$ also vanish if one or more of the v_i are $\leq \eta_0$. For simplicity we shall assume that the $\sigma_r(v_1, \ldots, v_r)$ are continuous for $\eta_0 \leq v_i \leq \alpha$,[11] this is no essential restriction since we could otherwise approximate from above (for the purpose of considering upper bounds) or from below (for getting lower bounds) with continuous σ_r.[12]

For a Λ which satisfies (3.10) and for which $R_u(\Lambda) \leq z$, we shall use the notation

$$\Lambda_u^+(\mathcal{P}, \sigma, z)$$

and for a Λ which satisfies (3.10) and for which $\lambda_d = 0$ for $d > z$ we write

$$\Lambda^+[\mathcal{P}, \sigma, z].$$

Similarly for Λ that satisfy (3.10′) and $R_u(\Lambda) \leq z$ or $\lambda_d = 0$ for $d > z$ we write

$$\Lambda_u^-(\mathcal{P}, \sigma, z)$$

and

$$\Lambda^-[\mathcal{P}, \sigma, z]$$

respectively.

We may now introduce in analogy with (8.1), (8.1′), (8.1″), and (8.1‴)

$$\liminf_{x\to\infty} T_u(\Lambda^+) = T_{g,\sigma}^+(\alpha)\,, \tag{9.2}$$

where $\Lambda^+ = \Lambda_u^+(\mathcal{P}(x^\alpha), \sigma, x)$, and

$$\limsup_{x\to\infty} T_u(\Lambda^-) = T_{g,\sigma}^-(\alpha)\,, \tag{9.2′}$$

where $\Lambda^- = \Lambda_u^-(\mathcal{P}(x^\alpha), \sigma, x)$.

Similarly for $0 < \eta < \alpha < 1$,

$$\liminf_{x\to\infty} T_u(\Lambda^+) = T_{g,\sigma}^+(\eta, \alpha)\,, \tag{9.2″}$$

where $\Lambda^+ = \Lambda_u^+(\mathcal{P}(x^\eta, x^\alpha), \sigma, x)$, and

$$\limsup_{x\to\infty} T_u(\Lambda^-) = T_{g,\sigma}^-(\eta, \alpha)\,, \tag{9.2‴}$$

where $\Lambda^+ = \Lambda_u^+(\mathcal{P}(x^\eta, x^\alpha), \sigma, x)$.

Here, as we did in Section 8, we consider as $x \to \infty$, all possible sequences of $u(p)$ that satisfy (4.2) and (4.4) and $0 \leq u(p) < p$, and furthermore that g and g' satisfies the requirements specified in Section 4.

That the definitions above define finite quantities, is rather clear in case of (9.2″) and (9.2‴). In case of (9.2′) it is also clear, since we are not tied to the normalization $\lambda_1 = 1$ for these sieves, the sieve with all $\lambda_d = 0$ is permissible

[11] More precisely, continuous to the right at η_0 and to the left at α.

[12] Bounds for the errors so committed are easily obtained.

for the lower bound problem so it is clear that $T^-_{g,\sigma}(\alpha) \geq 0$. As to (9.2), if C is an upper bound for the weights σ_0, $\sigma_r(v_1, \ldots, v_r)$ for $1 \leq r \leq r_0$ and $\eta_0 \leq v_i \leq \alpha$, then the sieve $C\,\Lambda^+[\mathcal{P}(x^{\eta_0}),\, x^{1-\delta}]$ would be permissible so that we would have

$$T^+_{g,\sigma}(\alpha) \leq C\,T^+_g(\eta_0)\,\exp(L\,(\eta_0, \alpha))\,,$$

and therefore bounded.

We now can prove the analog of Lemma 1,

Lemma 4. *We have*

(i) $T^+_{g,\sigma}(\alpha)$ *is continuous in* α *for* $0 \leq \alpha < 1$.

(ii) $T^-_{g,\sigma}(\alpha)$ *is continuous for* $0 \leq \alpha < 1$.

(iii) $T^+_{g,\sigma}(\eta, \alpha)$ *is continuous in both* η *and* α *for* $0 < \eta \leq \alpha < 1$, *and*

(iv) $T^-_{g,\sigma}(\eta, \alpha)$ *is continuous in both* η *and* α *for* $0 < \eta \leq \alpha < 1$.

We sketch the proof for the first two statements, the last two can be proved along similar lines and are actually simpler.

Let $0 < \alpha_1 < \alpha_2 < 1$. From the definition of $T^+_{g,\alpha}(\alpha_2)$ it follows that there is a sequence of x and associated with each x in this sequence a set of $u(p)$ satisfying our conditions and a $\Lambda^+_u(\mathcal{P}(x^{\alpha_2}), \sigma, x)$, such that when $x \to \infty$ through this sequence

$$\lim T_u(\Lambda^+_u(\mathcal{P}(x^{\alpha_2}), \sigma, x)) = T^+_{g,\sigma}(\alpha_2)\,.$$

If we for such x take the same set of $u(p)$ and restrict the sieve $\Lambda^+_u(\mathcal{P}(x^{\alpha_2}), \sigma, x)$ to the range $\mathcal{P}(x^{\alpha_1})$ we have clearly

$$T_u(\Lambda^+_u(\mathcal{P}(x^{\alpha_1}), \sigma, x)) \leq T_u(\Lambda^+_u(\mathcal{P}(x^{\alpha_2}), \sigma, x))\,,$$

from which we get

$$T^+_{g,\sigma}(\alpha_1) \leq T^+_{g,\sigma}(\alpha_2)\,. \tag{9.3}$$

Again from the definition of $T^+_{g,\sigma}(\alpha_1)$, we have that there exists a sequence of x tending to ∞, and with each such x associated a set of $u(p)$ satisfying our conditions and a $\Lambda^+_u(\mathcal{P}(x^{\alpha_1}), \sigma, x)$ such that

$$\lim T_u(\Lambda^+_u(\mathcal{P}(x^{\alpha_1}), \sigma, x)) = T^+_{g,\sigma}(\alpha_1)\,.$$

We now consider the sequence $x' = 2\,x$, where x runs through our sequence above, and we wish to modify the sieve $\Lambda^+_u(\mathcal{P}(x^{\alpha_1}), \sigma, x)$ so that it can be used on the range $\mathcal{P}(x'^{\alpha_2})$. From (7.11) or (7.20) we see that given η_0 and $\alpha_2 < 1$ we can find an $\eta > 0$ and $\leq \eta_0$ such that there is a sieve $\Lambda^+[\mathcal{P}(x^{\eta}),\, x^{(1-\alpha_2)/2}]$ with $T_u(\Lambda^+[\mathcal{P}(x^{\eta}),\, x^{(1-\alpha_2)/2}]) < 2$ for x large enough. Let us refer to the λ in the first sieve (the $\Lambda^+_u(\mathcal{P}(x^{\alpha_1}, \sigma, x))$) by $\lambda^{(1)}_d$ and in the second sieve by $\lambda^{(2)}_d$, if we construct a new sieve

$$\Lambda^+_u(\mathcal{P}(x'^{\alpha_2}), \sigma)$$

as follows: we define for d in $(\mathcal{P}(x'^{\alpha_2}))$

$$\lambda_d = (1+\varepsilon)\,\lambda_d^{(1)} + C \sum_{x^{\alpha_1}\le p \le x'^{\alpha_2}} \lambda_{d/p}^{(2)}, \tag{9.4}$$

where $0 < \varepsilon < 1/2$ and C again is a constant which majorizes the σ_r. It is easily seen that the new sieve is a

$$\Lambda_u^+(\mathcal{P}(x'^{\alpha_2}), \sigma, x'),$$

for x' large enough. Furthermore, we get for x' large enough

$$\begin{aligned} T_u(\Lambda_u^+(\mathcal{P}(x'^{\alpha_2}),\sigma,x')) &\le (1+\varepsilon)\,T_u^+(\Lambda_u^+(\mathcal{P}(x^{\alpha_1}),\sigma,x))\,\frac{e_u(x^{\alpha_1})}{e_u(x'^{\alpha_2})} \\ &\quad + 2\,C\,\frac{e_u(x^{\eta})}{e_u(x'^{\alpha_2})} \sum_{x^{\alpha_1}\le p<x'^{\alpha_2}} \frac{u(p)}{p}, \end{aligned}$$

or by letting x' go to ∞ in our sequence

$$T_{g,\sigma}^+(\alpha_2) \le (1+\varepsilon)\,T_{g,\sigma}^+(\alpha_1)\,\exp(L\,(\alpha_1,\alpha_2)) + 2\,C\,\exp(L\,(\eta,\alpha_2))\,L(\alpha_1,\alpha_2),$$

or since we may choose ε arbitrarily close to zero

$$T_{g,\sigma}^+(\alpha_2) \le T_{g,\sigma}^+(\alpha_1)\,\exp(L\,(\alpha_1,\alpha_2)) + 2\,C\,\exp(L\,(\eta,\alpha_2))\,L(\alpha_1,\alpha_2). \tag{9.5}$$

(9.3) and (9.5) clearly prove (i).

To prove (ii) we proceed in a similar way, one gets

$$T_{g,\sigma}^-(\alpha_1) \ge T_{g,\sigma}^-(\alpha_2) - C\,\exp(L\,(\eta_0,\alpha_2))\,L(\alpha_1,\alpha_2), \tag{9.6}$$

in a similar manner as (9.3) was obtained. We then construct from a $\Lambda^{(1)} = \Lambda_u^-(\mathcal{P}(x^{\alpha_1}),\sigma,x)$ and a $\Lambda^{(2)} = \Lambda^+[\mathcal{P}(x^{\eta}), x^{(1-\alpha_2)/2}]$ a $\Lambda_u^-(\mathcal{P}(x'^{\alpha_2}),\sigma,x')$ by a construction similar to (9.4), taking for d in $(\mathcal{P}(x'^{\alpha_2}))$

$$\lambda_d = (1-\varepsilon)\,\lambda_d^{(1)} - C \sum_{x^{\alpha_1}\le p\le x'^{\alpha_2}} \lambda_{d/p}^{(2)},$$

and obtain in a similar way

$$T_{g,\sigma}^-(\alpha_2) \ge (1-\varepsilon)\,T_{g,\sigma}^-(\alpha_1)\,\exp(L\,(\alpha_1,\alpha_2)) - 2\,C\,\exp(L\,(\eta,\alpha_2))\,L(\alpha_1,\alpha_2),$$

or since ε can be chosen arbitrarily small,

$$T_{g,\sigma}^-(\alpha_2) \ge T_{g,\sigma}^-(\alpha_1)\,\exp(L\,(\alpha_1,\alpha_2)) - 2\,C\,\exp(L\,(\eta,\alpha_2))\,L(\alpha_1,\alpha_2).$$

This combined with (9.6) give (ii). We need next an analog of Lemma 2. If we have a sequence F_0, $F_1(v_1), \ldots$, $F_r(v_1,\ldots,v_r), \ldots$, where F_0 is a constant and F_r a function of r variables $0 < v_i \le \alpha$ or $\eta \le v_i \le \alpha$ for $i = 1,\ldots,r$, and such that $F_r = 0$ if $v_1 + \ldots + v_r > 1 - \delta$, and such that the Λ system with $\lambda_1 = F_0$,

$$\lambda_d = F_r\left(\frac{\log p_1}{\log x}, \ldots, \frac{\log p_r}{\log x}\right)$$

for $d = p_1 \ldots p_r$, does satisfy the inequalities (3.10), we shall denote it by $\Lambda^+_{F,\sigma}[\alpha;\, 1-\delta]$ or $\Lambda^+_{F,\sigma}[\eta, \alpha; 1-\delta]$ depending on the range. Similarly, if the Λ satisfies the inequalities (3.10′) we write $\Lambda^-_{F,\sigma}[\alpha;\, 1-\delta]$ or $\Lambda^-_{F,\sigma}[\eta, \alpha;\, 1-\delta]$.

Lemma 5. *Let $0 < \eta < \eta_0 < \alpha < 1$ and let η' and α' be defined by*

$$L(\eta, \eta') = L(\alpha', \alpha) = \frac{2}{N+2}\, L(\eta, \alpha)$$

and

$$\delta = \frac{L(\eta, \alpha)}{(N+2)\,\bar{g}'(\eta, \alpha)}$$

then there exists a $\Lambda^+_{F,\sigma}[\eta, \alpha';\, e^{-\delta}]$ such that for a given $\varepsilon > 0$ and N large enough

$$\lim_{x\to\infty} T_u(\Lambda^+_{F,\sigma}[\eta, \alpha';\, e^{-\delta}]) \le T^+_{g,\sigma}(\eta', \alpha) + \varepsilon\, T^+_g(\eta, \eta_0)\, \exp(L\,(\eta_0, \alpha'))\,. \tag{9.7}$$

Furthermore, on the same assumptions, there exists a $\Lambda^-_{F,\sigma}[\eta, \alpha';\, e^{-\delta}]$ such that for $\varepsilon > 0$ and N large enough

$$\lim_{x\to\infty} T_u(\Lambda^-_{F,\sigma}[\eta, \alpha';\, e^{-\delta}]) \ge T^+_{g,\sigma}(\eta', \alpha) - \varepsilon\, T^+_g(\eta, \eta_0)\, \exp(L\,(\eta_0, \alpha'))\,. \tag{9.8}$$

In (9.7) and (9.8) x may go to infinity through any sequence.

We shall sketch the proof for the case that $g(t)$ is strictly increasing between η_0 and α, and later briefly indicate what modifications have to be made if $g(t)$ is constant in any subinterval of (η_0, α). We go back to the division of the interval (η, α) used in the proof of Lemma 2. If $g(t)$ is strictly increasing in (η_0, α), we choose our N so large that if $\eta_0 \le v_i < v_i' \le \alpha$, for $i = 1, \ldots, r$, $r \le r_0$, and

$$L(v_i, v_i') \le \frac{3\,L(\eta, \alpha)}{N+2}\,, \tag{9.9}$$

then

$$\sigma_r(v_1, \ldots, v_r) \le \sigma_r(v_1', \ldots, v_r') + \frac{\varepsilon}{2}\,. \tag{9.10}$$

It is also evident from the construction of the intervals $\Delta_i = (t_{i-1}, t_i)$ and $\Delta_i' = (t_{i+1}, t_{i+2})$ for $i = 1, \ldots, N$, that if v is in Δ_i and v' is in Δ_i', then

$$L(v, v') \le \frac{3\,L(\eta, \alpha)}{N+2}\,.$$

If N is chosen large enough there is a $\Lambda_F^+[\eta, \eta_0;\, e^{-\delta}]$ such that

$$\lim_{x\to\infty} T_u(\Lambda_F^+[\eta,\eta_0;\, e^{-\delta}]) \le \frac{3}{2}\, T_g^+(\eta,\eta_0)\,.$$

If we now consider a sequence of x_j tending to infinity and such that with each x_j there is associated a $\Lambda_u^+(\mathcal{P}(x_j^{\eta'},\, x_j^{\alpha}), \sigma, x_j)$ such that

$$\lim_{j\to\infty} T_u(\Lambda_u^+(\mathcal{P}(x_j^{\eta'}, x_j^{\alpha}), \sigma;\, x_j)) = T_{g,\sigma}^+(\eta',\alpha)\,,$$

and form ratios like in (8.21), it is easily shown again, as it was in the former situation, that we can pick out a subsequence of the x_j, such that these ratios tend to zero if $\sum_{\nu=1}^r t_{i_\nu+1} > 1$, and to finite limits $\ell_j(i_1,\ldots,i_r)$ for $\sum_{\nu=1}^r t_{i_\nu+1} \le 1$. It also is easily seen that if we now define a sieve $\Lambda_{F,\sigma}^+[\eta,\alpha';\, e^{-\delta}]$ as follows for $d = p_1 \ldots p_r$, we write

$$\lambda_d = \ell(i_1,\ldots,i_r) + \frac{\varepsilon}{2}\,\lambda_d'\,, \tag{9.11}$$

where the λ_d' is taken from the $\Lambda_F^+[\eta,\eta_0;\, e^{-\delta}]$ mentioned above and $\log p_\nu/\log x$ lies in Δ_{i_ν} for $\nu = 1,\ldots,r$. One then gets easily

$$\lim_{x\to\infty} T_u(\Lambda_{F,\sigma}^+[\eta,\alpha';\, e^{-\delta}]) \le T_{g,\sigma}^+(\eta',\alpha) + \frac{3}{4}\,\varepsilon\, T_g^+(\eta,\eta_0)\, \exp(L\,(\eta_0,\alpha'))\,,$$

which proves (9.7) (and with a little bit to spare!).

If $g(t)$ is not strictly increasing in (η_0,α), there are intervals where $g(t)$ is constant. This means that some of the intervals Δ_i in (η_0,α) may not become small as N is chosen large, thus we cannot always use the inequality (9.10) as points in Δ_i and Δ_i' are not necessarily close together. We call an interval in Δ_i exceptional if $t_{i+2} - t_{i-1} > \Delta(\varepsilon)$, where $\Delta(\varepsilon)$ is the upper bound of the Δ such that (9.10) holds whenever $0 \le v_i' - v_i \le \Delta$ for all $i = 1,\ldots,r$; $\eta_0 \le v_i < v_i' \le \alpha$. Clearly there are less than $3\,(\alpha-\eta_0)/\Delta(\varepsilon)$ exceptional Δ_i in (η_0,α) and for large x the $\sum u(p)/p$ taken over these exceptional intervals is clearly $< (1+\varepsilon_x)\,\frac{3\,(\alpha-\eta_0)}{\Delta(\varepsilon)}\,\frac{L(\eta,\alpha)}{N+2}$, where $\varepsilon_x \to 0$ as $x \to \infty$. We assume that δ is small enough that $e^{-\delta} - \alpha > (1-\alpha)/2$, and can then construct a $\Lambda_F^+[\eta,\eta_0;\, e^{-\delta}-\alpha]$ such that for x large enough

$$T_u(\Lambda_F^+[\eta,\eta_0;\, e^{-\delta}-\alpha]) < C\,(g;\eta_0,\alpha)\,.$$

Let now C' be a constant which is a majorant of the σ_r, $r = 1,\ldots,r_0$, and denote the λ in this $\Lambda_F^+[\eta,\eta_0;\, e^{-\delta}-\alpha]$ by $\lambda_d^{(2)}$, and call p exceptional if $\log x/\log p$ belongs to one of the exceptional Δ_i in (η_0,α); we then modify the construction (9.11) by adding on the right hand side the term

$$C' \sum_{\substack{x^\eta \le p \le x^{\alpha'} \\ p \text{ exceptional}}} \lambda_{d/p}^{(2)}\,. \tag{9.12}$$

This gives us a new $\Lambda^+_{F,\sigma}[\eta,\alpha';\, e^{-\delta}]$ and we find now

$$\lim_{x\to\infty} T_u(\Lambda^+_{F,\sigma}[\eta,\alpha';\, e^{-\delta}]) \le T^+_{g,\sigma}(\eta',\alpha) + \frac{3}{4}\,\varepsilon\, T^+_g(\eta,\eta_0)\, \exp(L\,(\eta_0,\alpha')) + \frac{L\,(\eta,\alpha)}{N+2}\,\frac{3\,(\alpha-\eta_0)}{\Delta(\varepsilon)}\, C\,(g;\,\eta_0,\alpha)\, \exp(L\,(\eta_0,\alpha'))\,.$$

For N large enough the third term on the right hand side is less than one third of the second term, and so we get

$$\lim_{x\to\infty} T_u(\Lambda^+_{F,\sigma}[\eta,\alpha';\, e^{-\delta}]) \le T^+_{g,\sigma}(\eta',\alpha) + \varepsilon\, T^+_g(\eta,\eta_0)\, \exp(L\,(\eta_0,\alpha'))\,.$$

This proves (9.7) in the general case. The proof of (9.8) can be carried out along quite similar lines. In (9.10) we would reverse the inequality and put $-\varepsilon/2$ on the right hand side. In (9.11) we reverse the sign of $\frac{\varepsilon}{2}\,\lambda'_d$, and we subtract instead of adding the expression (9.12) to complete our construction of a $\Lambda^-_{F,\sigma}[\eta,\alpha';\, e^{-\delta}]$ for which (9.8) holds.

We finally need an analog of Lemma 3, for this we again need to assume (8.26) or some similar restriction on $g(t)$.

Lemma 6. *If* (8.26) *holds, we have for* $\alpha < 1$

(i) $$\lim_{\eta\to 0} T^+_{g,\sigma}(\eta,\alpha) = T^+_{g,\sigma}(\alpha)\,,$$

and

(ii) $$\lim_{\eta\to 0} T^-_{g,\sigma}(\eta,\alpha) = T^-_{g,\sigma}(\alpha)\,.$$

Furthermore, if the function $g'(t)$ is effectively known, the differences $T^+_{g,\sigma}(\alpha) - T^+_{g,\sigma}(\eta,\alpha)$ and $T^-_{g,\sigma}(\eta,\alpha) - T^-_{g\sigma}(\alpha)$ can be effectively estimated.

To prove the first statement we proceed much as in the proof of the first part of Lemma 3. We combine according to Principle I in Section 6 a $\Lambda^2[x^\eta, x^{\sqrt{\eta}}]$ with

$$T_u(\Lambda^2\,[x^\eta, x^{\sqrt{\eta}}]) < (1-e^{-1/\sqrt{\eta}})^{-1}\,,$$

with the $\Lambda^+_{F,\sigma}[\eta,\alpha';\, e^{-\delta}]$ from Lemma 5 this is seen to give a

$$\Lambda^+[\mathcal{P}(x^{\alpha'}),\sigma, x^{e^{-\delta}+\sqrt{\eta}}],$$

where $e^{-\delta}+\sqrt{\eta} < 1$. It should be remarked that Principle I works in this more general situation because of our assumption that the $\sigma_r(v_1,\ldots,v_r) = 0$ if one or more of the v_i are $< \eta_0$ and $\eta < \eta_0$. It remains to complete the range by adding the range $\mathcal{P}(x^{\alpha'}, x^\alpha)$. This can be easily done along the lines used to prove (i) in Lemma 4. We construct in this way a $\Lambda^+[\mathcal{P}(x^\alpha),\sigma,x^{1-\varepsilon}]$ such that

$$\lim_{x\to\infty} T_u(\Lambda^+[\mathcal{P}(x^\alpha),\sigma,x^{1-\varepsilon}])$$

can come arbitrarily close to the limit on the left hand side of (i), this proves the first statement of the Lemma.

The second part is proved following the pattern used to prove the second part of Lemma 3. Again, we see we can use Principle I with the construction (6.5) to add the range $\mathcal{P}(x^\eta)$ to the range $\mathcal{P}(x^\eta, x^{\alpha'})$ of the sieve obtained in the second part of Lemma 5. To complete the range by adding the range $\mathcal{P}(x^{\alpha'}, x^\alpha)$ we follow the pattern used to prove (ii) in Lemma 4. In this way a $\Lambda^-[\mathcal{P}(x^\alpha), \sigma, x^{1-\varepsilon}]$ is constructed such that

$$\lim_{x\to\infty} T_u(\Lambda^-[\mathcal{P}(x^\alpha), \sigma, x^{1-\varepsilon}])$$

can be made arbitrarily close to the limit on the left hand side of (ii) which proves the second part of the Lemma.

The part about the effective estimation of the differences $T^+_{g,\sigma}(\alpha) - T^+_{g,\sigma}(\eta, \alpha)$ and $T^-_{g,\sigma}(\eta, \alpha) - T^-_{g,\alpha}(\alpha)$ can be deduced from the steps in the above proofs. This is somewhat messier than in the corresponding statement in Lemma 3, but fairly simple if we have a positive lower bound for $g'(t)$ in (η_0, α). We can now prove, in a manner quite similar to the way Theorem 1 was proved,

Theorem 3. *We have under our assumptions about g, g' and σ for $\alpha < 1$,*

$$\lim_{x\to\infty} \frac{M_u^+(W_x, \mathcal{P}(x^\alpha), \sigma)}{E_u(W_x, \mathcal{P}(x^\alpha), \sigma)} = T^+_{g,\sigma}(\alpha)\,,$$

and

$$\lim_{x\to\infty} \frac{M_u^-(W_x, \mathcal{P}(x^\alpha), \sigma)}{E_u(W_x, \mathcal{P}(x^\alpha), \sigma)} = T^-_{g,\sigma}(\alpha)\,.$$

Similar results hold if we replace the range $\mathcal{P}(x^\alpha)$ by $\mathcal{P}(x^\eta, x^\alpha)$ and $T^\pm_{g,\sigma}(\alpha)$ by $T^\pm_{g,\sigma}(\eta, \alpha)$. These limits are effectively computable.

We can of course also give an analog of Theorem 2 for this more general situation, but shall not bother to formulate it here, the form is of course quite obvious.

We made the requirement of our "weights" $\sigma_r(v_1, \ldots, v_r)$ that they vanish if any of the arguments v_i were less than some positive bound η_0. When in practice sifting with weights this requirement is usually made with some quite small η_0. It is natural to ask what happens if we give up this requirement. If we want the expected value of the sifting for the range $\mathcal{P}(x^\alpha)$ to be of the order $x\, e_u(\mathcal{P}(x^\alpha))$ we must have

$$\sum_d \frac{\sigma(d)}{f'(d)}$$

bounded as $x \to \infty$. This requires that for $1 \le r \le r_0$

$$\int \cdots \int_{0<v_i\le\alpha} \sigma_r(v_1, \ldots, v_r)\, \frac{dg(v_1)\ldots dg(v_r)}{v_1 \ldots v_r} < \infty\,,$$

but it also requires stronger assumptions about the $u(p)$ with $\log p = o(\log x)$, since (4.2) says nothing about these. If we assume for instance that $S_u(\xi) < C \log \xi$ for $\xi > 1$ and some constant C, which would imply $g(t) \le C\,t$, we could for instance permit $\sigma_r(v_1, \dots, v_r)$ which are majorized by some constant times v^δ where $\delta > 0$ and v is the smallest of the v_i, $i = 1, \dots, r$. Weaker assumptions about $S_u(\xi)$ and $g(t)$ would require stronger restrictions on the $\sigma_r(v_1, \dots, v_r)$.

It is possible in the case where we assume $S_u(\xi) < C \log \xi$, and $\sigma_r(v_1, \dots, v_r) < C\, v^\delta$, where $v = \min(v_1, \dots, v_r)$ for $1 \le r \le r_0$ to prove analogs of Lemmas 4, 5, and 6. An analog of Theorem 3, as well as an analog of Theorem 2 can also be proved.

The main new difficulty comes in adding the beginning range $\mathcal{P}(x^\eta)$ to the range $\mathcal{P}(x^\eta, x^{\alpha'})$ in the proof of the analog of Lemma 6, since we can't use the simple upper bound sieve used before on the range $\mathcal{P}(x^\eta)$ but must construct a more complicated upper bound sieve with weights for $\mathcal{P}(x^\eta)$. This can be done by modifying the simple upper bound sieve suitably. Combining the two ranges is also trickier to affect for the upper bound case, we cannot just use Principle I but must do further modifications. The lower bound case is actually simpler for this situation since we there actually may work with the same simple sieves for the early range $\mathcal{P}(x^\eta)$ as used before, since to replace the $\sigma(d)$, where d contains a prime in $\mathcal{P}(x^\eta)$, with zero can only lower our bounds.

In closing the treatment of sifting with weights it should be remarked that while, as we have seen, one can develop the general theory and prove existence theorems in a manner rather similar to what could be done for ordinary sifting, when it comes to the actual construction of effective sieves the situation is quite different. We simply have no direct way of constructing sieves with weight, nothing like the combinatorial Principle V or like the $\Lambda^2\,\Lambda^-$ sieves mentioned in Section 6.

Sieves for sifting with weights have so far always been constructed or put together from the sieves developed for ordinary sifting.

10. Some asymptotic formulas relating to the Λ^2 and $\Lambda^2\,\Lambda^-$ sieves

From our treatment in Section 7 of the Λ^2 and $\Lambda^2\,\Lambda^-$ sieves, it is evident that the estimation of sums of the type

$$\sum_{\substack{d \in (\mathcal{P}(\xi)) \\ d < z}} \frac{1}{f'(d)}, \tag{10.1}$$

or more generally

$$\sum_{\substack{d\in(\mathcal{P}(\xi))\\ d<z}} \frac{\sigma\left(\dfrac{\log d}{\log z}\right)}{f'(d)}, \tag{10.1'}$$

where $\sigma(v)$ is a continuous function defined on the interval $(0, 1)$, play an important role. We wish to investigate the asymptotic behavior of such sums if $\xi = x^v$; $z = x^w$ as x goes to infinity.

We write for brevity

$$\Sigma_x(v; w) = \sum_{\substack{d\in(\mathcal{P}(x^\eta))\\ d<x^w}} \frac{1}{f'(d)},$$

and also

$$\Sigma_x(\eta, v; w) = \sum_{\substack{d\in(\mathcal{P}(x^\eta, x^v))\\ d<x^w}} \frac{1}{f'(d)}.$$

For the "complete" sums, where we remove the restriction $d \le x^w$, we write $\Sigma_x(v)$ and $\Sigma_x(\eta, v)$. We have of course $\Sigma_x(v) = e_u^{-1}(x^v)$ and $\Sigma_x(\eta, v) = e_u^{-1}(x^\eta, x^v)$.

For $\delta > 0$ we have clearly

$$\Sigma_x(\eta; \delta)\, \Sigma_x(\eta, v, w-\delta) < \Sigma_x(v; w) < \Sigma_x(\eta)\, \Sigma_x(\eta, v; w). \tag{10.2}$$

We wish to study the ratio

$$\frac{\Sigma_x(v; w)}{\Sigma_x(v)} = r_x(v, w), \tag{10.3}$$

as $x \to \infty$, and to prove that it tends to a limit function depending on $g(t)$ and v and w. Since $\Sigma_x(v) = \Sigma_x(\eta)\, \Sigma_x(\eta, v)$, we get from (10.2) that

$$r_x(v, w) < \frac{\Sigma_x(\eta, v;\, w)}{\Sigma_x(\eta, v)}, \tag{10.4}$$

and also

$$r_x(v, w) > \frac{\Sigma_x(\eta; \delta)}{\Sigma_x(\eta)}\ \frac{\Sigma_x(\eta, v; w-\delta)}{\Sigma_x(\eta, v)}. \tag{10.4'}$$

From (10.4) we see that as $x \to \infty$, we have

$$\begin{aligned}\limsup_{x\to\infty} r_x(v, w) &\le \exp(-L(\eta, v))\\ &\times \left\{1 + \sum_{r=1}^{\infty} \frac{1}{r!} \int\!\!\cdots\!\!\int_{\substack{\eta\le t_i<v\\ t_1+\ldots+t_r\le w}} \frac{dg(t_1)\ldots dg(t_r)}{t_1\ldots t_r}\right\}.\end{aligned} \tag{10.5}$$

Looking at the right hand side of (10.4′), we see that the first factor is the inverse of the right hand side of (7.8) with $\xi = x^{\eta}$ and $z = x^{\delta}$, and from (7.11) we see that

$$\liminf_{x\to\infty} \frac{\Sigma_x(\eta;\delta)}{\Sigma_x(\eta)} \geq 1 - \exp\left(-\frac{\delta}{\eta}\log\frac{\delta}{e\,g(\eta)} - \frac{g(\eta)}{\eta}\right), \tag{10.6}$$

while the second factor on the right hand side of (10.4′) tends to the limit

$$\exp(-L(\eta,v))\left\{1+\sum_{r=1}^{\infty}\frac{1}{r!}\int\cdots\int_{\substack{\eta\leq t_i<v\\ t_1+\ldots+t_r\leq w-\delta}}\frac{dg(t_1)\ldots dg(t_r)}{t_1\ldots t_r}\right\},$$

thus if we choose η small enough in relation to δ that $e^2\,g(\eta) < \delta$, we get from this (10.4′) and (10.6), that

$$\begin{aligned}\liminf_{x\to\infty} r_x(v,w) &\geq (1-e^{-\delta/\eta})\exp(-L(\eta,v))\\ &\times\left\{1+\sum_{r=1}^{\infty}\frac{1}{r!}\int\cdots\int_{\substack{\eta\leq t_i<v\\ t_1+\ldots+t_r\leq w-\delta}}\frac{dg(t_1)\ldots dg(t_r)}{t_1\ldots t_r}\right\}.\end{aligned}\tag{10.7}$$

Consider the first sum in the curly bracket on the right hand side of (10.5). We introduce the function $\rho_\delta(w)$ defined as follows

$$\rho_\delta(w) = \begin{cases} 1 & \text{for } w\geq 0,\\ 1+w/\delta & \text{for } 0\geq w\geq -\delta,\\ 0 & \text{for } w\leq -\delta.\end{cases}\tag{10.8}$$

It is easy to see that $\rho_\delta(w)$ can be expressed as

$$\rho_\delta(w) = \frac{1}{2\pi i}\int_{-i\infty}^{i\infty}\frac{e^{\delta s}-1}{\delta s}\frac{e^{ws}}{s}\,ds, \tag{10.9}$$

where we integrate along the imaginary axis, but pass around $s=0$ on the right. Clearly the sum we consider is less than

$$1+\sum_{r=1}^{\infty}\frac{1}{r!}\int_{\eta}^{v}\cdots\int_{\eta}^{v}\rho_\delta(w-t_1\ldots-t_r)\frac{dg(t_1)\ldots dg(t_r)}{t_1\ldots t_r}.$$

If we here insert the expression (10.9) for ρ_δ we see that the resulting expression converges even if we take absolute values in each term under the integral sign, so we may move the integration with respect to s outside on the summation and get

$$\begin{aligned}&\frac{1}{2\pi i}\int_{-i\infty}^{\infty}\frac{e^{\delta s}-1}{\delta s}\frac{e^{ws}}{s}\,ds\left\{1+\sum_{r=1}^{\infty}\frac{1}{r!}\int_{\eta}^{v}\cdots\int_{\eta}^{v}e^{-s(t_1+\ldots t_r)}\frac{dg(t_1)\ldots dg(t_r)}{t_1\ldots t_r}\right\}\\ &=\frac{1}{2\pi i}\int_{-i\infty}^{i\infty}\frac{e^{\delta s}-1}{\delta s}\frac{e^{ws}}{s}\exp\left(\int_{\eta}^{v}e^{-st}\frac{dg(t)}{t}\right)ds.\end{aligned}$$

From this and (10.5) we now get

$$\limsup_{x\to\infty} r_x(v,w) < \exp\left(-\int_\eta^v \frac{dg(t)}{t}\right)$$

$$\times \frac{1}{2\pi i}\int_{-i\infty}^{i\infty} \frac{e^{\delta s}-1}{\delta s}\,\frac{e^{ws}}{s}\,\exp\left(-\int_\eta^v e^{-st}\,\frac{dg(t)}{t}\right)ds$$

$$= \frac{1}{2\pi i}\int_{-i\infty}^{i\infty} \frac{e^{\delta s}-1}{\delta s}\,\exp\left(ws-\int_\eta^v \frac{1-e^{-st}}{t}\,dg(t)\right)\frac{ds}{s}.$$

The expression on the right hand side exists as a function of η also for $\eta = 0$ and is continuous there, so we get letting $\eta \to 0$ that
(10.10)

$$\limsup_{x\to\infty} r_x(v,w) \le \frac{1}{2\pi i}\int_{-i\infty}^{i\infty} \frac{e^{\delta s}-1}{\delta s}\,\exp\left(ws-\int_0^v \frac{1-e^{-st}}{t}\,dg(t)\right)\frac{ds}{s}.$$

Next we look at the sums in the curly bracket on the right hand side of (10.7). We see in a similar way that this is greater than

$$1+\sum_{r=1}^{\infty}\int_\eta^v\cdots\int_\eta^v \rho_\delta(w-2\,\delta-t_1-\ldots-t_r)\,\frac{dg(t_1)\ldots dg(t_r)}{t_1\ldots t_r}$$

$$= \frac{1}{2\pi i}\int_{-i\infty}^{i\infty} \frac{e^{-\delta s}(1-e^{-\delta s})}{\delta s}\,\exp\left(ws+\int_\eta^v \frac{e^{-st}}{t}\,dg(t)\right)\frac{ds}{s},$$

and so we get from (10.7) in a similar way

$$\liminf_{x\to\infty} r_x(v,w) \ge (1-e^{-\delta/\eta})$$

$$\times \frac{1}{2\pi i}\int_{-i\infty}^{i\infty} \frac{e^{-\delta s}(1-e^{-\delta s})}{\delta s}\,\exp\left(ws-\int_\eta^v \frac{1-e^{-st}}{t}\,dg(t)\right)\frac{ds}{s}.$$

We may again let $\eta \to 0$ and obtain

(10.10′)
$$\liminf_{x\to\infty} r_x(v,w) \ge \frac{1}{2\pi i}\int_{-i\infty}^{i\infty} \frac{e^{-\delta s}(1-e^{-\delta s})}{\delta s}\,\exp\left(ws-\int_0^v \frac{1-e^{-st}}{t}\,dg(t)\right)\frac{ds}{s}.$$

From the way the right hand sides of (10.10) and (10.10′) arose, it is clear that the right hand side of (10.10) is decreasing as δ decreases, while the right hand side of (10.10′) increases when δ decreases, when $\delta \to 0$ both integrals converge to

$$\frac{1}{2\pi i}\int_{-i\infty}^{i\infty} \exp\left(w\,s-\int_0^v \frac{1-e^{-st}}{t}\,dg(t)\right)\frac{ds}{s},$$

this can be shown by dividing the integrals into a central part $(-iT, iT)$ for large T and two tail parts $(-i\infty, iT)$ and $(iT, i\infty)$, for $w > 0$, one can show that the tail parts remain uniformly bounded (and small if T is large) as $\delta \to 0$, while the central parts of the integrals in (10.10) and (10.10′) respectively converge to the central part of the integral above. For the cases one is mostly interested in when $g'(t)$ is bounded away from zero for t small the integrals in question actually converge uniformly in δ for $0 \le \delta \le \delta_0$ even when absolute values are taken of the integrands, so the passage to the limit needs no justification. At any rate we get now

$$\lim_{x\to\infty} r_x(v,w) = \frac{1}{2\pi i}\int_{-i\infty}^{i\infty} \exp\left(w\,s - \int_0^v \frac{1-e^{-st}}{t}\,dg(t)\right)\frac{ds}{s}.$$

If we introduce the notation

$$h(v,w) = \frac{1}{2\pi i}\int_{-i\infty}^{i\infty} \exp\left(w\,s - \int_0^v \frac{1-e^{-st}}{t}\,dg(t)\right)\frac{ds}{s}, \tag{10.11}$$

we get now from (10.3) remembering the definition of $\Sigma_x(v;w)$ and the fact that $\Sigma_x(v) = e_u^{-1}(x^v)$

Lemma 7. *As $x \to \infty$ we have the asymptotic relation*

$$\sum_{\substack{d\in(\mathcal{P}(x^v))\\ d\le x^w}} \frac{1}{f'(d)} \sim h(v,w)\,e_u^{-1}(x^v),$$

where $h(v,w)$ is given by (10.11).

As a function of w, $h(v,w)$ is clearly an increasing function with $h(v,0) = 0$[13] and $h(v,\infty) = 1$. As a function of w it behaves better if $g'(t)$ is large for t near zero. It is easy to show that if $g'(t) \to \infty$ as $t \to 0$, then $h(v,w)$ has derivatives of any order with respect to w.

We can now also determine the asymptotic behavior of the sum (10.1′) for $\xi = x^v$, $z = x^w$ when $x \to \infty$. Writing the sum as a Stieltjes integral involving the sum (10.1) we easily get

Lemma 7′. *As $x \to \infty$ we have the asymptotic relation*

$$\sum_{\substack{d\in(\mathcal{P}(x^v))\\ d\le x^w}} \frac{\rho(\log d/(w\,\log x))}{f'(d)} \sim e_u^{-1}(x^v)\int_{t=0}^{t=w} \rho(t/w)\,dh(v,t).$$

---

[13] We are here disregarding the uninteresting case when $e_u(x)$ does not tend to zero as $x \to \infty$.

The asymptotic evaluation of these sums essentially concludes the treatment of the case of the upper bound for variable sifting density, by giving the main term asymptotically as (7.8) shows. As for the remainder term $R_u(\Lambda^2)$, estimates can be gotten directly from (5.4). One can also estimate the remainder term by noting that from (7.3′) and (7.5) it is obvious that the resulting λ_d are of absolute value ≤ 1, since if d and ρ denote numbers in $(\mathcal{P})$,

$$|\lambda_d| = \sum_{\delta/d} \frac{1}{f'(\delta)} \sum_{\substack{\rho\le z/d\\(\rho,d)=1}} \frac{1}{f'(\rho)} \left(\sum_{\rho\le z} \frac{1}{f'(\rho)}\right)^{-1} .$$

Thus in particular, if $u(p)$ is either ≥ 1 or equal to 0 (which usually is the case in sifting), we have

$$R_u(\Lambda^2) \le \left(\sum_{d\le z} u(d)\right)^2 . \tag{10.12}$$

Another form for the remainder term which is valid if $u(p)$ is either ≥ 1 or equal to 0, can be derived from (7.3′) and (7.3″), as follows: We have from $R_u(\Lambda^2) \le \left(\sum_{d\le z} u(d)\,|\lambda_d|\right)^2$ and (7.3′)

$$u(d)\,\lambda_d = \mu(d)\,d \sum_{d\,|\,\rho} \frac{y_\rho}{f'(\rho)} ,$$

that with the $y_\rho \ge 0$, we get

$$\sum_{d\le z} u(d)\,|\lambda_d| \le \sum_{\rho} \frac{y_\rho}{f'(\rho)} \sum_{d|\rho} d = \sum_{\rho} \frac{\sigma_1(\rho)\,y_\rho}{f'(\rho)} ,$$

where $\sigma_1(\rho)$ denotes the sum of the divisors of ρ. Since $\sum_\rho y_\rho/f'(\rho) = 1$, and $y_\rho = 0$ for $\rho > z$, we see that

$$\sum_{d\le z} u(d)\,\lambda_d \le \max_{\rho\le z} \sigma_1(\rho) \le z \max_{\rho\le z} \frac{\sigma_1(\rho)}{\rho} .$$

For large z, it is easily shown that the last factor on the right hand side is asymptotic to $\frac{6}{\pi^2}\,e^\gamma \log\log z$, where γ is Euler's constant and so $< \frac{12}{11} \log\log z$. We get from this for large z

$$R_u(\Lambda^2) < 1.2\,z^2(\log\log z)^2 . \tag{10.13}$$

For the $\Lambda^2\,\Lambda^-$ sieve we had developed two forms of the main term (7.13) and (7.13′), the last form given for the special choice of the Λ^-, where $\lambda'_1 = 1$, $\lambda'_p = -1$ for $p \le \xi$ and $\lambda'_d = 0$ for all other d. The maximizing y_ρ are not so easily obtained, but one way of making a good choice is to put

$$y_\rho = \sigma(\log\rho/\log z) , \tag{10.14}$$

where $\sigma(t)$ is a nonnegative function defined for $0 \le t \le 1$. The resulting sums can then be evaluated asymptotically using Lemma 7′ and one is left with a problem in the calculus of variations in order to choose $\sigma(t)$ in the best possible way. Usually this problem cannot be solved explicitly and one has to proceed by successive approximations.

If one does not specialize the Λ^- to the simple form chosen to get (7.13′), one may still use the form (10.14) and transform the problem, but this general form is difficult to work with unless the Λ^- has been somewhat specialized, and in particular the Λ^- chosen so that the λ'_d vanish if d has more than a certain number of prime factors.

We shall return to this and make use of these results in a later section where we shall be concerned with the case of constant sifting density, where $g(t) = k\,t$ and k is a constant. They will be particularly useful in considering the sifting problem for large constant sifting density.

Since we may choose $z = x^{1/2-\varepsilon}$ with $\varepsilon > 0$ in our Λ^2 upper bound sieve, we clearly have

$$T_g^+(\alpha) \le \frac{1}{h(\alpha, 1/2-\varepsilon)},$$

letting here $\varepsilon \to 0$, we obtain

Theorem 8. *Under the assumptions (4.2) and (4.4), we have*

$$\limsup_{x\to\infty} \frac{M_u^+(W_x, \mathcal{P}(x^\alpha))}{E_u(W_x, \mathcal{P}(x^\alpha))} \le \frac{1}{h(\alpha, 1/2)}.$$

In particular we have

$$T_g^+(\alpha) \le \frac{1}{h(\alpha, 1/2)}.$$

We shall see later that there is at least one important case (apart from the trivial case of $\alpha = 0$) when equality holds in the above statements.

For the lower bound one cannot give any similar simple result, even if we restrict ourselves to a $\Lambda^2\,\Lambda^-$ where Λ^- is the one we choose to get the forms (7.13′) and (7.14). If we in (7.14) take $\xi = x^\alpha$, $z = x^u$, with $\alpha + 2\,u < 1$, and choose the y_ρ to be of the form $\sigma(\log\rho/\log z)$, where $\sigma(t) \ge 0$ has $\sigma(0) = 1$ and $\sigma(t) = 0$ for $t > 1$ and is continuous except for a possible discontinuity at $t = 1$, we are led (using Lemma 7′ for the sums involving ρ, and taking $v = 1$) to the expression

(10.15)

$$\int_0^u \sigma^2\left(\frac{w}{u}\right) h'_w(\alpha, w)\,dw - \int\limits_{w=0}^{w=u}\int\limits_{t=0}^{t=\alpha} \left(\sigma\left(\frac{w}{u}\right) - \sigma\left(\frac{w+t}{u}\right)\right)^2 h'_w(\alpha, w)\,\frac{g'(t)}{t}\,dw dt$$

divided by

$$\left(\int_0^u \sigma\left(\frac{w}{u}\right) h'_w(\alpha, w)\,dw\right)^2. \tag{10.15′}$$

To make a favorable choice of $\sigma(t)$ in general seems not feasible, since as a problem in the calculus of variations this is rather intractable even for specific simple $g(t)$. What one can do is to start out with a choice of $\sigma(t)$ which equals 1 for a certain stretch $0 \le t \le t_0 < 1$ and then tapers linearly off to zero at $t = 1$, and one tries to choose t_0 favorably, next one adds small adjustment terms to $\sigma(t)$ and tries to choose their coefficients in the optimal way. The result is of course only of interest if the numerator comes out positive.

We shall return to this in the case of large constant sifting density in a later section.

From Lemma 7 and Lemma 7′ it is comparatively easy to see that the sieves Λ^2 and $\Lambda^2\,\Lambda^-$ constructed in Section 7 to obtain (7.20) and (7.20′) can be replaced by Λ_F sieves in the sense explained in Section 8. For the λ_d in the Λ^2 used to get (7.20) we see that as $x \to \infty$, we have for $\xi = x^\alpha$, $z = x^w$

$$\lambda_d \sim \mu(d)\,\frac{h(\alpha, w - \log d/\log x)}{h(\alpha, w)}\,, \tag{10.16}$$

corresponding to the set of functions for Λ

$$F_r(v_1, \dots, v_r) = (-1)^r\,\frac{h(\alpha, w - v_1 - \dots - v_r)}{h(\alpha, w)}\,. \tag{10.16′}$$

This leads to another set of functions $F_r^*(v_1, \dots, v_r)$ for the Λ^2. For the Λ in the $\Lambda^2\,\Lambda^-$ we can also get asymptotic formulas using Lemma 7′ which shows that the λ_d in the Λ are again

$$\sim (-1)^r\,F\left(\frac{\log p_1}{\log x}, \dots, \frac{\log p_r}{\log x}\right)$$

where F is a certain function. Since in the Λ^- we had $F_0 = 1$; $F_1(v) = -1$ for $0 \le v \le \log\xi/\log x$, and $F_1(v) = 0$ for $v > \log\xi/\log x$, $F_r(v_1, \dots, v_r) = 0$ for $r > 1$, we see that the $\Lambda^2\,\Lambda^-$ is also a Λ_{F^*} with a more complicated set of functions F_r^*.

It may be mentioned that these methods can be developed within the formalism of the Λ_F systems if one so desires.

We shall in the following as an abbreviation often write V for an aggregate of variables v, $F(V)$ will, if the aggregate contains r arguments v, stand for $F_r(V)$, we also introduce the symbol $\mu(V) = (-1)^r$ if V contains r arguments v. Taking as our point of departure the last expression in (8.40″) for $0 < \eta < \alpha$ and assuming that $F^+ = F^*$ has arisen from another set of functions F in the way

$$F^*(V) = \sum_{V' \cup V'' = V} F(V')\,F(V'')\,, \tag{10.17}$$

it is not difficult to see that if we introduce the new set of functions $Y(V)$ by the transformation

$$\mu(V)\,Y(V) = F(V) + \int_\eta^\alpha F(V,v)\,\frac{dg(v)}{v} + \dots$$

$$(10.18) \qquad + \frac{1}{r!}\int_\eta^\alpha \cdots \int_\eta^\alpha F(V,v_1,\dots,v_r)\,\frac{dg(v_1)\dots dg(v_r)}{v_1\dots v_r} + \dots,$$

which is actually an involution, so that

$$\mu(V)\,F(V) = Y(V) + \int_\eta^\alpha Y(V,v)\,\frac{dg(v)}{v} + \dots$$

$$(10.18') \qquad + \frac{1}{r!}\int_\eta^\alpha \cdots \int_\eta^\alpha Y(V,v_1,\dots,v_r)\,\frac{dg(v_1)\dots dg(v_r)}{v_1\dots v_r} + \dots,$$

then the identity holds
(10.19)

$$F_0^* + \int_\eta^\alpha F^*(v)\,\frac{dg(v)}{v} + \dots + \frac{1}{r!}\int_\eta^\alpha \cdots \int_\eta^\alpha F^*(v_1,\dots,v_r)\,\frac{dg(v_1)\dots dg(v_r)}{v_1\dots v_r} + \dots$$

$$= Y_0^2 + \int_\eta^\alpha (Y(v))^2\,\frac{dg(v)}{v} + \dots + \frac{1}{r!}\int_\eta^\alpha \cdots \int_\eta^\alpha (Y(v_1,\dots,v_r))^2\,\frac{dg(v_1)\dots dg(v_r)}{v_1\dots v_r} +$$

In order that $F_0^* = F_0^2 = 1$, we need that
(10.20)

$$1 = Y_0 + \int_\eta^\alpha Y(v)\,\frac{dg(v)}{v} + \dots + \frac{1}{r!}\int_\eta^\alpha \cdots \int_\eta^\alpha Y(v_1,\dots,v_r)\,\frac{dg(v_1)\dots dg(v_r)}{v_1\dots v_r} + \dots$$

The condition $F_r(v_1,\dots,v_r) = 0$ for $v_1 + \dots + v_r > 1/2$ which implies $F_r^*(v_1,\dots,v_r) = 0$ for $v_1 + \dots + v_r > 1$, clearly implies $Y(v_1,\dots,v_r) = 0$ for $v_1 + \dots + v_r > 1/2$, and obviously also $Y(v_1,\dots,v_r) = 0$ for $v_1 + \dots + v_r > 1/2$ implies $F_r(v_1,\dots,v_r) = 0$ for $v_1 + \dots + v_r > 1/2$.

It is clear that minimizing the right hand side of (10.19) with the side condition (10.20) leads to a $Y(V)$ that is a constant for the sum of the arguments $v \le 1/2$ and zero otherwise. This constant is of course determined by (10.20). Going back to (8.40″) this leads to

$$T_g^+(\eta,\alpha) \le \exp(L(\eta,\alpha)) \left\{ 1 + \sum_{r=1}^\infty \frac{1}{r!} \underset{\substack{\eta\le v_i\le\alpha \\ v_1+\dots+v_r\le 1/2}}{\int\cdots\int} \frac{dg(v_1)\dots dg(v_r)}{v_1\dots v_r} \right\}^{-1}.$$

The reciprocal of the expression on the right hand side can also be expressed as

$$\frac{1}{2\pi i}\int_{-i\infty}^{i\infty} \exp\left(\frac{s}{2} - \int_\eta^\alpha \frac{1-e^{-st}}{t}\,dg(t)\right)\frac{ds}{s},$$

which as $\eta \to 0$ tends to $h(\alpha, 1/2)$. The form of $F(V)$ is of course easily determined from (10.18′), as $\eta \to 0$ we recover the formula (10.16′).

Let us finally mention a slight modification of the set Λ obtained by the Λ^2 method, from (7.3′) and (7.5) one has

$$\mu(d)\,\lambda_d = f(d) \sum_{\substack{d\,|\,\rho \\ \rho \le z}} \frac{1}{f'(\rho)} \left(\sum_{\rho \le z} \frac{1}{f'(\rho)} \right)^{-1}. \tag{10.21}$$

Here ρ runs over the numbers of $(\mathcal{P}(x^\alpha))$. Let us define f also for non squarefree numbers as a totally multiplicative function so that $f(\rho_1)\,f(\rho_2) = f(\rho_1\,\rho_2)$ also when $(\rho_1, \rho_2) > 1$. We shall use the symbol $\sum^*$ to indicate that summation is made over all positive integers composed of primes from $\mathcal{P}(x^\alpha)$ whether squarefree or not and $\sum_\rho^*$ to indicate that summation is made only over positive integers all of whose prime factors divide ρ. We now replace our λ_d given by (10.21), by

$$\mu(d)\,\lambda'_d = \sum_{\rho \le z/d}^{*} \frac{1}{f(\rho)} \left(\sum_{\rho \le z}^{*} \frac{1}{f(\rho)} \right)^{-1}. \tag{10.22}$$

It is easily seen that we have now

$$\mu(\rho) \sum_{\rho\,|\,d} \frac{\lambda'_d}{f(d)} = \frac{1}{f(\rho)} \sum_{\delta \le z/\rho}{}_{\rho}^{*} \frac{1}{f(\delta)} \left(\sum_{\rho \le z}^{*} \frac{1}{f(\rho)} \right)^{-1} \le \frac{1}{f'(\rho)} \left(\sum_{\rho \le z}^{*} \frac{1}{f(\rho)} \right)^{-1}.$$

But the left hand sides here are the $y_\rho / f'(\rho)$ that correspond to the λ'_d, they give us when inserted in the expression

$$\sum_\rho \frac{y_\rho^2}{f'(\rho)}$$

the bound

$$\sum_{\rho \le z} \frac{1}{f'(\rho)} \left(\sum_{\rho \le z}^{*} \frac{1}{f(\rho)} \right)^{-2}, \tag{10.23}$$

instead of

$$\left(\sum_{\rho \le z} \frac{1}{f'(\rho)} \right)^{-1}, \tag{10.23′}$$

which corresponded to the λ_d in (10.21). But it is easy to see that as $x \to \infty$ the two expressions

$$\sum_{\rho \le z}^{*} \frac{1}{f(\rho)} \quad \text{and} \quad \sum_{\rho \le z} \frac{1}{f'(\rho)}$$

are asymptotically equal since the first terms can be in handled much in the same way we dealt with (10.1) or the second sum, and the end formula turns out to be the same. So for asymptotic results the λ'_d are as good as the λ_d and it is obvious their absolute values depend only on the size of z/d and decrease monotonically as d increases. Their asymptotic behavior as $x \to \infty$ is of course the same as for the λ_d.

The sieves Λ^2 and $\Lambda^2\,\Lambda^-$ are, except in a few cases, capable of improvements by means of some of the principles mentioned in Section 6.

11. The combinatorial sieve as developed by Brun, Buchstab and Rosser

We described in Section 6 a very general principle for constructing sieves, Principle V of that section.

This in essence goes back to Viggo Brun who developed the first effective sieve method ca. seventy years ago, though neither he nor his successors described it in terms of Λ systems. The sieves that arise from Principle V we shall refer to as combinatorial sieves, a characteristic feature of them is that the λ_d are either equal to $\mu(d)$ or to zero.

In trying to construct a sieve using Principle V, one encounters two basic problems: (a) One has to give some definite rules according to which the choice between the two alternatives $\lambda_{p_1 d} = -\lambda_d$ or $\lambda_{p_1 d} = 0$ is made, and made in a way that is advantageous rather than the opposite. (b) These rules must also be of such a nature that we are able to estimate the sum

$$\sum_d \frac{\lambda_d}{f(d)},$$

well.

Brun and his early successors considered essentially the case of sifting an interval, and only the case of constant sifting density. They operated directly with quantities like the $M_u^+(W_x, \mathcal{P})$ and $M_u^-(W_x, \mathcal{P})$ (though using other symbols) and their notation to some extent served to obscure some facets of the problems.

The way that eventually led to the optimal rules of choice in application of Principle V can be briefly described as follows:

We introduce the notation $W_x(p)$ to denote the weighted set formed by the w_n for which $p \times n$. If p does not divide d we have then

$$N_d(W_x(p)) = N_{dp}(W_x)\,. \tag{11.1}$$

Also if $\alpha > \beta \geq 0$, it is obvious that

$$M(W_x, \mathcal{P}(x^\alpha)) = M(W_x, \mathcal{P}(x^\beta)) - \sum_{x^\beta \leq p < x^\alpha} M(W_x(p), \mathcal{P}(p))\,. \tag{11.2}$$

From this follows of course[14]

$$(11.3)\qquad M_u^+(W_x,\mathcal{P}(x^\alpha)) \le M_u^+(W_x,\mathcal{P}(x^\beta)) - \sum_{x^\beta\le p<x^\alpha} M_u^-(W_x(p),\mathcal{P}(p))\,,$$

and

$$(11.3')\qquad M_u^-(W_x,\mathcal{P}(x^\alpha)) \ge M_u^-(W_x,\mathcal{P}(x^\beta)) - \sum_{x^\beta\le p<x^\alpha} M_u^+(W_x(p),\mathcal{P}(p))\,.$$

While Brun in principle had these inequalities, in his form they were obscured by his notation which actually carried irrelevant information so he did not realize that the terms in the sums on the right hand side of (11.3) and (11.3′) refer to a problem identical with the original one but with the parameter x/p instead of x. Basically Brun did not observe that sifting a section of an arithmetic progression where the difference is relatively prime to the primes in the sifting range, is really the same as sifting an interval; had he done so, he might likely have anticipated Buchstab and perhaps Rosser's work.

At any rate, Buchstab was the first to realize the true significance of these inequalities, and used them to improve the earlier bounds obtained by Brun and his earliest successors by repeated use of these inequalities starting with these bounds and iterating the use of (11.3) and (11.3′).

In such iteration of the use of (11.3) and (11.3′), it is rather obvious where one at each stage would stop: In using (11.3) we would stop when α is so large that for the largest $p \le x^\alpha$, the best bound we have for $M_u^-(W_x(p),\mathcal{P}(p))$ becomes trivial, that is zero. In using (11.3′) we would stop where the right hand side is about to become zero or negative.

Since, when we are dealing with constant sifting density, say $g'(t) = k$ so that $g(t) = k\,t$, the sifting problem remains the same if we keep the same $u(p)$ sequence and replace x by some x^a where $0 < a < \infty$,[15] and since we have for $d \in (\mathcal{P}(p))$

$$N_d(W_x(p)) = \frac{u(p)\,u(d)}{p\,d}\,x + R_{p\,d}(x)\,,$$

with

$$|R_{p\,d}(x)| \le u(p)\,u(d)\,,$$

we see that if we divide by $u(p)$, we have precisely the same assumptions satisfied for $\frac{1}{u(p)}\,W_x(p)$ as for W_x, but with x/p instead of x. It seems natural, since we know that for $g(t) = k\,t$ we either must have $T_g^-(\alpha)$, which we now denote simply by $T_k^-(\alpha)$, positive for all $\alpha < 1$ or there is an α_k, $0 < \alpha_k < 1$

[14] In terms of Λ^+ and Λ^- sieves, (11.3) is what one gets by taking the optimal $\Lambda^+(\mathcal{P}(x^\beta), x)$ and extending the range using Principle III. Similarly (11.3′) can be gotten from Principle IV by extending the range of the optimal lower bound $\mathcal{P}(x^\beta)$ sieve.

[15] In general this leads to replacing $g(t)$ with $g(a\,t)/a$, and so, unless $g(t) = k\,t$, one gets a different density function.

such that $T_k^-(\alpha_k) = 0$ and $T_k^-(\alpha) > 0$ for $\alpha < \alpha_k$, to assume that when we iterate the use of the inequalities (11.3) and (11.3′) there is some constant $\beta_k \le 1$, and $\beta_k \le \alpha_k$, when α_k exists, such that we break off in (11.3) when $\log p/\log(x/p)$ reaches β_k, which means where $\alpha/(1-\alpha) = \beta_k$, and in the use of (11.3′) where $\alpha = \beta_k$. Since we do not know what this β_k might be it seems well motivated to choose some parameter β and to construct a sieve Λ^+ and one Λ^- which are such that it corresponds to carrying out the iterations and breaking off in (11.3) at $\alpha/(1-\alpha) = \beta$ and (11.3′) when $\alpha = \beta$. One then could later see where the most favorable choice of β would be.

Such a sieve was first described by Rosser in around 1950 or so, and he succeeded in developing its theory for the case $k = 1$, in which case he could show that his sieve gave the optimal results. I had occasion to see a manuscript (quite long, it was never published) of his in the early fifties, it also contained some results for $k = 2$, but he had not determined theoretically the limits of what the method could give in that case, only numerical results were given for $k = 2$.

In the middle to late fifties I developed the theory of Rosser's sieve (or as I have referred to it in my Stony Brook lectures the Buchstab–Rosser sieve, and in later years as the $B^2 R$ sieve, acknowledging the basic contributions of Brun and Buchstab to the development of the combinatorial sieve), for general k.[16] It was briefly mentioned in my Stony Brook lectures which however were primarily designed to give the general theory and not specific sieve methods.

Later Henryk Iwaniec independently took up this problem and treated it also quite fully in Iwaniec [1], [2], and [3]. His methods were rather different from mine, the results however are of course identical.

In what follows I shall give my own version of the theory of the $B^2 R$ sieve.

We shall not be working with a parameter $0 < \beta \le 1$, as would be indicated by my introduction. Instead we shall use the reciprocal of this as our parameter and call that β. From certain points of view this is simpler, and it is also the way I originally did it. This means that what above was referred to as β_k will be the reciprocal of the optimal choice of β.

We can define the $B^2 R$ sieves as Λ systems as is done by the inequalities (5.3) and (5.4) in my Stony Brook lectures, I shall however here prefer to define two systems of functions $F_r^{\pm}$ as follows:

We have a parameter $\beta \ge 1$ which we keep fixed. We define two sets of inequalities for variables $v_1 \le v_2 \le \ldots \le v_r$ which lie in the interval

[16] With the exception of one omission due to an error of calculation for $k = 3$, this was discovered and corrected when I lectured extensively on this at Pennsylvania State University in the Spring of 1974.

$0 \leq v \leq 1$, first

$$\begin{aligned} (\beta+1)\, v_r &\leq 1\,, \\ (\beta+1)\, v_{r-2} + v_{r-1} + v_r &\leq 1\,, \\ &\vdots \\ (\beta+1)\, v_{r-2t} + v_{r-2t+1} + \ldots + v_r &\leq 1\,, \\ &\vdots \end{aligned} \tag{11.4}$$

and the second set

$$\begin{aligned} \beta\, v_r &\leq 1\,, \\ (\beta+1)\, v_{r-1} + v_r &\leq 1\,, \\ &\vdots \\ (\beta+1)\, v_{r-2t+1} + v_{r-2t+2} + \ldots + v_r &\leq 1\,, \\ &\vdots \end{aligned} \tag{11.4$'$}$$

Our functions $F_r^+(v_1, \ldots, v_r)$ and $F_r^-(v_1, \ldots, v_r)$ are now defined in the following manner: We have $F_0^+ = F_0^- = 1$, and for $r \geq 1$,

$$F_r^+(v_1, \ldots, v_r) = \begin{cases} (-1)^r\,, & \text{if all inequalities (11.4) are satisfied,} \\ 0 & \text{otherwise,} \end{cases} \tag{11.5}$$

and[17]

$$F_r^-(v_1, \ldots, v_r) = \begin{cases} (-1)^r\,, & \text{if all inequalities (11.4$'$) are satisfied,} \\ 0 & \text{otherwise.} \end{cases} \tag{11.5$'$}$$

It should be noted that the inequalities (11.4) and (11.4′) (more specifically the last of them with $t = [(r-1)/2]$ in case of (11.4) and $t = [r/2]$ in case of (11.4′)) imply that $v_1 + \ldots + v_r \leq 1$.

It is easily seen that this definition is equivalent to the following recursive definition

$$F_r^+(v_1, \ldots, v_r) = \begin{cases} 0, & \text{for } (\beta+1)\, v_r > 1, \\ -F_{r-1}^-\left(\frac{v_1}{1-v_r}, \ldots, \frac{v_{r-1}}{1-v_r}\right) & \text{for } (\beta+1)\, v_r \leq 1, \end{cases} \tag{11.6}$$

and

$$F_r^-(v_1, \ldots, v_r) = \begin{cases} 0, & \text{for } \beta\, v_r > 1, \\ -F_{r-1}^+\left(\frac{v_1}{1-v_r}, \ldots, \frac{v_{r-1}}{1-v_r}\right) & \text{for } \beta\, v_r \leq 1, \end{cases} \tag{11.6$'$}$$

[17] As we have defined F^- here we see that it only permits construction of a $\Lambda^-(\mathcal{P}(x^\alpha))$ if $\alpha \leq 1/\beta$. We could modify this definition so as to allow such construction for any $\alpha \leq 1$, while leaving the case $\alpha \leq 1/\beta$ unchanged. It is however simpler to restrict α for the time being, whenever expressions like $t_k^-(\eta, \alpha)$ and $t_k^-(\alpha)$ occurs we shall assume $\alpha \leq 1/\beta$, until at the end of our development of the theory of this sieve, when we shall consider $t_k^-(\alpha)$ also for $1/\beta < \alpha \leq 1$.

This follows if we in (11.4) and (11.4′) subtract v_r on both sides of the inequalities below the first one in each set, and then divide through by $1 - v_r$, this gives us precisely the inequalities which define the right hand side of (11.6) and (11.6′) respectively. We also observe that (11.4) shows that if r is even

$$F_r^+(v_1, v_2, \ldots, v_r) = -F_{r-1}^+(v_2, \ldots, v_r)$$

since the set of inequalities is the same for the two F, while if r is odd there is one extra inequality that has to be satisfied to make $F_r^+(v_1, \ldots, v_r)$ different from zero, so that for odd r we have

$$F_r^+(v_1, \ldots, v_r) = \begin{cases} 0, & \text{or} \\ -F_{r-1}^+(v_2, \ldots, v_r). \end{cases}$$

This is precisely the relationship in a combinatorial upper bound sieve. We see similarly from (11.4′) that if r is odd

$$F_r^-(v_1, v_2, \ldots, v_r) = -F_{r-1}^-(v_2, \ldots, v_r),$$

while if r is even

$$F_r^-(v_1, \ldots, v_r) = \begin{cases} 0, & \text{or} \\ -F_{r-1}^-(v_2, \ldots, v_r). \end{cases}$$

Which, as we remember, is the relationship needed for a combinatorial lower bound sieve.

We shall need to make use of (8.42), and give therefore first complete the argument to prove the equality of the two sides of that equation. Since the proof is the same whether we are dealing with a F_r^+ or a F_r^- set, we omit the $\pm$ symbol attached to the F's. On the left hand side of (8.42), we may pick out from the $\theta_F(v_1, \ldots, v_{r'})$ with $r' \geq r$ the combinations $F_r(v_1', \ldots, v_r') + F_{r-1}(v_2', \ldots, v_r')$ where v_1' is the smallest v_i in θ_F and $v_1' \leq v_2' \leq \ldots \leq v_r'$, the v in θ_F that do not occur among the $v_1', \ldots, v_r'$ we rename $v_{r+1}', \ldots, v_{r+s}'$ with $r + s = r'$, for these we do not specify ordering, they are of course all in the interval (v_1', α), apart from that they can run freely independent of each other and of the $v_2', \ldots, v_r'$.

We see now that the contribution of our combination $F_r(v_1', \ldots, v_r') + F_{r-1}(v_2', \ldots, v_r')$ to the sum on the left hand side of (8.42) becomes

$$\int \cdots \int_{\eta \leq v_1' \leq \ldots \leq v_r' \leq \alpha} (F_r(v_1', \ldots, v_r') + F_{r-1}(v_2', \ldots, v_r')) \frac{dg(v_1') \ldots dg(v_r')}{v_1' \ldots v_r'}$$
$$\times \left(1 + \sum_{s=1}^{\infty} \frac{1}{s!} \int_{v_1'}^{\alpha} \cdots \int_{v_1'}^{\alpha} \frac{dg(v_{r+1}') \ldots dg(v_{r+s}')}{v_{r+1}' \ldots v_{r+s}'} \right)$$

here the sum in the bracket is seen to be

$$\sum_{s=0}^{\infty} \frac{1}{s!} \left(\int_{v_1'}^{\alpha} \frac{dg(v)}{v} \right)^s = \exp(L(v_1', \alpha)).$$

Inserting this in the above and dropping the dashes on the v's, we get the general term for $r \geq 1$ of the series on the right hand side of (8.42). The first term on the left hand side of (8.42) which was so far not involved in our consideration, gives us of course the first term on the right hand side. In the case we are currently considering when $g(t) = k\,t$ we see that $\exp(L(v,\alpha)) = (\alpha/v)^k$, so (8.42) becomes

(11.7)

$$1 + k \int_\eta^\alpha \theta_F(v)\frac{dv}{v} + \ldots + \frac{k^r}{r!} \int_\eta^\alpha \cdots \int_\eta^\alpha \theta_F(v_1,\ldots,v_r)\frac{dv_1\ldots dv_r}{v_1\ldots v_r} + \ldots$$

$$= 1 + \alpha^k \sum_{r=1}^\infty k^r \int_{\eta\le v_1\le\ldots\le v_r\le\alpha}\cdots\int (F_r(v_1,\ldots,v_r) + F_{r-1}(v_2,\ldots,v_r))\frac{dv_1\ldots dv_r}{v_1^{k+1}\ldots v_r}\,.$$

We now turn our attention to the expression (11.7) when we insert our F_r^+ and F_r^- systems constructed above and let $\eta \to 0$. We denote the expression (11.7) by $t_k^+(\eta,\alpha)$ if we insert the F_r^+ and by $t_k^-(\eta,\alpha)$ if we insert the F_r^-.

From the definition of F_r^+ by (11.4) and (11.5), we see that

$$F_r^+(v_1,\ldots,v_r) + F_{r-1}^+(v_2,\ldots,v_r) = 0\,,$$

unless r is odd and the following set of inequalities hold

$$\begin{gathered}(\beta+1)\,v_r \le 1\,,\\ (\beta+1)\,v_{r-2} + v_{r-1} + v_r \le 1\,,\\ \vdots\\ (\beta+1)\,v_3 + v_4 + \ldots + v_r \le 1\,,\\ (\beta+1)\,v_1 + v_2 + \ldots + v_r > 1\,,\\ \vdots\end{gathered} \tag{11.8}$$

we refer to this set of inequalities as $\mathcal{V}_r^+$ and note that for $r = 1$ it is to be interpreted as the single inequality $(\beta+1)\,v_1 > 1$.

Similarly we see from (11.4′) and (11.5′) that

$$F_r^-(v_1,\ldots,v_r) + F_{r-1}^-(v_2,\ldots,v_r) = 0\,,$$

unless r is even and the following set of inequalities hold

$$\begin{gathered}\beta\,v_r \le 1\,,\\ (\beta+1)\,v_{r-1} + v_r \le 1\,,\\ (\beta+1)\,v_{r-3} + v_{r-2} + \ldots + v_r \le 1\,,\\ \vdots\\ (\beta+1)\,v_3 + v_4 + \ldots + v_r \le 1\,,\\ (\beta+1)\,v_1 + v_2 + \ldots + v_r > 1\,,\end{gathered} \tag{11.8'}$$

We refer to this set of inequalities as $\mathcal{V}_r^-$, for $r = 2$ it consists just of $\beta\, v_2 \le 1$, $(\beta+1)\, v_1 + v_2 > 1$. It is also clear that in the cases where $F_r^+ + F_{r-1}^+ \ne 0$ it has the value 1, and when $F_r^- + F_{r-1}^- \ne 0$ it has the value -1. We shall in what follows simply write $v \in \mathcal{V}_r^+$ or $v \in \mathcal{V}_r^-$ to indicate that $v_1 \le v_2 \le \ldots \le v_r$ satisfies the system (11.8) or (11.8′). Since we intend to let $\eta \to 0$, it is no restriction to assume that $\eta < 1/(\beta+1)$. We get then for $t_k^+(\eta,\alpha)$ from the right hand side of (11.7)[18]

$$t_k(\eta,\alpha) = 1 + k\,\alpha^k \int\limits_{1/(\beta+1)}^{\alpha} \frac{dv}{v^{k+1}} + \ldots \tag{11.9}$$
$$+\, \alpha^k\, k^{2r+1} \underset{\substack{\eta \le v_1 \le \cdots \le v_{r+1} \le \alpha \\ v \in \mathcal{V}_{2r+1}^+}}{\int \cdots \int} \frac{dv_1 \ldots dv_{2r+1}}{v_1^{k+1} v_2 \ldots v_{2r+1}} + \ldots,$$

and similarly

$$t_k^-(\eta,\alpha) = 1 - \alpha^2\, k^2 \underset{\substack{\eta \le v_1 \le v_2 \le \alpha \\ v \in \mathcal{V}_2^-}}{\int \cdots \int} \frac{dv_1 dv_2}{v_1^{k+1} v_2} - \ldots \tag{11.9′}$$
$$- \ldots - \alpha^k\, k^{2r} \underset{\substack{\eta \le v_1 \le \cdots \le v_{2r} \le \alpha \\ v \in \mathcal{V}_{2r}^-}}{\int \cdots \int} \frac{dv_1 \ldots dv_{2r}}{v_1^{k+1} v_2 \ldots v_{2r}} - \ldots,$$

We wish first to show that if we put $\eta = 0$ on the right hand side (11.9) and (11.9′), the resulting series converge for all $\beta \ge 1$ if $k < 1/2$, and for any $k > 0$ if $\beta \ge \beta_0(k)$. We write for the integral in the general term in (11.9) with $\eta = 0$,

$$\mathfrak{I}_{2r+1}^+ = \underset{\substack{0 \le v_1 \le \cdots \le v_{2r+1} \le \alpha \\ v \in \mathcal{V}_{2r+1}^+}}{\int \cdots \int} \frac{dv_1 \ldots dv_{2r+1}}{v_1^{k+1} v_2 \ldots v_{2r+1}}, \tag{11.10}$$

and similarly for the integral in the general term in (11.9′) with $\eta = 0$,

$$\mathfrak{I}_{2r}^- = \underset{\substack{0 \le v_1 \le \cdots \le v_{2r} \le \alpha \\ v \in \mathcal{V}_{2r}^-}}{\int \cdots \int} \frac{dv_1 \ldots dv_{2r}}{v_1^{k+1} v_2 \ldots v_{2r}}. \tag{11.10′}$$

Looking at the last inequality in (11.8) or (11.8′), we see that in case of $\mathfrak{I}_{2r+1}^+$ we have, since v_{2r+1} is the largest v, that

$$(\beta + 2r + 1)\, v_{2r+1} > 1. \tag{11.11}$$

Similarly in $\mathfrak{I}_{2r}^-$ we have

$$(\beta + 2r)\, v_{2r} > 1. \tag{11.11′}$$

[18] if $\alpha < 1/(\beta+1)$, the second term on the right hand side is taken as zero.

The two last inequalities in (11.8) show that the v_{2r+1} must lie in an interval of length not more than $(\beta+1)\,v_1+v_2-\beta\,v_3 \le 2\,v_1$, and the two last inequalities in (11.8′) show that the same holds for the v_{2r} in $\mathfrak{I}_{2r}^-$. Thus we see that the integration over v_{2r+1} in $\mathfrak{I}_{2r+1}^+$ yields less than $2\,(\beta+2r+1)\,v_1$, and that over v_{2r} in $\mathfrak{I}_{2r}^-$ less than $2\,(\beta+2r)\,v_1$.

Inserting this and assuming $k<1$ we get

$$\text{(11.12)}\qquad \mathfrak{I}_{2r+1}^+ < 2\,(\beta+2r+1)\int\cdots\int_{0\le v_1\le\ldots\le v_{2r}\le\alpha}\frac{dv_1\ldots dv_{2r}}{v_1^k v_2\ldots v_{2r}} = 2\,(\beta+2r+1)\,\frac{\alpha^{1-k}}{(1-k)^{2r}}\,,$$

and similarly

$$\text{(11.12′)}\qquad \mathfrak{I}_{2r}^- < 2\,(\beta+2r)\,\frac{\alpha^{1-k}}{(1-k)^{2r-1}}\,.$$

If we now insert these upper bounds in (11.9) and (11.9′) we see that the series on the right hand side converge uniformly if $k/(1-k)\le 1-\delta$, with $\delta>0$, so the passage to the limit $\eta\to 0$ is justified for $k<1/2$. We denote $\lim_{\eta\to0} t_k^+(\eta,\alpha)$ by $t_k^+(\alpha)$ and similarly use the notation $t_k^-(\alpha)$. For $k<1/2-\delta$, we wish to see what happens to $t_k^+(\alpha)$ and $t_k^-(\alpha)$ as $\alpha\to 0$. For α very small it is clear that after the first term the terms in the series

$$t_k^+(\alpha) = 1+\ldots+\alpha^k\,k^{2r+1}\int\cdots\int_{\substack{0\le v_1\le\cdots\le v_{2r+1}\le\alpha\\ v\in\mathcal{V}_{2r+1}^+}}\frac{dv_1\ldots dv_{2r+1}}{v_1^{k+1}v_2\ldots v_{2r+1}}+\ldots$$

vanish as long as

$$(\beta+2r+1)\,\alpha\le 1\,.$$

Using this and the bound (11.12) for the larger r for which $(\beta+2r+1)\,\alpha>1$, we easily see that

$$\text{(11.13)}\qquad t_k^+(\alpha) = 1+O(e^{-c(\delta)/\alpha})\,,\qquad \text{as } \alpha\to 0\,.$$

Similarly one proves

$$\text{(11.13′)}\qquad t_k^-(\alpha) = 1-O(e^{-c(\delta)/\alpha})\,,\qquad \text{as } \alpha\to 0\,.$$

Now let us consider the case that $\beta>1$. From (11.8) we have

$$(\beta+1)\,v_1+v_2+\ldots+v_{2r+1} > 1 \ge (\beta+1)\,v_{2t+1}+v_{2t+2}+\ldots+v_{2r+1}\,,$$

for all $1\le t\le r$. This gives

$$(\beta+1)\,v_1+v_2+\ldots+v_{2t} > \beta\,v_{2t+1}\,,$$

and a fortiori,

$$(\beta+1)\,v_1 + 2\,v_3 + \ldots + 2\,v_{2t-1} > (\beta-1)\,v_{2t+1}\,, \tag{11.14}$$

for $t=1$, this gives

$$v_3 < \frac{\beta+1}{\beta-1}\,v_1\,,$$

and by induction from (11.14) follows

$$v_{2t+1} < \left(\frac{\beta+1}{\beta-1}\right)^t v_1 \qquad \text{for } 1 \le t \le r\,. \tag{11.15}$$

Inserting this in the inequality

$$(\beta+1)\,v_1 + v_2 + \ldots + v_{2r+1} > 1$$

we get

$$(\beta+1)\,v_1 + 2\,\frac{\beta+1}{\beta-1}\,v_1 + \ldots + 2\left(\frac{\beta+1}{\beta-1}\right)^r v_1 > 1\,,$$

or

$$\frac{(\beta+1)^{r+1}}{(\beta-1)^r}\,v_1 > 1\,,$$

so we get

$$v_1 > \frac{(\beta-1)^r}{(\beta+1)^{r+1}}\,. \tag{11.16}$$

We can argue in a similar way from the inequalities (11.8′) and get

$$v_{2t+1} < \left(\frac{\beta+1}{\beta-1}\right)^t v_1\,, \qquad \text{for } 1 \le t \le r-1\,, \tag{11.17}$$

and in addition the inequality

$$v_{2r} < \left(\frac{\beta+1}{\beta-1}\right)^r v_1\,. \tag{11.17′}$$

From the inequality $(\beta+1)\,v_1 + v_2 + \ldots + v_{2r} > 1$, one gets in this case

$$(\beta+1)\,v_1 + 2\,\frac{\beta+1}{\beta-1}\,v_1 + \ldots + 2\left(\frac{\beta+1}{\beta-1}\right)^{r-1} v_1 + \left(\frac{\beta+1}{\beta-1}\right)^r v_1 > 1\,,$$

or $\beta\left(\frac{\beta+1}{\beta-1}\right)^r v_1 > 1$, so in this case we have

$$v_1 > \frac{1}{\beta}\left(\frac{\beta-1}{\beta+1}\right)^r\,. \tag{11.18}$$

For $\beta > 1$ we can now give new upper bounds for the integrals $\mathfrak{I}^+_{2r+1}$ and $\mathfrak{I}^-_{2r}$. In (11.10) we replace the factor $1/v_1^k$ by $(\beta+1)^{(r+1)k}/(\beta-1)^{rk}$ which according to (11.16) is an upper bound for it. In the remaining integral we replace the integration domain by

$$\frac{(\beta-1)^r}{(\beta+1)^r} < v_1 \le v_2 \le \ldots \le v_{2r+1} \le \alpha\,,$$

(which according to the above contains the original domain) we from this easily get

$$\mathfrak{I}^+_{2r+1} < \frac{1}{(2r+1)!}\,\frac{(\beta+1)^{(r+1)\,k}}{(\beta-1)^{rk}}\left(\log\frac{\alpha\,(\beta+1)^{r+1}}{(\beta-1)^r}\right)^{2r+1}.$$

We see from this that the $(2r+1)$th root of the corresponding term in the series for $t_k^+(\alpha)$ is less than

$$\alpha^{k/(2r+1)}\,k\,((2r+1)!)^{-1/(2r+1)}\,\frac{(\beta+1)^{(r+1)k/(2r+1)}}{(\beta-1)^{rk/(2r+1)}}\,\log\frac{\alpha\,(\beta+1)^{r+1}}{(\beta-1)^r}$$
$$\sim k\,\frac{e}{2r}\left(\frac{\beta+1}{\beta-1}\right)^{k/2} r\,\log\frac{\beta+1}{\beta-1} = \frac{k\,e}{2}\left(\frac{\beta+1}{\beta-1}\right)^{k/2}\log\frac{\beta+1}{\beta-1} < 1$$

for β sufficiently large. Actually $\beta \ge 4k+1$ is easily seen to suffice. For the series for $t_k^-(\alpha)$ we obtain a similar result by using the lower bound (11.18) for v_1 in $\mathfrak{I}^-_{2r}$. This proves the convergence.

Lemma 8. *The series*
(11.19)

$$t_k^+(\alpha) = 1 + k\,\alpha^k\int\limits_{\frac{1}{\beta+1}}^{\alpha}\frac{dv}{v^{k+1}} + \ldots + \alpha^k\,k^{2r+1}\underset{\substack{0\le v_1\le\cdots\le v_{2r+1}\le\alpha\\ v\in\mathcal{V}^+_{2r+1}}}{\int\cdots\int}\frac{dv_1\ldots dv_{2r+1}}{v_1\ldots v_{2r+1}} + \ldots$$
$$= 1 + k\int\limits_0^{\alpha}\frac{\theta_{F^+}(v)}{v}\,dv + \ldots + \frac{k^r}{r!}\int\limits_0^{\alpha}\cdots\int\limits_0^{\alpha}\theta_{F^+}(v_1,\ldots,v_r)\,\frac{dv_1\ldots dv_r}{v_1\ldots v_r} + \ldots$$

converge absolutely and uniformly for $k \le 1/2-\delta$, $1 \le \beta \le B$, $0 \le \alpha \le A$, where δ is any fixed positive number, A and B any large positive constants. The same holds true for $0 < k \le K$, $1+4k \le \beta \le 1+BK$, $0 \le \alpha \le A$, where K is an arbitrary large positive constant. The same results hold if in addition $\alpha \le 1/\beta$, for the series

(11.19′)

$$t_k^-(\alpha) = 1 - \alpha^2 k^2 \iint\limits_{\substack{0\le v_1\le v_2\le\alpha \\ v\in\mathcal{V}_2^-}} \frac{dv_1 dv_2}{v_1^{k+1} v_2} - \dots$$

$$- \alpha^k k^{2r} \int\limits_{\substack{0\le v_1\le\dots\le v_{2r}\le\alpha \\ v\in\mathcal{V}_{2r}^-}} \cdots \int \frac{dv_1 \dots dv_{2r}}{v_1^{k+1} v_2 \dots v_{2r}} - \dots$$

$$= 1 + k \int_0^\alpha \frac{\theta_{F^-}(v)}{v}\, dv + \dots + \frac{k^r}{r!} \int_0^\alpha \cdots \int_0^\alpha \theta_{F^-}(v_1, \dots, v_r) \frac{dv_1 \dots dv_r}{v_1 \dots v_r} + \dots$$

Our argument earlier proves this for the first series that occurs in (11.19) and (11.19′). The conclusion concerning the second series ocurring in the two equations is quite obvious from (11.7) since for our F^+ and F^- we are dealing in each case with infinite series with only positive terms, or in the case of F^-, all but the first term are negative. This makes the passage to limit $\eta \to 0$ simple. The second series in (11.19) and (11.19′), being power series in k with real positive (or, in case of (11.19′), negative) coefficients, of course converge uniformly in k for $k \le k_0$ if they converge for k_0. By a well known theorem these power series will converge up to the nearest singularity on the positive real axis. Thus, if we can find other analytic expressions for $t_k^+(\alpha)$ and $t_k^-(\alpha)$ as functions of k which are of such a nature as to tell us about the nature and location of the singularities, we will be able to describe the precise region of convergence of all four series that occur in (11.19) and (11.19′), since it is clear that in (11.19) for instance the two series converge or diverge together. Our next goal is therefore to find such expressions.

It is clearly enough if we can arrive at such expressions assuming for instance that $k < 1/2$. Since we are actually dealing with functions that are analytic in k, the new expressions would represent $t_k^+(\alpha)$ and $t_k^-(\alpha)$ as long as the power series converges.

We first prove a kind of recursive relationship between $t_k^+(\alpha)$ and $t_k^-(\alpha)$.

Lemma 9. *We have*

$$\frac{d\,\alpha^{-k}\, t_k^+(\alpha)}{d\,\alpha} = \begin{cases} 0, & \text{for } \alpha > 1/(\beta+1), \\ -k\,\alpha^{-k-1}\, t_k^-(\alpha/(1-\alpha)), & \text{for } \alpha < 1/(\beta+1), \end{cases}$$

and for $\alpha < 1/\beta$,

$$\frac{d\,\alpha^{-k}\, t_k^-(\alpha)}{d\,\alpha} = -k\,\alpha^{-k-1}\, t_k^+(\alpha/(1-\alpha)).$$

The proof of this Lemma can be based on either of the two series for $t_k^+(\alpha)$ and $t_k^-(\alpha)$ given in (11.19) and (11.19′). We shall here prove the first part using the first series in (11.19) and the second part using the second series in (11.19′). One could as easily do it with the opposite choice.

First if $\alpha \geq 1/(\beta+1)$, we see if we integrate the second term in the first series in (11.19) that the formula becomes

$$t_k^+(\alpha) = (\beta+1)^k\,\alpha^k + \sum_{r=1}^{\infty} \alpha^k k^{2r+1} \underset{\substack{0\leq v_1\leq\ldots\leq v_{2r+1}\\ v\in\mathcal{V}_{2r+1}^+}}{\int\cdots\int} \frac{dv_1\ldots dv_{2r+1}}{v_1^{k+1}\,v_2\ldots v_{2r+1}}\,.$$

The definition of $\mathcal{V}_{2r+1}^+$ for $r\geq 1$ already implies $v_{2r+1}\leq 1/(\beta+1)\leq\alpha$, so we may drop the α in the description of the integration domains. Thus the right hand side is simply α^k times a quantity independent of α, which shows that

$$\frac{d}{d\alpha}\left(\alpha^{-k}\,t_k^+(\alpha)\right) = 0$$

for $\alpha > 1/(\beta+1)$. For $\alpha = 1/(\beta+1)$ we see that the right derivative exists and is zero.

Next assume $\alpha \leq 1/(\beta+1)$. We observe that for $r\geq 1$, if $v_{2r+1}\leq\alpha$, $v_1\leq v_2\leq\ldots\leq v_{2r+1}$ are in $\mathcal{V}_{2r+1}^+$, then

$$\frac{v_1}{1-v_{2r+1}} \leq \frac{v_2}{1-v_{2r+1}} \leq\ldots\leq \frac{v_{2r}}{1-v_{2r+1}}$$

are in $\mathcal{V}_{2r}^-$ and bounded by $\frac{v_{2r+1}}{1-v_{2r+1}} \leq \frac{\alpha}{1-\alpha} \leq \frac{1}{\beta}$, and conversely. Thus if we introduce new variables in our integrals as follows, $v_i' = v_i/(1-v_{2r+1})$ for $i=1,\ldots,2r$ and $v = v_{2r+1}$, we can write the first series in (11.19) as

$$t_k^+(\alpha) = 1 + k\alpha^k\int_0^\alpha \frac{dv}{v^{k+1}}\left\{\sum_{r=1}^{\infty}\left(\frac{v}{1-v}\right)^k k^{2r} \underset{\substack{0\leq v_1'\leq\ldots\leq v_{2r}'\leq\frac{v}{1-v}\\ v'\in\mathcal{V}_{2r}^-}}{\int\cdots\int} \frac{dv_1'\ldots dv_{2r}'}{v_1'^{\,k+1}\,v_2'\ldots v_{2r}'}\right\}$$

$$= 1 + k\,\alpha^k\int_0^\alpha \frac{dv}{v^{k+1}}\left\{1 - t_k^-\left(\frac{v}{1-v}\right)\right\}.$$

If we here divide by α^k and differentiate we get

$$\frac{d}{d\alpha}\left(\alpha^{-k}\,t_k^+(\alpha)\right) = -k\,\alpha^{-k-1} t_k^-\left(\frac{\alpha}{1-\alpha}\right).$$

Finally let $\alpha\leq 1/\beta$, and consider the second series in (11.19′). The general term here may also be written as

$$k^r \underset{0\leq v_1\leq\ldots\leq v_r\leq\alpha}{\int\cdots\int} \theta_{F^-}(v_1,\ldots,v_r)\,\frac{dv_1\ldots dv_r}{v_1\ldots v_r}\,,$$

we can therefore rewrite the formula as
(11.20)

$$t_k^-(\alpha) = 1 + k\int_0^\alpha \frac{dv}{v}\left\{\sum_{r=0}^\infty k^r \underset{0\le v_1\le\ldots\le v_r\le v}{\int\cdots\int} \theta_{F^-}(v_1,\ldots,v_r,v)\,\frac{dv_1\ldots dv_r}{v_1\ldots v_r}\right\}.$$

Here we may write

$$\theta_{F^-}(v_1,\ldots,v_r,v) = \theta_{F^-}(v_1,\ldots,v_r) + \theta_{F^-}^{(v)}(v_1,\ldots,v_r),$$

where we have designated the aggregate of terms in $\theta_{F^-}(v_1,\ldots,v_r,v)$ that contain v as the last argument by $\theta_{F^-}^{(v)}(v_1,\ldots,v_r)$. Thus for every term

$$F_s^-(v_{i_1},\ldots,v_{i_s})$$

with $1\le i_1\le\ldots\le i_s\le r$ occurring in $\theta_{F^-}(v_1,\ldots,v_r)$ there is a term

$$F_{s+1}^-(v_{i_1},\ldots,v_{i_s},v)$$

in $\theta_{F^-}^{(v)}(v_1,\ldots,v_r)$, but since $v\le\alpha\le 1/\beta$, we have by (11.6′)

$$F_{s+1}^-(v_{i_1},\ldots,v_{i_s},v) = -F_s^+\left(\frac{v_{i_1}}{1-v},\ldots,\frac{v_{i_s}}{1-v}\right).$$

Thus we see that

$$\theta_{F^-}^{(v)}(v_1,\ldots,v_r) = -\theta_{F^+}\left(\frac{v_1}{1-v},\ldots,\frac{v_r}{1-v}\right),$$

and consequently

$$\theta_{F^-}(v_1,\ldots,v_r,v) = \theta_{F^-}(v_1,\ldots,v_r) - \theta_{F^+}\left(\frac{v_1}{1-v},\ldots,\frac{v_r}{1-v}\right).$$

Inserting this in (11.20), we get

$$t_k^-(\alpha) = 1 + k\int_0^\alpha \frac{dv}{v}\left(t_k^-(v) - t_k^+\left(\frac{v}{1-v}\right)\right).$$

Differentiating with respect to α we get

$$\frac{dt_k^-(\alpha)}{d\alpha} = \frac{k}{\alpha}\left(t_k^-(\alpha) - t_k^+\left(\frac{\alpha}{1-\alpha}\right)\right),$$

which is easily seen to be equivalent to the last statement of Lemma 9. It is also true that $t_k^+(\alpha)$ has a left derivative at $1/(\beta+1)$, and $t_k^-(\alpha)$ a left derivative at $1/\beta$ and the values of these are given by the expressions valid for $\alpha < 1/(\beta+1)$ and $\alpha < 1/\beta$ respectively.

For the further analysis it will be convenient to introduce the new variable $u = 1/\alpha$ (which brings the advantage of replacing the operation $\alpha \mapsto \alpha/(1-\alpha)$ by $u \mapsto u-1$), and instead of t_k^+ and t_k^- the two functions

$$\Delta_k(u) = t_k^+(\alpha) - t_k^-(\alpha)\,, \tag{11.21}$$

and

$$\sigma_k(u) = t_k^+(\alpha) + t_k^-(\alpha) - 2\,. \tag{11.21}$$

Both functions are defined for $u \geq \beta$. From (11.13) and (11.13′) we see that for $k \leq 1/2 - \delta$

$$\Delta_k(u) = O(e^{-c(\delta)\,u}), \qquad \sigma_k(u) = O(e^{-c(\delta)\,u})\,, \tag{11.22}$$

as $u \to \infty$. For $u \geq \beta + 1$, the results of Lemma 9 give easily

$$(u^k\,\Delta_k(u))' = -k\,u^{k-1}\,\Delta_k(u-1)\,, \tag{11.23}$$

and

$$(u^k\,\sigma_k(u))' = k\,u^{k-1}\,\sigma_k(u-1)\,. \tag{11.23′}$$

If we put

$$A_k = (\beta+1)^k\,t_k^+\left(\frac{1}{\beta+1}\right)\,, \tag{11.24}$$

and

$$B_k = t_k^-\left(\frac{1}{\beta}\right)\,, \tag{11.24′}$$

we find for $\beta \leq u \leq \beta + 1$,

$$(u^k\,\Delta_k(u))' = -k\,u^{k-1}\,t_k^+\left(\frac{1}{u-1}\right) = -kA_k\,\frac{u^{k-1}}{(u-1)^k}\,, \tag{11.25}$$

and

$$(u^k\,\sigma_k(u))' = k\,A_k\,\frac{u^{k-1}}{(u-1)^k} - 2k\,u^{k-1}\,. \tag{11.25′}$$

In addition we have

$$\Delta_k(\beta) = A_k\,\beta^{-k} - B_k\,, \tag{11.26}$$

and

$$\sigma_k(\beta) = A_k\beta^{-k} + B_k - 2\,. \tag{11.26′}$$

Now consider the Laplace transforms

$$D_k(z) = \int_\beta^\infty \Delta_k(u)\, e^{-uz} du\,, \tag{11.27}$$

and

$$S_k(z) = \int_\beta^\infty \sigma_k(u)\, e^{-uz} du\,. \tag{11.27'}$$

For $k \le 1/2 - \delta$, we see from (11.22) that $D_k(z)$ and $S_k(z)$ are regular and analytic in the half plane $\mathrm{Re}\,(z) > -c(\delta)$.

For convenience we shall for a while drop the subscripts k, and rewrite (11.23) as

$$u\,\Delta'(u) + k\,\Delta(u) + k\,\Delta(u-1) = 0\,,$$

for $u \ge \beta + 1$, and (11.25) as

$$u\,\Delta'(u) + k\,\Delta(u) = -k\,A\,(u-1)^{-k}\,,$$

for $\beta \le u \le \beta + 1$. From this follows

$$\begin{aligned} &\int_\beta^\infty u\,\Delta'(u)\, e^{-uz} du + k \int_\beta^\infty \Delta(u)\, e^{-uz} du + k \int_{\beta+1}^\infty \Delta(u-1)\, e^{-uz} du \\ &= -k\,A \int_\beta^{\beta+1} \frac{e^{-uz}}{(u-1)^k}\, du\,. \end{aligned} \tag{11.28}$$

Here

$$\int_\beta^\infty u\,\Delta'(u)\, e^{-uz} du = -\frac{d}{dz} \int_\beta^\infty \Delta'(u)\, e^{-uz} du\,,$$

and by integration by parts

$$\int_\beta^\infty \Delta'(u)\, e^{-uz} du = -\Delta(\beta)\, e^{-\beta z} + z \int_\beta^\infty \Delta(u)\, e^{-uz}\, du = z\, D(z) - (A\,\beta^{-k} - B)\, e^{-\beta z}$$

so that

$$\int_\beta^\infty u\,\Delta'(u)\, e^{uz} du = -(z\,D(z))' - (A\beta^{1-k} - B\,\beta)\, e^{-\beta z}\,.$$

Also

$$k \int_\beta^\infty \Delta(u)\, e^{-uz} du + k \int_{\beta+1}^\infty \Delta(u-1)\, e^{-uz} du = k\,(1 + e^{-z})\, D(z)\,.$$

Thus (11.28) becomes

$$-(z\,D(z))' + k(1 + e^{-z})\, D(z) = -k\,A \int_\beta^{\beta+1} \frac{e^{-uz}}{(u-1)^k}\, du + (A\,\beta^{1-k} - B\,\beta)\, e^{-\beta z}\,,$$

or

$$(z\,D(z))' - k\,(1 + e^{-z})\, D(z) = U(z)\,, \tag{11.29}$$

where

$$U(z) = U_k(z) = k\,A \int_{\beta}^{\beta+1} \frac{e^{-uz}}{(u-1)^k}du - (A\,\beta^{1-k} - B\,\beta)\,e^{-\beta z}\,. \tag{11.30}$$

In a quite similar manner one finds

$$(z\,S(z))' - k\,(1-e^{-z})\,S(z) = V(z)\,, \tag{11.31}$$

where $V(z) = V_k(z)$ is given by

$$V(z) = \int_{\beta}^{\beta+1} \left(\frac{kA}{(u-1)^k} - 2k\right) e^{-uz}du + \beta\,(A\,\beta^{-k} + B - 2)\,e^{-\beta z}\,. \tag{11.32}$$

The equations (11.29) and (11.31) can be solved by standard methods. Since the solutions we seek go strongly to zero as the real part of z goes to infinity, we find

$$z\,D(z) = -z^{2k}\,e^{-k\,\omega(z)} \int_z^{\infty} \frac{U(\zeta)\,e^{k\,\omega(\zeta)}}{\zeta^{2k}}d\zeta\,, \tag{11.33}$$

where

$$\omega(z) = \int_0^z \frac{1-e^{-\zeta}}{\zeta}\,d\zeta\,, \tag{11.34}$$

and similarly

$$z\,S(z) = -e^{k\,\omega(z)} \int_z^{\infty} V(\zeta)\,e^{-k\,\omega(\zeta)}d\zeta\,. \tag{11.35}$$

The general solutions of the equations differ from (11.33) and (11.35) respectively by

$$C\,z^{2k}\,e^{-k\,\omega(z)}\,,$$

and

$$C\,e^{k\,\omega(z)}\,,$$

but both of these expressions grow like z^k as $z \to \infty$, since we have

$$\omega(z) = \gamma + \log z + \int_z^{\infty} \frac{e^{-\zeta}}{\zeta}d\zeta\,,$$

where γ is Euler's constant.

Since $S(z)$ clearly is regular at $z=0$, we get from (11.35) by taking $z=0$, that

$$\int_0^{\infty} V_k(\zeta)\,e^{-k\,\omega(\zeta)}d\zeta = 0\,, \tag{11.36}$$

where we now have introduced the subscripts k. The fact that $D(z)$ is single valued for $\operatorname{Re} z > -c(\delta)$, shows that the integral on the right hand side of

(11.33) is independent of the path of integration as long as it remains in that half plane, thus, in particular, we must have

$$\frac{1}{2\pi i}\int_L \frac{U_k(\zeta)\, e^{k\,\omega(\zeta)}}{(-\zeta)^{2k}}\, d\zeta = 0\,, \tag{11.37}$$

where L is a path from $+\infty$ which loops around the origin and returns to $+\infty$ as illustrated, starting with $\arg(-\zeta) = \pi$.

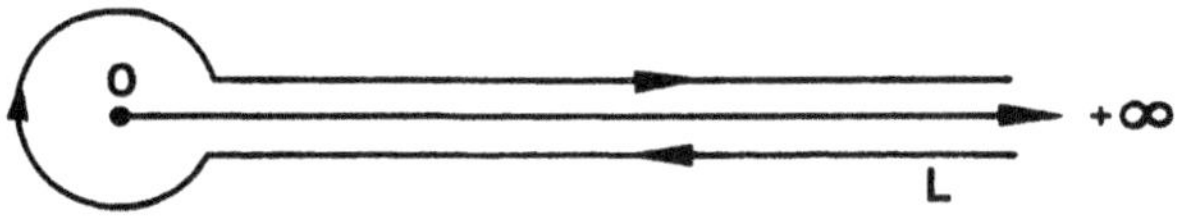

(11.36) and (11.37) now give us two linear equations to determine A_k and B_k. (11.37) is a homogeneous equation in A_k and B_k, while (11.36) has a term independent of A_k and B_k. For $\beta > 1$ it is immediately apparent that the coefficients of A_k and B_k in both equations, as well as the term independent of A_k and B_k in (11.36), are integral functions of k, so that if they determine A_k and B_k at all (which we shortly shall see that they do), they show that A_k and B_k are meromorphic functions of k. For $\beta = 1$, the situation is not quite so obvious because of the presence of the term $k\int_\beta^{\beta+1} \frac{e^{-uz}}{(u-1)^k}\, du$ in $U_k(z)$ and $V_k(z)$ which clearly has a singularity at $k = 1$ if $\beta = 1$.

We shall need to simplify the expressions that occur as coefficients of A_k, B_k and the term independent of A_k and B_k in these equations.

Consider first (11.36), the "constant term" is given by

$$\begin{aligned}
(11.38)\quad & -\int_0^\infty e^{-k\,\omega(z)} \left(2k\int_\beta^{\beta+1} e^{-uz} du + 2\beta\, e^{-\beta z}\right) dz \\
&= -2\int_0^\infty e^{-k\omega(z)} \left(k\, e^{-\beta z}\,\frac{1-e^{-z}}{z} + \beta\, e^{-\beta z}\right) dz \\
&= -2\beta\int_0^\infty e^{-k\omega(z)-\beta z} dz + 2\int_0^\infty e^{-\beta z} de^{-k\omega(z)} \\
&= -2\beta\int_0^\infty e^{-k\omega(z)-\beta z} dz + 2\Big[\ e^{-\beta z - k\omega(z)}\Big]_0^\infty + 2\beta\int_0^\infty e^{-k\omega(z)-\beta z} dz \\
&= -2.
\end{aligned}$$

The coefficient of B_k is

$$\beta\int_0^\infty e^{-k\omega(z)-\beta z} dz = \beta\, J_{k,\beta}\,. \tag{11.39}$$

For the coefficient of A_k we get, using again the notation

$$\int_0^\infty e^{-k\omega(z)-\beta z} dz = J_{k,\beta}\,, \tag{11.40}$$

that the coefficient of A_k is

$$\beta^{1-k} J_{k,\beta} + k\int_0^\infty e^{-k\omega(z)} \int_\beta^{\beta+1} \frac{e^{-uz}}{(u-1)^k} du dz\,.$$

In the last integral we write $u = 1 + v/z$ and $\omega(z) = \gamma + \log z + \omega_1(z)$, where

$$\omega_1(z) = \int_z^\infty \frac{e^{-\zeta}}{\zeta} d\zeta ,$$

and it becomes for $\beta > 1$,

$$\begin{aligned} &k\, e^{-\gamma k} \int_0^\infty \frac{e^{-z-k\,\omega_1(z)}}{z} \int_{(\beta-1)\,z}^{\beta z} \frac{e^{-v}}{v^k} dv dz = e^{-k\gamma} \int_0^\infty \int_{(\beta-1)\,z}^{\beta z} \frac{e^{-v}}{v^k} dv de^{-k\omega_1(z)} \\ &= e^{-k\gamma} \left[e^{-k\omega_1(z)} \int_{(\beta-1)\,z}^{\beta z} \frac{e^{-v}}{v^k} \right]_0^\infty \\ &+ e^{-k\gamma} \int_0^\infty e^{-k\omega_1(z)} \left((\beta-1)^{1-k} e^{-(\beta-1)z} - \beta^{1-k} e^{-\beta z} \right) \frac{dz}{z^k} \qquad (11.41) \\ &= (\beta-1)^{1-k} \int_0^\infty e^{-k\omega(z)-(\beta-1)z} dz - \beta^{1-k} \int_0^\infty e^{-k\omega(z)} e^{-\beta z} dz \\ &= (\beta-1)^{1-k} J_{k,\beta-1} - \beta^{1-k} J_{k,\beta} . \end{aligned}$$

Combining this with the previous expression, we get for the coefficient of A_k,

$$(\beta-1)^{1-k} J_{k,\beta-1} ,$$

in the case $\beta > 1$. For $\beta = 1$, (11.41) is modified after the first line to

$$\begin{aligned} &e^{-k\gamma} \left[e^{-k\omega_1(z)} \int_0^z \frac{e^{-v}}{v^k} \right]_0^\infty - e^{-k\gamma} \int_0^\infty e^{-k\omega_1(z)} e^{-z} \frac{dz}{z^k} \\ &= e^{-k\gamma}\Gamma(1-k) - \int_0^\infty e^{-z-k\omega(z)} dz = e^{-k\gamma}\Gamma(1-k) - J_{k,1} . \end{aligned}$$

Thus for $\beta = 1$, the coefficient of A_k becomes

$$e^{-k\gamma}\Gamma(1-k) .$$

Together with (11.38) and (11.39), this shows that (11.36) takes the form

$$(\beta-1)^{1-k} J_{k,\beta-1} A_k + \beta\, J_{k,\beta} B_k = 2 , \tag{11.42}$$

for $\beta > 1$, while for $\beta = 1$ it becomes

$$e^{-k\gamma}\Gamma(1-k)\, A_k + J_{k,1} B_k = 2 . \tag{11.42$'$}$$

It should be remarked that for $k < 1$ it is easily shown that

$$\lim_{\beta\to 1} (\beta-1)^{1-k} J_{k,\beta-1} = e^{-\gamma k}\, \Gamma(1-k) ,$$

since writing $\beta = 1 + \delta$, we have

$$\begin{aligned} \delta^{1-k} J_{k,\delta} &= \delta^{1-k} \int_0^\infty e^{-\delta z - k\omega(z)} dz = e^{-k\gamma} \int_0^\infty e^{-\delta z - k\omega_1(z)} \frac{d\delta z}{(\delta z)^k} \\ &= e^{-k\gamma} \int_0^\infty e^{-v-k\omega_1(v/\delta)} \frac{dv}{v^k} . \end{aligned}$$

Here $\omega_1(v/\delta)$ tends uniformly to zero as $\delta \to 0$ for $0 < \varepsilon \le v < \infty$, thus the integral tends to

$$\int_0^\infty e^{-v} \frac{dv}{v^k} = \Gamma(1-k)\,.$$

Next we turn our attention to (11.37), we shall write

$$\mathcal{I}_{k,\beta} = \frac{1}{2\pi\, i} \int_L e^{-\beta z + k\omega(z)} \frac{dz}{(-z)^{2k}}\,. \tag{11.43}$$

The coefficient of B_k in (11.37) is then seen to be

$$\beta\, \mathcal{I}_{k,\beta}\,. \tag{11.44}$$

As in the case of (11.36) the coefficient of A_k is more complicated, it equals

$$-\beta^{1-k} \mathcal{I}_{k,\beta} + \frac{k}{2\pi\, i} \int_L \frac{e^{k\omega(z)}}{(-z)^{2k}} \int_\beta^{\beta+1} \frac{e^{uz}}{(u-1)^k} du dz\,.$$

In the second term we put $u = 1 + v/z$ and write $\omega(z) = \gamma + \log z + \omega_1(z)$, we get then for this term, assuming $\beta > 1$,

$$\begin{aligned}
(11.45)\quad & \frac{e^{-2k\pi\, i}}{2\pi\, i} e^{\gamma k} k \int_L \frac{e^{k\omega_1(z) - z}}{z} \int_{(\beta-1)\, z}^{\beta z} \frac{e^{-v}}{v^k} dv dz \\
&= -\frac{e^{-2k\pi\, i}}{2\pi\, i} e^{\gamma k} \int_L \int_{(\beta-1)\, z}^{\beta z} \frac{e^{-v}}{v^k}\, dv de^{k\omega_1(z)} \\
&= -\frac{e^{-2k\pi\, i}}{2\pi\, i} e^{\gamma k} \left[e^{k\omega_1(z)} \int_{(\beta-1)\, z}^{\beta z} \frac{e^{-v}}{v^k}\, dv \right]_L \\
&\qquad + \frac{e^{-2k\pi\, i}}{2\pi\, i} e^{\gamma k} \int_L \frac{e^{k\omega_1(z)}}{z^k} (\beta^{1-k}\, e^{-\beta z} - (\beta - 1)^{1-k} e^{-(\beta-1)\, z}) dz \\
&= \frac{e^{-2k\pi\, i}}{2\pi\, i} \left(\beta^{1-k} \int_L \frac{e^{k\omega(z) - \beta z}}{z^{2k}}\, dz - (\beta-1)^{1-k} \int_L \frac{e^{k\omega(z) - (\beta-1) z}}{z^{2k}} dz \right) \\
&= \frac{1}{2\pi\, i} \left(\beta^{1-k} \int_L \frac{e^{k\omega(z) - \beta z}}{(-z)^{2k}}\, dz - (\beta-1)^{1-k} \int_L \frac{e^{k\omega(z) - (\beta-1) z}}{(-z)^{2k}} dz \right) \\
&= \beta^{1-k}\, \mathcal{I}_{k,\beta} - (\beta - 1)^{1-k} \mathcal{I}_{k,\beta-1}\,.
\end{aligned}$$

Combining this with the previous expression, we get that for $\beta > 1$ the coefficient of A_k is

$$-(\beta-1)^{1-k} \mathcal{I}_{k,\beta-1}\,.$$

For $\beta = 1$, (11.45) is modified after the second line to

$$\begin{aligned}
(11.45')\quad & -\frac{e^{-2k\pi\, i}}{2\pi\, i} e^{\gamma k} \left[e^{k\omega_1(z)} \int_0^z \frac{e^{-v}}{v^k}\, dv \right]_L + \frac{e^{-2k\pi\, i}}{2\pi\, i} e^{\gamma k} \int_L \frac{e^{k\omega_1(z) - z}}{z^k} dz \\
&= -\frac{e^{-2k\pi\, i}}{2\pi\, i} e^{\gamma k} \left[e^{k\omega_1(z)} \int_0^z \frac{e^{-v}}{v^k} dv \right]_L + \frac{1}{2\pi\, i} \int_L \frac{e^{k\omega(z) - z}}{(-z)^{2k}}\, dz\,,
\end{aligned}$$

here the last term is $\mathcal{I}_{k,1}$. The expression in the $[\ ]_L$ bracket can be rewritten as

$$\frac{e^{k(\omega(z)-\gamma)}}{z^{2k}}\, z^k \int_0^z \frac{e^{-v}}{v^k}dv\,,$$

we see this is single valued (actually an integral function of z) apart from the factor z^{-2k}. In our L integrals our convention is to start with argument of $-z$ equal to π, or after taking out the factor $e^{\pi i}$ as we have done, the argument of z equal to 0 at the start and equal to -2π at the end of the loop L. This gives for the $[\ ]_L$ in (11.45′)

$$(e^{4k\pi i}-1)\,\Gamma(1-k)\,,$$

and (11.45′) finally becomes

$$-\frac{e^{2k\pi i}-e^{-2k\pi i}}{2\pi i}\, e^{\gamma k}\,\Gamma(1-k)+\mathcal{I}_{k,1} = -\frac{\sin 2k\pi}{\pi}\, e^{\gamma k}\,\Gamma(1-k)+\mathcal{I}_{k,1}\,.$$

Thus for $\beta=1$ the coefficient of A_k is

$$-\frac{\sin 2k\pi}{\pi}\, e^{\gamma k}\,\Gamma(1-k)\,.$$

For $\beta>1$, (11.37) becomes

$$-(\beta-1)^{1-k}\mathcal{I}_{k,\beta-1}\,A_k+\beta\,\mathcal{I}_{k,\beta}\,B_k=0\,, \qquad (11.46)$$

while for $\beta=1$ it has the form

$$-\frac{\sin 2k\pi}{\pi}\, e^{\gamma k}\,\Gamma(1-k)\,A_k+\mathcal{I}_{k,1}\,B_k=0\,. \qquad (11.46')$$

These equations combined with (11.42) and (11.42′) determine A_k and B_k as meromorphic functions of k in the complex k–plane. For real positive k this again determines $U_k(z)$ and $V_k(z)$ and thus via (11.33) and (11.35)[19] the functions $D_k(z)$ and $S_k(z)$. From this it is possible to recover the functions $\Delta_k(u)$ and $\sigma_k(u)$ by a Mellin transform and finally the functions $t_k^+(\alpha)$ and $t_k^-(\alpha)$.

It is however much simpler, once A_k and B_k are found, to use (11.24) and (11.24′) which at once gives us $t_k^+(1/(\beta+1))$ and $t_k^-(1/\beta)$, and then use Lemma 9 to recover $t_k^+(\alpha)$ and $t_k^-(\alpha)$ recursively for $\alpha<1/(\beta+1)$ and $\alpha<1/\beta$ respectively. For $\alpha>1/(\beta+1)$ we have $t_k^+(\alpha)=A_k\,\alpha^k$. For $1/\beta<\alpha\le 1$, we have earlier not defined $t_k^-(\alpha)$. It is however natural to define it by the formula for the derivative of $\alpha^{-k}\,t_k^-(\alpha)$ given in Lemma 9. This corresponds to extending the range from $\mathcal{P}(x^{1/\beta})$ to $\mathcal{P}(x^{\alpha})$ using Buchstab's inequality

[19] These formulas can also be simplified by a process similar to that used to simplify the coefficients of A_k in (11.42) and (11.46) or (11.42′) and (11.46′) depending on whether $\beta>1$ or $\beta=1$. We do not have need for these formulas however.

(11.3′) (for $\alpha = 1$ it leads to a finite $t_k^-(1)$ only when $k < 1$). From this we see that $t_k^+(\alpha)$ for $\alpha > 0$ and $t_k^-(\alpha)$ for $0 < \alpha < 1$, are also meromorphic in k with poles only where $t_k^+(1/(\beta+1))$ or $t_k^-(1/\beta)$ (that is to say: A_k and B_k) has poles. Since $t_k^+(\alpha)$ and $t_k^-(\alpha)$ according to Lemma 8 are given by power series in k where the coefficients of k^r with $r > 0$ are all of the same sign, these power series and therefore the B^2R sieves for upper and lower bounds themselves, converge for $k \le k(\beta)$ where $k(\beta)$ is the first pole of A_k or B_k on the positive real axis.[20]

From the two equations that determine A_k and B_k, we get for $\beta > 1$,

$$A_k = \frac{2\mathcal{I}_{k,\beta}}{(\beta-1)^{1-k}\,\mathcal{D}_{k,\beta}}, \tag{11.47}$$

and

$$B_k = \frac{2\mathcal{I}_{k,\beta-1}}{\beta\,\mathcal{D}_{k,\beta}}, \tag{11.47′}$$

where we have written

$$\mathcal{D}_{k,\beta} = \mathcal{I}_{k,\beta} J_{k,\beta-1} + \mathcal{I}_{k,\beta-1} J_{k,\beta}\,. \tag{11.48}$$

For $\beta = 1$, we have

$$A_k = \frac{2\,\mathcal{I}_{k,1}}{\Gamma(1-k)\left(e^{-\gamma k}\mathcal{I}_{k,1} + e^{\gamma k}\dfrac{\sin 2k\pi}{\pi}\,J_{k,1}\right)}, \tag{11.49}$$

and

$$B_k = \frac{2\,\sin 2k\pi}{\pi\left(e^{-\gamma k}\mathcal{I}_{k,1} + e^{\gamma k}\dfrac{\sin 2k\pi}{\pi}\,J_{k,1}\right)}. \tag{11.49′}$$

We now turn to a discussion of these formulas. We need first some results concerning the integrals $\mathcal{I}_{k,\beta}$ and $J_{k,\beta}$, namely

$$(\beta-1)\,\frac{d}{d\beta}\,\mathcal{I}_{k,\beta-1} = (k-1)\,\mathcal{I}_{k,\beta-1} + k\,\mathcal{I}_{k,\beta}\,, \tag{11.50}$$

and

$$(\beta-1)\,\frac{d}{d\beta}\,J_{k,\beta-1} = (k-1)\,J_{k,\beta-1} - k\,J_{k,\beta}\,, \tag{11.50′}$$

We prove the first formula with β instead of $\beta - 1$ as follows: We have using

[20] Both A_k and B_k have this pole. This is obvious for $\beta > 1$, for $\beta = 1$ it follows because one can show $k(1) < 1$.

integration by parts,

$$\begin{aligned}\beta\frac{d}{d\beta}\mathcal{I}_{k,\beta} &= \frac{1}{2\pi i}\int_L \frac{e^{k\omega(z)}\,\beta e^{-\beta z}}{(-z)^{2k-1}}\,dz = -\frac{1}{2\pi i}\int_L \frac{e^{k\omega(z)}}{(-z)^{2k-1}}\,de^{-\beta z}\\ &= \frac{1}{2\pi i}\int_L \frac{e^{k\omega(z)-\beta z}}{(-z)^{2k-1}}\left(k\omega'(z)-\frac{2k-1}{z}\right)dz\\ &= \frac{1}{2\pi i}\int_L \frac{e^{k\omega(z)-\beta(z)}}{(-z)^{2k-1}}\left(-\frac{k-1}{z}-\frac{k\,e^{-z}}{z}\right)dz\\ &= \frac{k-1}{2\pi i}\int_L \frac{e^{k\omega(z)-\beta z}}{(-z)^{2k}}\,dz + \frac{k}{2\pi i}\int_L \frac{e^{k\omega(z)-(\beta+1)\,z}}{(-z)^{2k}}\,dz\\ &= (k-1)\,\mathcal{I}_{k,\beta} + k\,\mathcal{I}_{k,\beta+1}\,.\end{aligned}$$

The second formula involving $J_{k,\beta}$ can be proved in a similar manner by partial integration.

It is easy to see that for fixed k, if β is large enough, then $\mathcal{I}_{k,\beta}$ is positive. One may for instance use the method of steepest descent to evaluate the integral defining $\mathcal{I}_{k,\beta}$ and find that

$$\mathcal{I}_{k,\beta} \sim \frac{1}{\Gamma(2k)}\,\beta^{2k-1}\,,$$

as $\beta \to \infty$ (for large β the integral is very similar to the integral

$$\frac{1}{2\pi\, i}\int_L \frac{e^{-\beta z}}{(-z)^{2k}}\,dz = \frac{1}{\Gamma(2k)}\,\beta^{2k-1}\,,$$

in the critical area near the origin).

As β decreases, $\mathcal{I}_{k,\beta}$ clearly decreases as long as $\mathcal{I}_{k,\beta} \geq 0$ and $\mathcal{I}_{k,\beta+1} > 0$. We thus have two possibilities: Either $\mathcal{I}_{k,\beta}$ stays positive for all $\beta > 0$, or we reach a β which we shall denote by $\tilde{\beta} = \tilde{\beta}(k) > 0$, such that $\mathcal{I}_{k,\tilde{\beta}} = 0$, while $\mathcal{I}_{k,\beta} > 0$ for $\beta > \tilde{\beta}$. For large k we may estimate $\tilde{\beta}$ by estimating $\mathcal{I}_{k,\beta}$ for large k and β by the method of steepest descent.

If we write $-z$ instead of z in the integral, the logarithm of the integrand is

$$\varphi(z) = -k\int_0^z \frac{e^t-1}{t}\,dt + \beta\, z - 2k\log z\,.$$

We find

$$\varphi'(z) = -k\,\frac{e^z+1}{z} + \beta\,,$$

and

$$\varphi''(z) = k\,\frac{e^z(1-z)+1}{z^2}\,.$$

$\varphi''(z)$ has one positive zero $z = 1 + c$, where c is the positive solution of $c\,e^{1+c} = 1$. One finds $c = 1/3.591\ldots$. For $\beta = \frac{1}{c}\,k$, $\varphi'(z)$ has a double zero at

$z = 1 + c$, for $\beta > \frac{1}{c}k$, there are two real positive zeroes of $\varphi'(z)$, while for $\beta < \frac{1}{c}k$ we have instead a pair of complex conjugate zeroes.

As a consequence, one finds for k large and $\beta \geq \lambda k$, where $\lambda > \frac{1}{c}$, that $\mathcal{I}_{k,\beta} > 0$ (and large), while for $\beta \leq \lambda' k$ where $\lambda' < \frac{1}{c}$, one finds that $\mathcal{I}_{k,\beta}$ is an oscillating function which changes sign fairly often. In particular one can deduce that $\tilde{\beta} \sim \frac{1}{c}k$ as $k \to \infty$.[21]

Returning to (11.50), we may also write it as

$$\frac{d}{d\beta}\beta^{1-k}\mathcal{I}_{k,\beta} = k\,\beta^{-k}\mathcal{I}_{k,\beta+1}\,. \tag{11.51}$$

From this we conclude that for $\tilde{\beta} - 1 \leq \beta \leq \tilde{\beta}$ (assuming $\tilde{\beta} - 1 > 0$, otherwise for $0 < \beta \leq \tilde{\beta}$), $\beta^{1-k}\mathcal{I}_{k,\beta}$ is decreasing as β decreases, and so certainly negative. Assuming $\tilde{\beta} \geq 1$, and looking at the expression

$$\mathcal{D}_{k,\beta} = \mathcal{I}_{k,\beta}J_{k,\beta-1} + \mathcal{I}_{k,\beta-1}J_{k,\beta}\,,$$

we see (since the $J_{k,\beta}$ are always positive for $\beta > 0$), that

$$\mathcal{D}_{k,\beta} > 0$$

for $\beta \geq \tilde{\beta} + 1$, while clearly $\mathcal{D}_{k,\tilde{\beta}} < 0$. Thus there must be a β which we call $\hat{\beta}$ and define as the largest β for which $\mathcal{D}_{k,\beta} = 0$. It is clear from the above that

$$\tilde{\beta} < \hat{\beta} < \tilde{\beta} + 1\,.$$

It is easy to show that for $k > 1/2$ there is always a $\tilde{\beta}(k) > 0$,[22] and that for $k \geq 1$ we have $\tilde{\beta}(k) \geq 1$ and so a $\hat{\beta}(k)$ certainly exists for $k \geq 1$, a closer analysis shows that a $\hat{\beta}(k)$ exists for $k > k_0$ where k_0 is the smallest positive k for which the denominator in (11.49) or (11.49′) becomes zero, it is obvious that $1/2 < k_0 < 1$.

From (11.50) and (11.50′) one can after some calculation verify the formulas

$$\begin{aligned}\beta\frac{d}{d\beta}\left(\beta(1-\beta)\right)^{1-k}\mathcal{D}_{k,\beta} &= k\left(\beta(1-\beta)\right)^{1-k}\left(\mathcal{I}_{k,\beta+1}J_{k,\beta-1} - \mathcal{I}_{k,\beta-1}J_{k,\beta+1}\right)\\ &= k\,\frac{\left(\beta(1-\beta)\right)^{1-k}}{J_{k,\beta}}\left(J_{k,\beta-1}\,\mathcal{D}_{k,\beta+1} - J_{k,\beta+1}\mathcal{D}_{k,\beta}\right).\end{aligned} \tag{11.52}$$

---

[21] This can be made much more precise; $\tilde{\beta} = \frac{1}{c}k + O(k^{1/3})$ this is fairly easy. Kai Man Tsang has given an asymptotic expansion for $\tilde{\beta}$

$$\tilde{\beta} = \frac{1}{c}k + c_1k^{1/3} + c_2 + c_3\,k^{-1/3} + \ldots + O(k^{-n/3})\,,$$

where n may be chosen arbitrarily large.

[22] And that as $k \to 1/2$, $\tilde{\beta}(k) \to 0$. For $0 < k \leq 1/2$, $\mathcal{I}_{k,\beta} > 0$ for all $\beta > 0$.

This shows that at $\beta = \hat{\beta}(k)$ we have

$$\frac{d}{d\beta}\,\mathcal{D}_{k,\beta} > 0\,,$$

and for $k \geq 1$, also that this inequality holds for $\beta \geq \hat{\beta}$. Thus $\mathcal{D}_{k,\beta}$ always changes sign as β passes through $\hat{\beta}(k)$ and is negative for β immediately below $\hat{\beta}$. Since $\hat{\beta} > \tilde{\beta}$, $\mathcal{I}_{k,\hat{\beta}} > 0$, and so the expression on the right hand side of (11.47) blows up at $\beta = \hat{\beta}$ and becomes negative for β immediately below $\hat{\beta}$.

Since we know from Lemma 8 that the series (11.19) converges for $\beta \geq 1 + 4k$, and has to "blow up" at the boundary of its domain of convergence (since it represents a function which is meromorphic in k and β for $\mathrm{Re}\,(\beta) > 1$), the series (11.19) (and of course also (11.19′) for $\alpha < 1$) must converge as long as $\beta > \hat{\beta}(k)$, and we see even that for a pair k_0, $\hat{\beta}(k_0)$, we must have convergence if both the inequalities $k \leq k_0$, $\beta \geq \hat{\beta}(k_0)$ hold, except in the case when both equality signs hold. Thus the set of points with coordinates $k, \hat{\beta}(k)$ represent a curve which separates the domain of k, β for which the B^{2}R sieve converges from a domain where we clearly have divergence (since $\mathcal{D}_{k,\beta}$ is negative and $\mathcal{I}_{k,\beta} > 0$ immediately on the other side of this curve, (11.47) would make $A_k = t_k^+(1)$ negative, clearly absurd). If we use a coordinate system with k as the horizontal axis, the domain of convergence is the region to the left of this curve (or between the curve and the vertical axis).

Since in the region of convergence clearly

$$\frac{d\,A_k}{dk} > 0$$

we get

$$\mathcal{D}_{k,\beta}\,\frac{d}{dk}(\beta-1)^{k-1}J_{k,\beta} - (\beta-1)^{k-1}\,J_{k,\beta}\,\frac{d}{dk}\,\mathcal{D}_{k,\beta} > 0$$

and so at a point on the curve, where $\mathcal{D}_{k,\beta} = 0$ and $J_{k,\beta} > 0$, we have

$$\frac{d}{dk}\,\mathcal{D}_{k,\beta} \leq 0\,.$$

Combining this with $\frac{d}{d\beta}\,\mathcal{D}_{k,\beta} > 0$ obtained above, we see that the tangent of the curve (which clearly cannot be vertical) might possibly be horizontal, but only at a point of inflection. Since the average slope of the curve approaches $1/c$ as $k \to \infty$,[23] one could probably, by a very thorough quantitative analysis of the integrals involved, prove that the slope is asymptotic to $1/c$ (and also that points of inflection do not occur), but we have no particular need of this.

Next comes the question: For a given k, what is the most favorable choice of β? Going back to the inequalities (11.3) and (11.3′) which inspired the construction of the B^{2}R sieve, and considering that it is best to replace

[23] c being again the positive solution of $c\,e^{c+1} = 1$, since $\tilde{\beta} \sim \frac{1}{c}\,k$ and $\tilde{\beta} < \hat{\beta} < \tilde{\beta} + 1$.

$M_u^-(W_x(p), \mathcal{P}(p))$ by zero at the point where we have no positive lower bound available and not before, one is naturally led to the conclusion that the most favorable choice is to take β as close to 1 as we can without making $t_k^-(1/\beta)$ or B_k negative. For $0 < k \le 1/2$, (11.49) shows that if we take $\beta = 1$, then $B_k > 0$ for $k < 1/2$ and $B_k = 0$ for $k = 1/2$, thus for $k \le 1/2$ $\beta = 1$ should be the optimal choice. For $k > 1/2$, we see from (11.47′) that we can make $B_k = 0$, if we choose β so that $\mathcal{I}_{k,\beta-1} = 0$, and since we need to take a β which will give convergence $\beta = \beta^* = \tilde{\beta} + 1$ would be the choice.

A more rigorous argument goes as follows: We seek the β which makes $t_k^+(1) = A_k$ as small as possible. We have from (11.47), and using (11.51) and (11.52), that

$$\begin{aligned}
\frac{d}{d\beta}\log A_k &= \frac{d}{d\beta}\log \beta^{1-k}\mathcal{I}_{k,\beta} - \frac{d}{d\beta}\log(\beta(\beta-1))^{1-k}\mathcal{D}_{k,\beta} \\
&= \frac{k}{\beta}\,\frac{\mathcal{I}_{k,\beta+1}}{\mathcal{I}_{k,\beta}} - \frac{k}{\beta}\,\frac{\mathcal{I}_{k,\beta+1}J_{k,\beta-1} - \mathcal{I}_{k,\beta-1}J_{k,\beta+1}}{\mathcal{I}_{k,\beta}J_{k,\beta-1} + \mathcal{I}_{k,\beta-1}J_{k,\beta}} \\
&= \frac{k}{\beta}\,\frac{\mathcal{I}_{k,\beta-1}}{\mathcal{I}_{k,\beta}}\left(\frac{\mathcal{I}_{k,\beta+1}J_{k,\beta} + \mathcal{I}_{k,\beta}J_{k,\beta+1}}{\mathcal{I}_{k,\beta}J_{k,\beta-1} + \mathcal{I}_{k,\beta-1}J_{k,\beta}}\right) \\
&= \frac{k}{\beta}\,\frac{\mathcal{I}_{k,\beta-1}}{\mathcal{I}_{k,\beta}}\,\frac{\mathcal{D}_{k,\beta+1}}{\mathcal{D}_{k,\beta}}\,.
\end{aligned}$$

Thus we see that $\frac{dA_k}{d\beta} > 0$ for $\beta > \beta^*$, $\frac{dA_k}{d\beta} < 0$ for $\hat{\beta} < \beta < \beta^*$, as long as $\beta^* > 1$ (which is to say, for $k > 1/2$), while, if $0 < k \le 1/2$ (so that there is no $\tilde{\beta} > 0$) we have $\frac{dA_k}{d\beta} > 0$ for all $\beta > 1$. This shows that for $k > 1/2$, $A_k(\beta)$ has a minimum for $\beta = \beta^*$, while for $0 < k \le 1/2$ we get the smallest $A_k(\beta)$ for $\beta = 1$. For $k > 1/2$, β^* is the largest β for which $\mathcal{I}_{k,\beta-1} = 0$; concerning the curve formed by the points with coordinates $k, \beta^*(k)$, for $k > 1/2$, we can show that its slope is positive and finite. We have from (11.50), that

$$\left[(\beta-1)\frac{d}{d\beta}\mathcal{I}_{k,\beta-1}\right]_{\beta=\beta^*} = k\,\mathcal{I}_{k,\beta^*} > 0\,,$$

so $\frac{d}{d\beta}\mathcal{I}_{k,\beta-1} > 0$ at $\beta = \beta^*$. Also since clearly

$$\frac{d}{dk}t_k^-\left(\frac{1}{\beta}\right) < 0\,,$$

we get, inserting the expression (11.47′) for $t_k^-(1/\beta)$, that

$$\mathcal{D}_{k,,\beta}\,\frac{d\mathcal{I}_{k,\beta-1}}{dk} - \mathcal{I}_{k,\beta-1}\,\frac{d}{dk}\mathcal{D}_{k,\beta} < 0\,,$$

putting $\beta = \beta^*$ so $\mathcal{I}_{k,\beta-1} = 0$, we get since $\mathcal{D}_{k,\beta^*} > 0$, that

$$\frac{d\mathcal{I}_{k,\beta-1}}{dk} < 0\,,$$

for $\beta = \beta^*(k)$.

The calculation of $\beta^*(k)$ for $k > 1/2$ is somewhat cumbersome for general k, since the integral that defines $\mathcal{I}_{k,\beta-1}$ is taken over the infinite complex loop L. If $2k$ is an integer however, the integrand is single valued, and we may replace the loop L by a closed curve around the origin. The condition $\mathcal{I}_{k,\beta-1} = 0$, then becomes

$$\left[\frac{d^{2k-1}}{dz^{2k-1}}\, e^{k\omega(z)-(\beta-1)z}\right]_{z=0} = 0\,,$$

since the left hand side is a polynomial in β of degree $2k-1$, we get an algebraic equation of which β^* is the largest root.[24] Thus when $2k$ is an integer β^* is always an algebraic number.

We can now formulate

Theorem 9. *The B^2R sieve and the series (11.19) and (11.19′) converge when the point with coordinates (k,β) lies in the region defined by $k \geq 0$, $\beta \geq 1$, and bounded by the curve described by the points $(k,\hat{\beta}(k))$ for $k > k_0$, where k_0 is the zero between 1/2 and 1 of the denominator on the right hand side of (11.49′). On the boundary of this region there is divergence at $(k_0,1)$ and along the curve $(k,\hat{\beta}(k))$. Outside this region the method diverges.*

For (k,β) in the region of convergence the values of $t_k^+(\alpha)$ and $t_k^-(\alpha)$ can be computed using the values for $A_k = (\beta+1)^k t_k^+(1/(\beta+1))$ and $B_k = t_k^-(1/\beta)$ given by (11.47) and (11.47′) or (11.49) and (11.49′), by applying the recursion formulas of Lemma 9.

The optimal choice of β, $\beta^*(k)$ for a given k, is $\beta^* = 1$ for $0 < k \leq 1/2$, and for $k > 1/2$ it is given by the largest zero of $\mathcal{I}_{k,\beta-1}$. The point (k,β^*) always lies in the region of convergence. For $2k$ an integer, $\beta^*(k)$ is always an algebraic number. For $k > 1/2$ we have $A_k(\beta^*) = \frac{2(\beta^*-1)^{k-1}}{J_{k,\beta^*-1}}$, and $B_k(\beta^*) = 0$.

---

[24] In the fifties when I first arrived at the condition in the form (11.37), I calculated the derivative of the function $U_k(z)\,e^{-k\omega(z)}$ (using the form (11.30) for $U_k(z)$ and putting $B_k = 0$) for $2k = 1,2,3,4,5$ and 6. In the last case I must have made an error and ended up with a term involving $\log\frac{\beta}{\beta-1}$. Partly as a consequence of this, I did not at that time think the integrals involved allowed much simplification and so did not try. When I gave (and when I wrote up) my lectures at Stony Brook in 1969, my main object was a different one, on the Buchstab–Rosser sieve I only included a little of what I remembered or could find old notes of, and without checking anything. Consequently on this point the account in my Stony Brook lectures contains inaccuracies and errors (the most mysterious being that I give an equation for $\beta^*(2)$ of degree 4 instead of degree 3, where I took it from I have no idea!). Only when I in the Spring of 1974 gave a course on Sieve Methods at the Pennsylvania State University, did I rework the material, see that the logarithmic term always drops out, and simplify several of the expressions. At that time I gave the material essentially the form it has in this exposition.

12. Remarks on the combinatorial sieve, possible generalizations

If one compares the combinatorial sieve, the B^2R sieve as I call it, described in the previous Section, with the Λ^2 and $\Lambda^2\Lambda^-$ sieves (for the present without comparing the results obtainable in various cases, we shall do that in a later section), one is struck by one significant feature: There is no real difference in complication between the upper and lower bound sieves! We may recall how much more complicated the extremal problem we were faced with in the $\Lambda^2\Lambda^-$ case is than that which arose in the Λ^2 case (and this even if we had restricted ourselves to the very simplest choice of the Λ^-!). This advantage, and the amazing effectivity of the B^2R sieve for small k (which we shall consider in a later section), are the great virtues of the combinatorial sieve. On the other hand, it is not very effective for large k, and it lacks a certain flexibility which is found in the Λ^2 and $\Lambda^2\Lambda^-$ sieves, as evidenced by the fact that they are so easily adapted to the situation of variable density.

If we try to adapt the combinatorial sieve to variable density, we see by going back to the inequalities (11.3) and (11.3′) which suggested the construction of our sieve in the case of constant density, that the sifting problems for the terms in the sum on the right hand side have a different density function $g'(t)$ from the original problem, namely if $p = x^v$ then $W_x(p)$ has the density function $g'(t(1-v))$ instead of $g'(t)$. This means that in our inequalities (11.4) and (11.4′) which defined our functions F_r^+ and F_r^+ instead of one fixed parameter β we have in (11.4) for instance to replace the β in the inequality

$$(\beta + 1)\, v_{r-2t} + v_{r-2t+1} + \ldots + v_r \leq 1,$$

by a β which is a function of $(1 - v_{r-2t+1} - \ldots - v_r)$, (more precisely β is determined by the function $g'((1 - v_{r-2t+1} - \ldots - v_r)t)$),[25] since we do not in advance know this function, we have to treat this as a problem in calculus of variations leaving the function free (except of course, it has to be ≥ 1). We may still form functions like $\Delta(u)$ and $\sigma(u)$ and consider their Laplace transform, but the problem of determining the optimal function β seems in general quite intractable. A general solution does not seem possible, though it may well be possible that there are certain special choices of $g(t)$ besides $g(t) = kt$, for which the problem may permit solution, but these $g(t)$ would not be likely to have a simple form and to try to identify them would probably be very difficult.

Returning to the inequalities (11.3) and (11.3′), we may consider again the case of constant density. If k is not so small (say, greater than 1), we may in (11.3′) replace $M_u^+(W_x(p), \mathcal{P}(p))$ with the upper bound obtainable by the Λ^2–method, whenever $\frac{\log p}{\log x/p} \geq 1/\beta'$, in a similar way as we earlier replaced

[25] This integer t from (11.4) that occurs in the subscript and the real variable t that occurs as a factor in the argument have of course no connection and should not be confused here.

$M_u^-(W_x(p), \mathcal{P}(p))$ by zero, whenever $\frac{\log p}{\log x/p} \geq 1/\beta$. This would lead us to the construction of a new sieve, which would now depend on two parameters β and β' to be chosen in the most favorable way later, and which could be defined recursively in the manner of the B^{2}R sieve that was defined by (11.6) and (11.6′), only we would this time have two alternatives in (11.6) also. Since the Λ^2 sieve depends on the sifting density k as well as the "cutoff point" α if the sifting range is $\mathcal{P}(x^\alpha)$, our $F^\pm$ will also be so dependent, thus we may write $F_r^\pm(v_1, \ldots, v_r; \alpha, k)$; if $L_r(v_1, \ldots, v_r; \alpha, k)$ is used to designate the F corresponding to the Λ^2 sieve, our recursion rules would be

(12.1)

$$F_r^+(v_1, \ldots, v_r; \alpha, k) = \begin{cases} L_r(v_1, \ldots, v_r; \alpha, k) & \text{for } \beta'\alpha \geq 1, \\ -F_{r-1}^-\left(\frac{v_1}{1-v_r}, \ldots, \frac{v_{r-1}}{1-v_r}; \frac{v_r}{1-v_r}, k\right) & \text{otherwise,} \end{cases}$$

and

(12.1′)

$$F_r^-(v_1, \ldots, v_r; \alpha, k) = \begin{cases} 0 & \text{for } \beta\alpha \geq 1, \\ -F_{r-1}^+\left(\frac{v_1}{1-v_r}, \ldots, \frac{v_{r-1}}{1-v_r}; \frac{v_r}{1-v_r}, k\right) & \text{otherwise.} \end{cases}$$

If we try to handle this according to the pattern used for the B^{2}R sieve, we meet the complication that since the $F_r^\pm$ depend on k, the coefficients in the power series in (11.19) and (11.19′) will also depend on k. Thus we could no longer appeal to the fact that a power series with coefficients of constant sign is convergent up to the first singularity on the positive real axis. To overcome this, we can replace the k in the definition of the sieve by (12.1) and (12.1′) by a k_0 which we consider fixed, develop the theory of this sieve for density k, where k is considered a variable, and only at the end put $k = k_0$. This is possible because the Λ^2 sieve which is optimal for density k_0 converges for k somewhat beyond k_0 (as is not hard to establish, it converges for $k < k_0+1/2$). The resulting formulas are rather more complicated than for the case considered in the previous section, and the conditions for optimizing the choice of the two parameters are rather messier. On the whole, the B^{2}RΛ^2 sieve (as we may call it) lacks the naturalness and inevitability of the B^{2}R sieve. There are several reasons that one may give for considering this sieve and its theory less interesting than the B^{2}R sieve. First of all, there is no indication that it leads to best possible results in any situation where the B^{2}R sieve does not. Only for $k > 1$ does it give better results and for large k the results we get for the lower bound are much inferior to what one obtains by the $\Lambda^2\Lambda^-$ sieve. Secondly, for numerical work in the "middle range" for k, it is clearly much simpler and faster to use the Buchstab iteration procedure starting with what is immediately obtainable by the Λ^2 sieve for the upper bound and the B^{2}R sieve for the lower bound and applying the inequalities (11.3) and (11.3′) repeatedly. In the limit case as $x \to \infty$, it is easy to see that (11.3) and (11.3′) become, for $0 < \beta < \alpha < 1$,

$$\alpha^{-k} T_k^+(\alpha) \leq \beta^{-k} T_k^+(\beta) - k \int_\beta^\alpha T_k^-\left(\frac{u}{1-u}\right) \frac{du}{u^{k+1}}, \tag{12.2}$$

and

$$(12.2') \qquad \alpha^{-k}T_k^-(\alpha) \geq \beta^{-k}T_k^-(\beta) - k\int_\beta^\alpha T_k^+\left(\frac{u}{1-u}\right)\frac{du}{u^{k+1}},$$

In (12.2) it is understood that we replace $T_k^-(u/(1-u))$ in the integrand on the right hand side by zero whenever it becomes negative. For very small α both the combinatorial sieves (even in Brun's original version!) and the Λ^2 and $\Lambda^2\Lambda^-$ sieves, give very good bounds for $T_k^\pm(\alpha)$, bounds which as $\alpha \to 0$ deviate from 1 by less than $O(e^{-c\frac{1}{\alpha}\log\frac{1}{\alpha}})$ with c a positive constant. We may of course in using (12.2) and (12.2′) appeal not only to the Λ^2 method for initial upper bounds, but also to the $\Lambda^2\Lambda^-$ method for initial lower bounds; if k is not small, this would lead to improved bounds for $T_k^+(\alpha)$ and $T_k^-(\alpha)$ for some range of α, but not for the α we are usually most interested in, namely the larger values, here we would not improve on what we can get directly from the Λ^2 and the $\Lambda^2\Lambda^-$ methods, however much we made use of (12.2) and (12.2′).[26]

Of course, once one combines the use of (12.2) and (12.2′) with the use of bounds obtained by the Λ^2 or the Λ^2 sieves, one no longer has to do with a true combinatorial sieve, the λ_d will no longer be restricted to the values $\mu(d)$ or zero.

13. A study of sieves in connection with a particular simple sifting problem

We have yet to study the $\Lambda^2\Lambda^-$ sieve more closely in the case of constant sifting density. Before we undertake this, however, it will be instructive to study in some detail a particular simple sifting problem, which can be analysed rather more completely.

Let R be a positive integer, and let us assume we have a sifting range $\mathcal{P}(x^\alpha, x^{\alpha'})$ with

$$(13.1) \qquad \frac{1}{R+1} < \alpha < \alpha' < \frac{1}{R}.$$

For simplicity, let us assume that within the range $\mathcal{P}$ the sifting density is constant[27] and such that

$$(13.2) \qquad v = \sum_{x^\alpha \leq p < x^{\alpha'}} \frac{u(p)}{p},$$

where R and v are kept fixed as $x \to \infty$.

[26] For this reason, what we called the $\mathrm{B}^2\mathrm{R}\Lambda^2$ sieve would for the larger α not improve on the upper bounds obtained directly by the Λ^2 sieve for $k > 1$.

[27] This assumption is purely for convenience, we might require only $\sum_{x^\alpha \leq p \leq x^{\alpha'}} u^2(p)/p^2 \to 0$, as $x \to \infty$, beside (13.1) and (13.2).

We observe that if d is a product of r primes from the range, then for $r \geq R+1$

$$d \geq x^{(R+1)\alpha} > x^{1+\eta},$$

holds with some fixed $\eta > 0$, while if $r \leq R$,

$$d < x^{R\alpha'} < x^{1-\eta},$$

with some fixed $\eta > 0$. This shows that for this sifting problem we may without loss restrict ourselves to sieves where $\lambda_d = 0$ if d has more than R prime factors.

But, beyond this, it is also easy to establish that we may, again without loss restrict ourselves to sieves where λ_d if $d = p_1 \dots p_r$, depends only on r, the number of prime factors of d. The proof is given along the lines followed in Section 8 to prove Lemma 2. Namely we form the ratios

$$\frac{\sum\limits_{p_i \in \mathcal{P}} \frac{\lambda_{p_1 \dots p_r}}{f(p_1) \dots f(p_r)}}{\sum\limits_{p_i \in \mathcal{P}} \frac{1}{f(p_1) \dots f(p_r)}}, \tag{13.3}$$ [28]

and show that as we let $x \to \infty$ and consider a sequence of optimal sieves, the ratio (13.3) tends to zero for $r > R$, and that we can take out a subsequence of x and of optimal sieves such that for $r \leq R$ each of the ratios (13.3) tend to a limit, which we denote by $\lambda(r)$. If we now define a new Λ by letting $\lambda_d = \lambda(r)$ if d contains r prime factors, then this Λ will be a sieve of the same type (upper or lower bound) as the sequence of sieves used in forming the ratios (13.3) and it will be as effective as $x \to \infty$, that is: the $T_u(\Lambda)$ will tend to the same limit as $x \to \infty$.

We shall now consider Λ systems with $\mathcal{P} = \mathcal{P}(x^\alpha, x^{\alpha'})$ where $\lambda_d = \lambda(r)$, that is, depends only on the number r of prime factors of d. One then easily sees that if $d = p_1 \dots p_r$, θ_d will also depend only on the number r of prime factors of d, since we have

$$\theta_d = \theta(r) = \sum_{\nu \leq R} \lambda(\nu) \binom{r}{\nu},$$

so $\theta(r)$ is actually a polynomial in r of degree at most R.

As $x \to \infty$, we see that the expression

$$\sum_d \frac{\lambda_d}{f(d)}$$

tends to the limit

$$\sum_{\nu \leq R} \frac{\lambda(\nu)}{\nu!} v^\nu = e^{-v} \sum_{r=0}^{\infty} \frac{\theta(r)}{r!} v^r. \tag{13.4}$$

[28] For $r = 0$ this ratio is understood as having the value 1.

The expression

$$(13.5)\qquad \sum_{r=0}^{\infty} \frac{\theta(r)}{r!} v^r ,$$

is actually the limit of $T_u(\Lambda)$ as $x \to \infty$, as one can easily see.

If our Λ is a Λ^+ the polynomial $\theta(r)$ must have the property that $\theta(0) = 1$, while $\theta(r) \geq 0$ for all positive integers r. Similarly if our Λ is a Λ^-, $\theta(r)$ must have $\theta(0) = 1$, while $\theta(r) \leq 0$ for all positive integers.

It is not difficult to deduce from this that the optimal value of the expression (13.5) is attained for a $\theta(r)$ polynomial of a very special form, namely, for the upper bound a $\theta(r)$ of the form

$$(13.6)\qquad \theta(r) = \prod_{i \leq [R/2]} \left(1 - \frac{r}{\nu_i}\right)\left(1 - \frac{r}{1+\nu_i}\right),$$

where the ν_i for $i = 1, 2, \ldots, [R/2]$ are positive integers such that $\nu_{i+1} \geq \nu_i + 2$. Similarly for the lower bound the optimal value is attained by a $\theta(r)$ of the form

$$(13.6')\qquad \theta(r) = (1-r) \prod_{i \leq [(R-1)/2]} \left(1 - \frac{r}{\nu_i}\right)\left(1 - \frac{r}{1+\nu_i}\right),$$

where the ν_i for $i = 1, 2, \ldots, [(R-1)/2]$ are positive integers such that $\nu_1 > 1$ and $\nu_{i+1} \geq \nu_i + 2$.

We shall briefly sketch a proof of this in the case of the upper bound. Since $\theta(0) = 1$ and the degree is at most R, the polynomial can be written as

$$\theta(r) = \prod (1 - \alpha_i r),$$

where there are at most R factors. We see first that complex α_i cannot occur since if $\alpha = \beta + i\gamma$

$$(1 - \alpha r)(1 - \bar{\alpha} r) \geq (1 - \beta r)^2,$$

with equality only for $r = 0$. Thus we would lower the value of $\theta(r)$ for all $r > 0$ if we replace the complex α's by their real parts, and without affecting the sign of $\theta(r)$. Negative α_i's could not occur, since we again would lower $\theta(r)$ for all $r > 0$, without affecting the sign, by omitting such factors. Similarly any $\alpha > 1$ may be replaced by 1. If all α now are real with $0 < \alpha \leq 1$ let α^* be one of these which is located either between the reciprocals of two integers $1/(\nu+1)$ and $1/\nu$ or at the reciprocal of an integer which is a multiple zero of $\theta(r)$. If we replace α^* by a variable μ, the expression (13.5) becomes a linear expression in μ. Clearly we may let μ change from α^* continuously without changing the sign of $\theta(r)$ for any positive integer r, if we only take care μ does not pass through the reciprocal of any integer n for which $\theta(n) \neq 0$. Clearly we can always choose to move μ in a direction that either decreases the expression (13.5) or leaves it unchanged. When we let μ move as far as it can under these rules it ends up as the reciprocal of an integer, which becomes a simple zero of the new $\theta(r)$. Continuing this process as long as possible, we will clearly end up with a $\theta(r)$ of the form

$$\prod_i \left(1-\frac{r}{\nu_i}\right)\left(1-\frac{r}{1+\nu_i}\right)$$

with all $\nu_i \geq 1$ and no overlapping of factors. In order to show that we have exactly $[R/2]$ such blocks of two factors so that the degree of $\theta(r)$ is exactly $2\,[R/2]$, one only has to show that if there are fewer terms in the product, we can always lower the value of (13.5) by adding a term $\left(1-\frac{r}{N}\right)\left(1-\frac{r}{N+1}\right)$ with N sufficiently large. Such a term decreases $\theta(r)$ for all $0 < r < 2N+1$. Since for $r > R$ it is easy to see that $|\theta(r)| < r^R/R!$, we can easily bound the "tail end" of the series (13.5), the part where $r \geq 2N+1$. For N sufficiently large one finds that the factor $\left(1-\frac{r}{N}\right)\left(1-\frac{r}{N+1}\right)$ reduces the value of the first non vanishing term in (13.5) with $r > 0$ by more than the estimate of the "tail end" and therefore reduces (13.5) as a whole. This shows that the smallest value of (13.5) for the upper bound is attained by some $\theta(r)$ of the form (13.6).

The proof for the lower bound case is very similar.

One can give bounds for how large the largest ν_i is in the expressions (13.6) and (13.6′) when the optimal values of (13.5) are attained. These are obtained by assuming the largest ν_i to be N and showing that (13.5) can be improved (that is: decreased in case of the upper bound, and increased in case of the lower bound) if N exceeds a certain bound depending on R and v, by replacing the factor

$$\left(1-\frac{r}{N}\right)\left(1-\frac{r}{N+1}\right) \tag{13.7}$$

with the factor

$$\left(1-\frac{r}{N-1}\right)\left(1-\frac{r}{N}\right) \tag{13.7′}$$

in the expression for $\theta(r)$. We can do this again by estimating the tail end of the series (13.5) where $r \geq N$ using the estimate $|\theta(r)| < r^R/R!$ valid for $r > R$, and showing that the first non vanishing term of (13.5) with $r > 0$ is reduced by more than the absolute value of the "tail" when we replace (13.7) with (13.7′). In this way we can show that

$$\nu_{\max} \leq \max(2R+2v, R+4v) \tag{13.8}$$

for instance. By more careful estimations these bounds can be somewhat improved.

This means that we only have a finite number of possible (13.6) and (13.6′) to check for a given R and v, so that the optimizing $\theta(r)$ can be found. In general it is unique, but for some situations two different $\theta(r)$ will both give the optimal value. In that case linear combinations $\alpha\,\theta_1(r) + (1-\alpha)\,\theta_2(r)$ with $0 < \alpha < 1$ will also be optimal although they are not of the form (13.6) or (13.6′). That this happens is clear, since when for fixed R, v increases continuously, the optimizing form (13.6) or (13.6′) of $\theta(r)$ cannot change continuously, nevertheless it is obvious that it has to change for certain values of v.[29] For these transition points v we then must have two different optimizing $\theta(r)$ of the form (13.6) or (13.6′). It seems likely that we never have three or more different optimizing $\theta(r)$ of the form (13.6) or (13.6′) for a given R and v.

It is instructive to see what the various special sieve methods give in this case. For the upper bound the combinatorial sieve gives the expression

$$e^v \sum_{\nu \le 2[R/2]} \frac{(-1)^\nu\, v^\nu}{\nu!}, \tag{13.9}$$

for (13.5). The polynomial $\theta(r)$ in this case is simply

$$\theta(r) = \prod_{1 \le \nu \le 2[R/2]} \left(1 - \frac{r}{\nu}\right). \tag{13.9′}$$

Similarly for the lower bound we get

$$e^v \sum_{\nu \le 2[(R-1)/2]+1} \frac{(-1)^\nu\, v^\nu}{\nu!}, \tag{13.10}$$

for (13.5), and the polynomial $\theta(r)$ in this case is

$$\theta(r) = \prod_{1 \le \nu \le 2[(R-1)/2]+1} \left(1 - \frac{r}{\nu}\right) \tag{13.10′}$$

Similarly we find that the Λ^2 method gives

$$\frac{e^v}{\displaystyle\sum_{\nu \le [R/2]} \frac{v^\nu}{\nu!}}, \tag{13.11}$$

for the expression (13.5). The resulting $\theta(r)$ is the square of a polynomial $p(r)$ of degree $[R/2]$, with $p(0) = 1$, and such that it minimizes (13.5). This polynomial $p(r)$ has real positive zeroes ≥ 1, and so $\theta(r)$ has the form

$$\theta(r) = \prod_{1 \le i \le [R/2]} \left(1 - \frac{r}{\mu_i}\right)^2, \tag{13.12}$$

[29] Except of course when R is so small that $\theta(r) = 1$ or $\theta(r) = 1 - r$ are the only permissible expressions (13.6) or (13.6′).

a form which seems well suited to approximate the behavior of an expression of the form (13.6) to some extent.

We may combine the combinatorial lower bound sieve with the inequality (11.3′) and bounds from the Λ^2 sieve; clearly using the Buchstab principle takes us outside the sieves where the λ_d depend only on the number of prime factors in d, but after completing the construction one can replace the resulting sieve by one where $\lambda_d = \lambda(r)$ and which gives the same result. We are in this way led to the expression

$$e^v \left\{ \sum_{\nu \le 2[(R-1)/2]+1} \frac{(-1)^\nu v_1^\nu}{\nu!} - \int_{v_1}^{v} \frac{dt}{\sum_{\nu \le [(R-1)/2]} \frac{t^\nu}{\nu!}} \right\}, \tag{13.13}$$

where $v_1 < v$ is to be chosen in the most favorable way.

Finally, the form (13.6′), makes it natural to try for the lower bound a $\theta(r)$ of the form

$$\theta(r) = (1-r)\, p^2(r)\,, \tag{13.14}$$

where $p(r)$ is a polynomial of degree $[(R-1)/2]$ such that $p(0) = 1$, and chosen so as to maximize (13.5). This, one sees, corresponds to a $\Lambda^2\Lambda^-$ sieve where the Λ^- has $\lambda_1 = 1$; $\lambda_p = -1$ for the primes in the range, and $\lambda_d = 0$ for all other d.

We shall write $R' = [(R-1)/2]$, and

$$p(r) = \sum_\nu \lambda(\nu) \binom{r}{\nu},$$

where $\lambda(0) = 1$ and $\lambda(\nu) = 0$ for $\nu > R'$, while the other $\lambda(\nu)$ are left to our disposal, and further write

$$y_r = (-1)^r \sum_\nu \frac{\lambda(r+\nu)}{\nu!}\, v^\nu, \tag{13.15}$$

so that

$$\lambda(r) = (-1)^r \sum_\nu \frac{y_{r+\nu}}{\nu!}\, v^\nu. \tag{13.15′}$$

This implies $y_r = 0$ for $r > R'$, while

$$\sum_\nu \frac{y_\nu}{\nu!}\, v^\nu = \lambda(0) = 1. \tag{13.15″}$$

We get from (7.13′) that the expression (13.5) becomes

$$e^v\, Q(y)\,, \tag{13.16}$$

where

$$(13.16') \qquad Q(y) = \sum_{r \le R'} \frac{v^r}{r!} y_r^2 - \sum_{r \le R'} \frac{v^{r+1}}{r!} (y_r - y_{r+1})^2 .$$

It is instructive to look at the following problem: For a given R, what is the least upper bound of those v for which the expression (13.5) can be made positive by a suitable choice of a $\theta(r)$ of the form (13.6′). We may call this number v_R. For a given R it can clearly be effectively determined as we have only a finite number of polynomials $\theta(r)$ of the form (13.6′) to test, since we can give an upper bound for v_R, namely R, and thus the largest ν_i in the form (13.6′) would be $\le 5R$.

To show that $v_R < R$,[30] we will show that if we put $v = R$, then (13.5) is negative with any choice of the polynomial (13.6′). Without loss we may assume R is odd, since R enters the expression (13.6′) only as $[(R-1)/2]$, we see that $v_R = v_{R-1}$ if R is even.

Let μ be the first positive integer for which $\theta(\mu) \ne 0$, clearly μ must be even, and if $\mu < R+1$, it is easily seen that the smallest absolute value $\theta(\mu)$ could have, is that taken at μ when $\theta(r)$ has the form

$$(1-r)\dots\left(1-\frac{r}{\mu-1}\right)\left(1-\frac{r}{\mu+1}\right)\dots\left(1-\frac{r}{R+1}\right).$$

The value of this at $r = \mu$ is

$$-\frac{\mu!\,(R-\mu+1)!}{(R+1)!},$$

and so the value of the term

$$\frac{\theta(\mu)}{\mu!} R^{\mu}$$

would be

$$-\frac{(R-\mu+1)!}{(R+1)!} R^{\mu}.$$

For $\mu = 2$ this expression is

$$-\frac{R}{R+1},$$

for all larger μ it is ≤ -1 and so (13.5) would be negative, the same is seen to be the case if $\mu = R+1$, when $\theta(r)$ would have the form

$$\left(1-\frac{r}{1}\right)\dots\left(1-\frac{r}{R}\right),$$

and we would have $\theta(R+1) = -1$.

[30] We assume $R > 1$, for $= 1$ we have $v_R = 1$.

It remains to consider further the case that $\mu = 2$. We then have two possibilities, either the next r for which $\theta(r) \neq 0$ is $r = R+2$, in which case we see that

$$\frac{\theta(R+2)}{(R+2)!} R^{R+2} = -\frac{2}{R} \frac{R^{R+2}}{(R+2)!},$$

or there is a $\mu' < R+2$, which is the second positive integer for which $\theta(r) \neq 0$. In that case the smallest absolute value for $\theta(2)$ is taken by the expression

$$(1-r)\left(1-\frac{r}{3}\right)\cdots\left(1-\frac{r}{\mu'-1}\right)\left(1-\frac{r}{\mu'+1}\right)\cdots\left(1-\frac{r}{R+2}\right)$$

it is seen that this expression also gives the smallest absolute value for $\theta(\mu')$. A simple computation shows that in both cases the contribution from the two first non–vanishing terms with $r > 0$ in (13.5) more than cancel the positive leading term 1. Thus for $v = R$, (13.5) is negative which proves that $v_R < R$.

The special sieves considered above will of course give us lower bounds for v_R, and it is not hard to estimate how they behave for large R.

For the combinatorial sieve we get from (13.10), assuming R odd that

$$e^v \sum_{\nu \leq R} \frac{(-1)^\nu v^\nu}{\nu!} = 1 - e^v \int_0^v \frac{(v-t)^R}{R!} e^{-t} dt. \tag{13.17}$$

Since for $0 < t < v$,

$$e^{-t} > \left(1-\frac{t}{v}\right)^v = v^{-v}(v-t)^v,$$

we have

$$\int_0^v (v-t)^R e^{-t} dt > v^{-v} \int_0^v (v-t)^{R+v} dv = \frac{v^{R+1}}{R+v+1}.$$

Again, for $0 < t < v$,

$$(v-t)^R < v^R e^{-\frac{R}{v}t},$$

we have

$$\int_0^v (v-t)^R e^{-t} dt < v^R \int_0^\infty e^{-(1+R/v)t} dt = \frac{v^{R+1}}{R+v}.$$

Thus we get from (13.17)

$$\sum_{\nu \leq R} \frac{(-1)^\nu v^\nu}{\nu!} = e^{-v} - \frac{v^{R+1}}{R!\,(R+v+\eta)}, \tag{13.18}$$

where $0 < \eta < 1$.

If we denote the positive root of the polynomial on the left hand side by v'_R, we see from (13.18) using Stirling's formula for $R!$ that

$$v'_R \sim c\,R, \tag{13.19}$$

where c is the constant defined by $c\,e^{1+c} = 1$, $c = \frac{1}{3.591\ldots}$ which we encountered in our treatment of the B^2R sieve.

Next we see what we can get from the expression (13.13), that is the combinatorial sieve combined with the Λ^2 sieve using Buchstab's inequality, we again assume as before that R is odd. We first rewrite the expression in the curly bracket in (13.13) as follows

$$(13.20)\qquad \begin{aligned} &e^{-v_1} - \frac{v_1^{R+1}}{R!\,(R+v_1+\eta)} - \int_{v_1}^{v}\left(\frac{1}{\sum_{\nu\le[(R-1)/2]}\frac{t^\nu}{\nu!}} - e^{-t}\right)dt - \int_{v_1}^{v} e^{-t}dt \\ &= e^{-v} - \frac{v_1^{R+1}}{R!(R+v_1+\eta)} - \int_{v_1}^{v}\frac{\sum_{\nu>[(R-1)/2]}\frac{t^\nu}{\nu!}}{e^t\left(e^t - \sum_{\nu>[(R-1)/2]}\frac{t^\nu}{\nu!}\right)}dt\,. \end{aligned}$$

Here if $t < \mu R$ with a fixed $\mu < 1/2$, we have

$$\sum_{\nu>[(R-1)/2]}\frac{t^\nu}{\nu!} = \omega\frac{t^{(R+1)/2}}{((R+1)/2)!}\,,$$

where $1 < \omega < 1/(1-2\mu)$, and also clearly

$$\sum_{\nu>[(R-1)/2]}\frac{t^\nu}{\nu!} < \frac{1}{2}\,e^t\,.$$

The last term in (13.20) can therefore, if $v \le \mu\, R$, be estimated as

$$\frac{\omega'}{((R+1)/2)!}\int_{v_1}^{v} t^{(R+1)/2}e^{-2t}dt\,,$$

where $1 < \omega' < 2/(1-2\mu)$. If we can choose $v_1 \ge (R+1)/4$, the largest value of the integrand in the last integral is at $t = v_1$, and the integral is then essentially of the magnitude

$$\frac{v_1^{(R+1)/2}}{((R+1)/2)!}\,e^{-2v_1}\,,$$

if we make this about equal to the term

$$\frac{v_1^{R+1}}{(R+1)!\,(R+v_1)}$$

we get that for large R, $v_1 \sim \lambda\, R$, where λ is the positive root of the equation $\lambda = 2\,e^{-1-4\lambda}$, and we see that the first term on the right hand side of (13.20), e^{-v} is larger than this if $v \le (c' - \varepsilon)\, R$ with $c' = 4\lambda - \log 2$, and ε fixed > 0 as $R \to \infty$. Thus we see that with the optimal choice of v_1, (13.13) changes sign for $v = v_R''$, where $v_R'' \sim c'\, R$ as $R \to \infty$. One finds that $c' = \frac{1}{2.882\ldots}$.

We turn finally to the $\Lambda^2\Lambda^-$ sieve and see that it is really enough to see for what v we can make the $Q(y)$ of (13.16′) positive with a suitable choice of y_r and for which v it is not possible with any choice of y_r. It is clear that we for this purpose can disregard the condition (13.55″). If we write[31]

$$y_r - y_{r+1} = \Delta_r\,,$$

we have for $R = 2R' + 1$

$$y_r = \sum_{j=r}^{R'} \Delta_j\,,$$

from which we get that

$$y_r^2 \le (R' + 1 - r) \sum_{j=r}^{R'} \Delta_j^2\,,$$

here equality holds only if all Δ_r are equal. If we insert this upper bound for y_r^2 in (13.16′), we get

$$\text{(13.21)} \qquad \begin{aligned} Q(y) &\le \sum_{r\le R'} \frac{v^r}{r!}(R'+1-r)\sum_{j=r}^{R'}\Delta_j^2 - \sum_{j\le R'}\frac{v^{j+1}}{j!}\Delta_j^2 \\ &= (R'+1-v)\sum_{j\le R'}\Delta_j^2 \sum_{r\le j}\frac{v^r}{r!}\,. \end{aligned}$$

From this we see that $Q(y)$ cannot be made positive for $v \ge R' + 1$ with any choice of y_r. On the other hand if $v < R' + 1$ we see that $Q(y)$ is positive if we choose all Δ_j equal, so that $y_r = (R' + 1 - r)\,\Delta$ for $0 \le r \le R'$, the value of Δ is of course determined by the condition (13.15″). Thus with the $\Lambda^2\Lambda^-$ method we get a positive lower bound for $v < v_R''' = [(R+1)/2]$.

Clearly we have

$$v_R' \le v_R'' \le v_R''' \le v_R\,,$$

for large R we have strict inequality everywhere. For $R = 1$, equality holds everywhere, for $R = 3$, we have equality only between v_R''' and v_R, and for the later values (not many, my computations are done without any electronic devices) v_R is larger than v_R''' but not drastically. I am inclined to conjecture that as R tends to infinity v_R is asymptotic to $\frac{1}{2}R$. We may summarize our results as

$$\text{(13.22)} \qquad v_R' \sim c\,R\,,$$

as $R \to \infty$ with $c = \frac{1}{3.591\ldots}$.

$$\text{(13.22′)} \qquad v_R'' \sim c'\,R\,,$$

[31] For $r = (R-1)/2 = R'$, $y_{R'+1}$ is understood to be zero.

as $R \to \infty$, with $c' = \frac{1}{2.882\ldots}$.

(13.22″)
$$v_R''' = \left[\frac{R+1}{2}\right] \sim \frac{1}{2}R,$$

as $R \to \infty$. Finally

(13.22‴)
$$\left[\frac{R+1}{2}\right] \le v_R < R.$$

It might be of interest to compute v_R for a number of larger values (mine do not go beyond single digits) of R, to see whether the ratio v_R/R seems to approach $1/2$ (from above, of course), or not.

14. The sifting limit for constant sifting density

We shall now look at a question which is very analogous to the one we have just discussed, but no doubt much more difficult and definitely of greater general interest. For a constant sifting density k, we define the sifting limit α_k as the lowest upper bound of the α for which $T_k^-(\alpha) > 0$. From our results on the B^{2}R sieve it is clear that $\alpha_k = 1$ for $k \le 1/2$. We may also speak of a sifting limit for sifting density k associated with a specific sieve procedure. For the B^{2}R sieve the sifting limit for sifting density k is the reciprocal of $\beta^*(k)$, we may denote this by α_k'. Clearly we must always have $\alpha_k' \le \alpha_k$. For $0 < k \le 1/2$ it is obvious that equality holds, since $\alpha_k' = 1$ and α_k could not be larger than 1. For $k = 1$ we find $\beta^*(1) = 2$, thus $\alpha_1' = 1/2$. We shall later see that α_1 also equals $1/2$. Whether there are any other k for which equality holds and $\alpha_k' = \alpha_k$ is not known at present. In Section 11 we saw that for $2k$ an integer $\beta^*(k)$ is the largest root of a polynomial of degree $2k-1$ with rational coefficients. Thus α_k' is an algebraic number when $2k$ is an integer and is easily computed. One finds

$$\alpha_{3/2}' = \frac{1}{3.366\ldots}, \qquad \alpha_2' = \frac{1}{4.834\ldots}, \qquad \alpha_{5/2}' = \frac{1}{6.358\ldots},$$

and so forth. For large k our results about the asymptotic behaviour of $\beta^*(k)$ gives that

(14.1)
$$\alpha_k' \sim \frac{c}{k},$$

where $c = \frac{1}{3.591\ldots}$ is again the positive solution of the equation $c\,e^{1+c} = 1$.

To see what the B^{2}RΛ^2 sieve gives (without having to develop the whole theory of this sieve method), we may from (12.2) and (12.2′) subtract the equation

$$\alpha^{-k} = \beta^{-k} - k\int_\alpha^\beta \frac{du}{u^{k+1}}.$$

Observing that as $\beta \to 0$, we have

$$\beta^{-k}\left(T_k^{\pm}(\beta) - 1\right) \to 0\,,$$

we get from (12.2) and (12.2′) respectively,

$$T_k^{+}(\alpha) \le 1 + k\,\alpha^k \int_0^{\alpha} \left(1 - T_k^{-}\left(\frac{u}{1-u}\right)\right) \frac{du}{u^{k+1}}\,, \tag{14.2}$$

and

$$T_k^{-}(\alpha) \ge 1 - k\,\alpha^k \int_0^{\alpha} \left(T_k^{+}\left(\frac{u}{1-u}\right) - 1\right) \frac{du}{u^{k+1}}\,. \tag{14.2′}$$

From Theorem 8 and (10.11) we get

$$T_k^{+}(\alpha) \le \frac{1}{h_k^{*}(\alpha)}\,, \tag{14.3}$$

where

$$h_k^{*}(\alpha) = \frac{1}{2\pi\, i} \int\limits_{-i\infty}^{i\infty} \exp\left(\frac{s}{2\alpha} - k \int_0^{s} \frac{1-e^{-t}}{t}\, dt\right) \frac{ds}{s} \tag{14.4}$$

the path of integration passing to the right of the pole at $s = 0$.

Assuming $\alpha < 1/(2k)$, we can write α as

$$\alpha = \frac{1}{2k}\,\frac{v}{e^v - 1}\,,$$

where v is real and positive. Shifting the path of integration in (14.4) to the left, so it passes through the point $s = -v$ on the real axis (where the derivative of the exponential in (14.4) becomes zero), taking account of the residue at $s = 0$ and estimating the remaining integral by the method of steepest descent, we get, using only a crude estimate for the integral, that[32]

$$h_k^{*}(\alpha) = 1 - \exp\left(-k\left(e^v - 1 - \int_0^{v} \frac{e^t - 1}{t}\, dt + o(1)\right)\right). \tag{14.5}$$

Thus for $u \le \mu/2k$, with fixed $\mu < 1$, (14.3) gives

$$T_k^{+}(u) - 1 \le \exp\left(-k\left(e^v - 1 - \int_0^{v} \frac{e^t - 1}{t}\, dt + o(1)\right)\right). \tag{14.6}$$

where now

$$2ku = \frac{v}{e^v - 1}\,. \tag{14.7}$$

---

[32] The term $o(1)$ can of course be estimated much sharper, in fact it is $O\left(\frac{\log k}{k}\right)$, or more accurately $\frac{a_1 \log k + a_2}{k} + o(1/k)$, where a_1 and a_2 are constants.

Writing u instead of $\frac{u}{1-u}$ in (14.2′), we can rewrite it as

$$T_k^-(\alpha) \geq 1 - k\,(2k\alpha)^k \int_0^{\frac{\alpha}{1-\alpha}} (T_k^+(u) - 1)\,\frac{(1+u)^{k-1}}{(2ku)^k}\,\frac{du}{u}\,. \tag{14.8}$$

For large k, it is the maximum of the expression $(T_u^+(u) - 1)\,(2ku)^{-k}$ that essentially determines the order of magnitude of the integral. Using (14.6) we get for this expression the upper bound

$$\exp\left(-k\left\{e^v - 1 - \int_0^v \frac{e^t - 1}{t}\,dt + \log\frac{v}{e^v - 1}\right\} + o(k)\right)\,. \tag{14.9}$$

The minimum of the expression in the curly bracket is seen to be attained for $v = \log 2$, which gives $u = \frac{\log 2}{2k}$. Thus for $\frac{\log 2}{2k} < \alpha < \frac{1}{2k}$ we get
(14.10)

$$T_k^-(\alpha) \geq 1 - \exp\left(k\left\{\log(2k\alpha) - 1 + \int_0^{\log 2}\frac{e^t - 1}{t}\,dt - \log\log 2 + o(1)\right\}\right)\,.$$

From this we see that we get a sifting limit α_k'' which has the asymptotic behavior

$$\alpha_k'' \sim \frac{c''}{k}\,, \tag{14.11}$$

where c'' is given by

$$c'' = \frac{1}{2}\exp\left(1 - \int_0^{\log 2}\frac{e^t - 1}{t}\,dt + \log\log 2\right)\,, \tag{14.12}$$ [33]

one finds

$$c'' = \frac{1}{2.445\ldots}\,.$$ [34]

Of course our argument above has not really used the full $\mathrm{B}^2\mathrm{R}\Lambda^2$ method, we only used (14.2′) once and (14.2) not at all. If we however carry out an iterated use of these inequalities, the further iterations do not seem to improve the bound for $T_k^+(u) - 1$ around $u = \frac{\log 2}{2k}$ (which is the crucial point) enough to make a real difference in the asymptotic behavior, so we still should end up with the constant (14.12) and the formula (14.11).

Our results in the previous section suggest that we should do rather better for large k with the $\Lambda^2\Lambda^-$ sieve than with the two previous sieves. This is borne out if we at first apply only the crude estimation (7.19) with $\xi = x^\alpha$ and $z^2\,\xi = x^{1-\varepsilon}$ with a small positive ε to be disposed of later. (7.19) then gives, letting $x \to \infty$, that

$$T_k^-(\alpha) \geq 1 - (1+k)\,\exp\left(-\frac{1 - 5\alpha - \varepsilon}{2\alpha}\,\log\frac{1 - 5\alpha}{2ke\alpha} - k\right)\,,$$

[33] This value was first given by N. Ankeny in [Ankeny 1].

[34] As a comparison it can be mentioned that if we had used (7.20) instead of Theorem 8 we should have obtained $\frac{1}{2.782\ldots}$ instead.

or letting $\varepsilon \to 0$,

$$T_k^-(\alpha) \geq 1 - (1+k)\exp\left(-\frac{1-5\alpha}{2\alpha}\log\frac{1-5\alpha}{2ke\alpha} - k\right), \tag{14.13}$$

which is valid for $\frac{1-5\alpha}{2k\alpha} > 1$. Since the right hand side of (14.13) is easily seen to be positive for large k if

$$\alpha = \frac{1}{2k + c\sqrt{k}\log(1+k)},$$

and $c > 2\sqrt{2}$, we get that

$$\alpha_k \geq \frac{1}{2k + c\sqrt{k}\log k}, \tag{14.14}$$

if $c > 2\sqrt{2}$ and k is greater than some $k_0(c)$.

We should of course expect to do somewhat better than this by making optimal (or close to optimal) choices for our y_ρ and by more precise estimations of the various sums entering. However, it is clear that by so doing we lose considerably in generality. (7.19) makes hardly any assumptions about the distribution of the sifting density, nor even about its existence, it only requires a sufficiently good estimate of $S_u(x^\alpha)$ for the specific α considered. This, besides being a stronger result for large k, distinguishes (14.14) from the previous results obtained by the $\mathrm{B}^2\mathrm{R}$ and $\mathrm{B}^2\mathrm{R}\Lambda^2$ methods. This generality is lost once we use the more precise estimations given in Section 10, these do depend on the constant sifting density. Also, it is rather clear that we cannot expect more than a quite small improvement on (14.14) in this way, since we should expect that it is easier by the Λ^2 method to obtain the estimate $T_k^+(\alpha) < 2$, than to obtain by the $\Lambda^2\Lambda^-$ method the estimate $T_k^-(\alpha) > 0$. (14.3) and (14.4) give, if we estimate $h_k(\alpha)$ asymptotically for α very near to $1/(2k)$ for large k, that $h_k^*(\alpha) = 1/2$ for some $\alpha = \frac{1}{2(k-1/9+o(1/\sqrt{k}))}$, thus we should not expect to be able to improve the right hand side of (14.14) beyond $1/(2(k-1/9))$ by the $\Lambda^2\Lambda^-$ method and it would seem doubtful whether even $1/(2k)$ could be attained.

We begin by going back to the form for the leading term as the ratio of (10.15) and (10.15′) given in Section 10. First we shall see how we make the derivation from (7.14) rigorous. In (7.14) we put $y_\rho = \sigma(\log\rho/\log z)$, where $\sigma(t)$ is a decreasing function with $\sigma(0) = 1$ and $\sigma(t) = 0$ for $t \geq 1$. We shall first assume $\sigma(t)$ to be not only continuous, but satisfying some Lipschitz condition

$$|\sigma(t) - \sigma(t')| < A\,|t - t'|^{1/2}. \tag{14.15}$$

We put $z = x^u$ and $\xi = x^\alpha$, with $2u + \alpha < 1$.[35]

[35] To avoid confusion the reader is reminded that when on this page and the next u occurs as a subscript, it refers to the sequence of $u(p)$, when not a subscript it refers to the number $u = \log z/\log x$.

The crucial part is of course the double sum in the numerator of (7.14), since the other part of the numerator as well as the denominator are easily handled using Lemma 7 or Lemma 7′. We divide (7.14) by $e_u(\xi)$ to get the $T_u(\Lambda^2\Lambda^-)$, and we do this by putting in a factor $e_u(\xi)$ in the numerator and a factor $(e_u(\xi))^2$ in the denominator; as $x \to \infty$ it is clear that the denominator tends to (10.15′), and also that the first term of the numerator tends to the first term of (10.15).

We divide the double sum into two parts by choosing a small ε, and dividing the summation range for p into $\mathcal{P}(x^\varepsilon)$ and $\mathcal{P}(x^\varepsilon, x^\alpha)$. In the summation range $\mathcal{P}(x^\varepsilon)$ we get from (14.15) that

$$\frac{1}{f(p)\,f'(\rho)}\left\{\sigma\left(\frac{\log\rho}{\log z}\right)-\sigma\left(\frac{\log\rho p}{\log z}\right)\right\}^2<\frac{A^2}{\log z}\,\frac{\log p}{f(p)\,f'(\rho)}\,,$$

so that

$$\begin{aligned}(14.16)\qquad & e_u(x^\alpha)\sum_{p<x^\varepsilon}\sum_{\rho\le z}\frac{1}{f'(p)\,f'(\rho)}\left\{\sigma\left(\frac{\log\rho}{\log z}\right)-\sigma\left(\frac{\log\rho p}{\log z}\right)\right\}^2\\ &<\frac{A^2\,e_u(x^\alpha)}{\log z}\sum_{p<x^\varepsilon}\frac{\log p}{f(p)}\sum_{\rho\le z}\frac{1}{f'(\rho)}<A^2\,\frac{S_u(x^\varepsilon)}{\log z}\\ &=\frac{A^2}{u}\,g(\varepsilon)+o(1)\,,\end{aligned}$$

as x goes to infinity.

The remaining double sum is handled by dividing the summation range into regions $I_{m,n}$ defined by:

$$x^{\varepsilon+\frac{m-1}{N}(\alpha-\varepsilon)}\le p<x^{\varepsilon+\frac{m}{N}(\alpha-\varepsilon)}\,,$$

and

$$x^{\frac{n-1}{N}u}\le\rho<x^{\frac{n}{N}u}$$

with $1 \le m, n \le N$, where N is large, say $N > \varepsilon^{-2}$. The general term in the double sum we write as

$$\frac{e_u(x^\alpha)\,\log p}{f(p)\,f'(\rho)}\,\frac{1}{\log p}\left\{\sigma\left(\frac{\log\rho}{\log z}\right)-\sigma\left(\frac{\log\rho p}{\log z}\right)\right\}^2.$$

In each $I_{m,n}$ we may replace the second factor first by its largest value and next by it smallest value, and thus enclose the sum over $I_{m,n}$ between these two bounds times

$$e_u(x^\alpha)\sum_{I_{m,n}}\frac{\log p}{f(p)\,f'(\rho)}\,.$$

For this last sum we get an asymptotic expression as $x \to \infty$ by using (4.2) and Lemma 7. Summing these upper and lower bounds over m and n, we get upper and lower bounds for the remaining part of our double sum. Letting now $N \to \infty$, we see that these bounds tend to the second term in (10.15), but with $t = \varepsilon$ as the lower bound for the integration over t. Letting now finally ε go to zero so that $g(\varepsilon) \to 0$ and using the estimation (14.16) we obtain (10.15). We now have that the ratio between (10.15) and (10.15′) is a lower bound for $T_g^-(\alpha)$. We may now get rid of the condition (14.15), since the resulting statement also holds if $\sigma(t)$ can be approximated arbitrarily well by $\sigma^*(t)$ which satisfy this condition so that the expressions (10.15) and (10.15′) formed with σ^* tend to those formed with σ, so we may now permit a discontinuity at $t = 1$. Finally we may also replace the condition $\alpha + 2u < 1$ with $\alpha + 2u \leq 1$.

In the case we are actually studying $g(t) = kt$ and

$$(14.17) \qquad h(\alpha, w) = \frac{1}{2\pi i} \int\limits_{-i\infty}^{i\infty} \exp\left(ws - k \int_0^{\alpha} \frac{1 - e^{-ts}}{t} dt \right) \frac{ds}{s}$$

$$\frac{1}{2\pi i} \int\limits_{-i\infty}^{i\infty} \exp\left(\frac{w}{\alpha} s - k \int_0^{s} \frac{1 - e^{-t}}{t} dt \right) \frac{ds}{s}$$

$$= h_k\left(\frac{w}{\alpha} \right),$$

where we have defined

$$h_k(v) = \int\limits_{-i\infty}^{i\infty} \exp\left(vs - k \int_0^{s} \frac{1 - e^{-t}}{t} dt \right) \frac{ds}{s},$$

the path of integration passing, as before, to the right of the pole at $s = 0$.

If we now in (10.15) and (10.15′) write $w = \alpha v$; $u = \alpha u'$ and $t = \alpha t'$, they become

$$(14.18) \quad \int_0^u \sigma^2\left(\frac{v}{u}\right) h_k'(v) dv - k \int_0^u h_k'(v) dv \int_0^1 \left(\sigma\left(\frac{v}{u}\right) - \sigma\left(\frac{v+t}{u}\right) \right)^2 \frac{dt}{t},$$

and

$$(14.18') \quad \left(\int_0^u \sigma\left(\frac{v}{u}\right) h_k'(v) dv \right)^2,$$

and the condition $2u + \alpha \leq 1$ is now replaced by $(2u + 1)\alpha \leq 1$. In trying to make the ratio positive by suitable choice of u and σ, we need of course not worry about the denominator (14.18′) which is positive in any case, and can concentrate on (14.18) and try to get it positive with as small a u as possible. We need some estimates concerning $h_k(v)$ for large k. We assume first that $|v - k| < 2\, k^{3/5}$ and shall estimate

$$(14.19) \qquad h_k'(v) = \frac{1}{2\pi i} \int\limits_{-i\infty}^{i\infty} \exp\left(vs - k \int_0^{s} \frac{1 - e^{-t}}{t} \right) ds \,.$$

We write $\Delta = k - v$, and

$$f(s) = vs - k\int_0^s \frac{1-e^{-t}}{t}\,dt = -\Delta\, s + k\left(s - \int_0^s \frac{1-e^{-t}}{t}dt\right).$$

We have

$$f'(s) = -\Delta + k\left(1 - \frac{1-e^{-s}}{s}\right),$$

which has a real zero s_0 with

$$s_0 = \frac{2\Delta}{k} + \frac{1}{3}\left(\frac{2\Delta}{k}\right)^2 + O\left(\left(\frac{2\Delta}{k}\right)^3\right).$$

We further find

$$f(s_0) = -\frac{\Delta^2}{k} - \frac{4}{9}\frac{\Delta^3}{k^2} + O\left(\frac{\Delta^4}{k^3}\right),$$

$$f''(s_0) = \frac{k}{2}\left(1 - \frac{4}{3}\frac{\Delta}{k} + O\left(\frac{\Delta^2}{k^2}\right)\right),$$

$$f'''(s_0) = -\frac{k}{3}\left(1 - \frac{3}{4}\frac{\Delta}{k} + O\left(\frac{\Delta^2}{k^2}\right)\right),$$

and

$$f^{IV}(s) = O(k) \qquad \text{for } s = O(1).$$

We choose as the path of integration in (14.19) the line segments joining the points $-i\infty$, $-i8k^{-2/5}$, $s_0 - i8k^{-2/5}$, $s_0 + i8k^{-2/5}$, $i8k^{-2/5}$ and $i\infty$.

On the line segments $|s| \geq 8\,k^{-2/5}$ along the imaginary axis, we have

$$\operatorname{Re} f(s) = -k\int_0^{|s|} \frac{1-\cos t}{t}\,dt,$$

from which we deduce that the integrand there is

$$O\left(\frac{e^{-10\,k^{1/5}}}{1+|s|^2}\right)$$

and so these parts contribute at most

$$O(e^{-10k^{1/5}}). \tag{14.20}$$

On the remaining parts of the integration path where $|s - s_0| < 9\,k^{-2/5}$, we write

$$f(s) = f(s_0) + \frac{(s-s_0)^2}{2}f''(s_0) + \frac{(s-s_0)^3}{6}f'''(s_0) + O(k\,|s-s_0|^4).$$

From this, inserting the values of $f(s_0)$, $f''(s_0)$ and $f'''(s_0)$ given above, we get also for the two horizontal line segments that their contribution to the integral is at most

$$(14.21') \qquad O(e^{-10\,k^{1/5}}).$$

We are left with the integral over the segment $(s_0 - i8k^{-2/5}, s_0 + i8k^{-2/5})$. Writing $s = s_0 + it$, the integrand takes the form

$$e^{f(s_0)-\frac{t^2}{2}f''(s)-\frac{it^3}{6}f'''(s_0)+O(k\,|t|^4)}$$
$$= e^{f(s_0)-\frac{t^2}{2}f''(s_0)}\left(1-\frac{i\,t^3}{6}f'''(s_0)+O(k\,|t|^4)+O(k^2\,t^6)\right).$$

If we add the stretches $|t| > 8\,k^{-2/5}$ to our integral, we change it by at most

$$(14.21'') \qquad O(e^{-10\,k^{1/5}}).$$

We are left with

$$\frac{1}{2\pi}\int_{-\infty}^{\infty} e^{f(s_0)-\frac{t^2}{2}f''(s_0)}\left(1-\frac{i\,t^3}{6}f'''(s_0)+O(k\,|t|^4)+O(k^2\,t^6)\right)dt$$

which gives

$$\frac{1}{2\pi}\sqrt{\frac{\pi}{f''(s_0)}}\,e^{f(s_0)}\left(1+O\left(\frac{k}{(f''(s_0))^2}+\frac{k^2}{(f''(s_0))^3}\right)\right).$$

Combining our results and inserting the values for $f(s_0)$ and $f''(s_0)$, we get

$$(14.22) \qquad h'_k(v) = \frac{e^{-\frac{\Delta^2}{k}-\frac{4}{9}\frac{\Delta^3}{k^2}+O\left(\frac{\Delta^4}{k^3}\right)}}{\sqrt{\pi k\left(1-\frac{4}{3}\frac{\Delta}{k}+O\left(\frac{\Delta^2}{k^2}\right)\right)}}\left(1+O\left(\frac{1}{k}\right)\right)$$
$$= \frac{1}{\sqrt{\pi k}}\,e^{-\frac{\Delta^2}{k}}\left(1+\frac{2}{3}\frac{\Delta}{k}-\frac{4}{9}\frac{\Delta^3}{k^2}+O\left(\frac{1}{k}+\frac{\Delta^6}{k^4}\right)\right).$$

We can make this expression more convenient for our later use by writing $k' = k - 1/3 - c$, where c is a constant to be chosen later. We also write $\Delta' = k' - v = \Delta - 1/3 - c$. With this notation the expression for $h'_k(v)$ takes the form

$$(14.23) \qquad h'_k(v) = \frac{1}{\sqrt{\pi k}}\,e^{-\frac{\Delta'^2}{k}}\left(1-2c\frac{\Delta'}{k}-\frac{4}{9}\frac{\Delta'^3}{k^2}+O\left(\frac{1}{k}+\frac{\Delta'^6}{k^4}\right)\right),$$

valid for $|\Delta'| \le k^{3/5}$.

We also need some estimate for use outside this region for v, namely for $v \leq k - k^{3/5}$. This could be obtained by using the integral (14.19) to estimate $h'_k(v)$, however, we shall use a more general elementary estimate for $h(\alpha, w)$, which can be simply derived without limiting ourselves to the case of constant sifting density. We shall show that for $g(\alpha) > w$, we have

$$h(\alpha, w) \leq e^{-\frac{(g(\alpha)-w)^2}{2\alpha g(\alpha)}}. \tag{14.24}$$

Our point of departure is the inequality, valid for $\eta > 0$,

$$\begin{aligned} e_u(\xi) \sum_{\substack{\rho \leq z \\ \rho \in (\overline{\mathcal{P}}(\xi))}} \frac{1}{f'(\rho)} &< e_u(\xi) z^\eta \sum_{\rho \in (\mathcal{P}(\xi))} \frac{\rho^{-\eta}}{f'(\rho)} \\ &= z^\eta \prod_{p \leq \xi} \left(1 + \frac{p^{-\eta}}{f'(p)}\right)\left(1 - \frac{1}{f(p)}\right) \\ &= z^\eta \prod_{p < \xi} \left(1 - \frac{1 - p^{-\eta}}{f(p)}\right) \\ &< \exp\left(\eta \log z - \sum_{p<\xi} \frac{1-p^{-\eta}}{f(p)}\right). \end{aligned} \tag{14.25}$$

Here we have

$$1 - p^{-\eta} \geq \eta \log p - \frac{\eta^2 \log^2 p}{2} \geq \eta \left(1 - \frac{\eta \log \xi}{2}\right) \log p.$$

This gives

$$e_u(\xi) \sum_{\substack{\rho \leq z \\ \rho \in (\overline{\mathcal{P}}(\xi))}} \frac{1}{f'(\rho)} < \exp\left(\eta\left(\log z - S_u(\xi)\right) + \frac{\eta^2}{2} S_u(\xi) \log \xi\right). \tag{14.26}$$

If $S_u(\xi) > \log z$ we may here put

$$\eta = \frac{S_u(\xi) - \log z}{S_u(\xi) \log z},$$

and get

$$e_u(\xi) \sum_{\substack{\rho \leq z \\ \rho \in (\overline{\mathcal{P}}(\xi))}} \frac{1}{f'(\rho)} < \exp\left(-\frac{(S_u(\xi) - \log z)^2}{2\, S_u(\xi) \log \xi}\right). \tag{14.26'}$$

Putting here $\xi = x^\alpha$, $z = x^w$ and letting $x \to \infty$, we get (14.24) using Lemma 7. In the case that $g(t) = kt$, we get from (14.24), using that in this case $h(\alpha, w) = h_k(w/\alpha)$, that

$$h_k\left(\frac{w}{\alpha}\right) \leq e^{-\frac{(k-w/\alpha)^2}{2k}},$$

if $w/\alpha < k$, or

$$h_k(v) \le e^{-\frac{(k-v)^2}{2k}}, \tag{14.27}$$

for $v < k$.[36]

(14.23) and (14.27) give us the means to evaluate asymptotically the integral (14.18) for large k with various choices of $\sigma(t)$. The experience with the problem of choosing the y_r in the previous section leads us to try first with $\sigma(t)$ a linear function between 0 and 1. We choose first $u = k'$, and put $\sigma(t) = k'(1-t) + \mu$, for $0 \le t \le 1$, which gives

$$\sigma\left(\frac{v}{u}\right) = k' - v + \mu, \qquad \text{for } 0 \le v \le u, \tag{14.28}$$

here μ is a constant to be chosen later. That this $\sigma(t)$ does not satisfy the normalization $\sigma(0) = 1$ is immaterial since the ratio between (10.15) and (10.15′) does not change if σ is multiplied by a constant. Since the denominator (10.15′) cannot be negative, it is enough to consider (10.15) and choose μ so that (10.15) comes out positive with a c as large as possible.

With our choices we can now rewrite (10.15) as

(14.29)

$$\begin{aligned}
&\int_0^{k'} (k'-v+\mu)^2\, h_k'(v)dv - k\int_0^{k'-1} h_k'(v)dv \int_0^1 tdt \\
&\qquad - k\int_{k'-1}^{k'} dv \int_0^{k'-v} h_k'(v)tdt - k\int_{k'-1}^{k'} dv \int_{k'-v}^{1} \frac{(k'-v+\mu)^2}{t}\, dt\, h_k'(v) \\
&= \int_0^{k'} (w+\mu)^2\, h_k'(k'-w)\, dw - \frac{k}{2}\int_1^{k'} h_k'(k'-w)\, dw \\
&\qquad - \frac{k}{2}\int_0^1 w^2 h_k'(k'-w)dw - k\int_0^1 (w+\mu)^2 \log\frac{1}{w}\, h_k'(k'-w)\, dw \\
&= \mathfrak{I}_1 - \mathfrak{I}_2 - \mathfrak{I}_3 - \mathfrak{I}_4 .
\end{aligned}$$

We first evaluate $\mathfrak{I}_1$. We divide the integration range into $0 \le w \le k^{3/5}$ and $k^{3/5} \le w \le k'$, and note that by (14.23) we have in the first part

$$h_k'(k'-w) = \frac{1}{\sqrt{\pi k}}\, e^{-\frac{w^2}{k}} \left(1 - 2c\frac{w}{k} - \frac{4}{9}\frac{w^3}{k^2} + O\left(\frac{1}{k} + \frac{w^6}{k^4}\right)\right),$$

thus we see that the integral over the first part of the range differs from

$$\frac{1}{\sqrt{\pi k}} \int_0^\infty (w+\mu)^2 e^{-\frac{w^2}{k}} \left(1 - 2c\frac{w}{k} - \frac{4}{9}\frac{w^3}{k^2}\right) dw$$

by an amount which is $O(1)$. Using partial integration and (14.27) to estimate the second part of $\mathfrak{I}_1$, we see that this tends rapidly to zero with $k \to \infty$ so

[36] For the case of constant density we could get rid of the 2 in the denominator of the exponent in (14.27) by using $\sum_{p\le\xi} \frac{\log^2 p}{f(p)} \sim \frac{k}{2}\log^2 \xi$ in deriving the result.

we have

$$(14.30)\qquad \mathfrak{I}_1 = \frac{1}{\sqrt{\pi k}} \int_0^\infty (w+\mu)^2 e^{-\frac{w^2}{k}} \left(1 - 2c\frac{w}{k} - \frac{4}{9}\frac{w^3}{k^2}\right) dw + O(1).$$

Next, we get, again using (14.23) and (14.27), that

$$\begin{aligned}(14.30')\mathfrak{I}_2 &= \frac{k}{2}\frac{1}{\sqrt{\pi k}} \int_1^\infty e^{-\frac{w^2}{k}} \left(1 - 2c\frac{w}{k} - \frac{4}{9}\frac{w^3}{k^2}\right) dw + O(1) \\ &= \frac{k}{2\sqrt{\pi k}} \left(\int_0^\infty e^{-\frac{w^2}{k}} \left(1 - 2c\frac{w}{k} - \frac{4}{9}\frac{w^3}{k^2}\right) dw - 1\right) + O(1).\end{aligned}$$

For $\mathfrak{I}_3$ we get

$$(14.30'')\qquad \mathfrak{I}_3 = \frac{k}{2}\frac{1}{\sqrt{\pi k}} \int_0^1 w^2 dw + O(1) = \frac{k}{6\sqrt{\pi k}} + O(1),$$

and finally for $\mathfrak{I}_4$

$$\begin{aligned}(14.30''')\qquad \mathfrak{I}_4 &= \frac{k}{\sqrt{\pi k}} \int_0^1 (w+\mu)^2 \log\frac{1}{w}\, dw + O(1) \\ &= \frac{k}{\sqrt{\pi k}} \left(\frac{1}{9} + \frac{\mu}{2} + \mu^2\right) + O(1).\end{aligned}$$

Observing that for $\lambda > -1$

$$\int_0^\infty w^\lambda e^{-w^2/k} dw = \frac{1}{2}\Gamma\left(\frac{\lambda+1}{2}\right) k^{(\lambda+1)/2},$$

we get from (14.30),...,(14.30'''), that

$$(14.31)\qquad \mathfrak{I}_1 - \mathfrak{I}_2 - \mathfrak{I}_3 - \mathfrak{I}_4 = \sqrt{\frac{k}{\pi}} \left(-\frac{c}{2} + \frac{\mu}{2} - \mu^2 - \frac{1}{9}\right) + O(1).$$

Thus we get (10.15) positive for large k if

$$-\frac{c}{2} + \frac{\mu}{2} - \mu^2 - \frac{1}{9} > 0,$$

or

$$c < \mu - 2\mu^2 - \frac{2}{9}.$$

The optimal choice of μ is seen to be $\mu = 1/4$ which gives

$$c < \frac{1}{8} - \frac{2}{9} = -\frac{7}{72},$$

more precisely, taking account of the remainder term $O(1)$ in (14.31) we get (10.15) positive for

$$(14.32)\qquad c < -\frac{7}{72} - \frac{a}{\sqrt{k}},$$

where a is some constant, which can be computed by keeping track of the remainder terms $O(1)$ in our formulas. It does turn out positive, but a more complicated choice of σ which involves a second degree term $\delta\,(k'-v)^2$ in addition to the earlier part with a small coefficient δ will give enough additional freedom to get (14.32) with a negative constant a (for large enough k).[37] Thus one gets (10.15) positive for

$$c=-\frac{7}{72}$$

in particular, for $k>k_0$. This gives

$$k'=k-\frac{1}{3}-c=k-\frac{17}{72}\,.$$

Since this shows that we get a positive lower bound for $\alpha=\frac{1}{2k'+1}=\frac{1}{2k+19/36}$, we get

$$\alpha_k>\frac{1}{2k+19/36}\,, \tag{14.33}$$

for $k>k_0$.

As it does not seem possible to improve the 19/36 by a modified choice of σ, we are led to try to use a $\Lambda^2\Lambda^-$ sieve where the Λ^- is somewhat more complicated than the simple sieve Λ^- so far used. We wish to retain the condition $\lambda'_d=0$ for $d>\xi=x^\alpha$, but to permit $\lambda'_d\neq 0$ for some composite d. This means going back to (7.13) and trying to develop it further in a similar way as we earlier did with the more specialized (7.13′). Making it into a dimensionless ratio in the y_ρ's as we earlier did with (7.13) to get (7.14), it becomes

$$\frac{\displaystyle\sum_{d\le\xi}\frac{\lambda'_d}{f(d)}\sum_{\substack{(\rho,d)=1\\ \rho\le z}}\frac{1}{f'(\rho)}\left\{\sum_{\delta\,|\,d}\mu(\delta)y_{\rho\delta}\right\}^2}{\left(\displaystyle\sum_{\rho\le z}\frac{y_\rho}{f'(\rho)}\right)^2}\,. \tag{14.34}$$

When we seek to get from this to a ratio between expressions similar to (but necessarily more complicated than) (10.15) and (10.15′), it is reasonable to choose a Λ^- such that we do not get too many new terms in the numerator and such that the λ'_d for the composite d that are allowed can be chosen small as k gets large. We make the following choice: $\lambda'_1=1$, $\lambda'_p=-1$ for $p<\xi$ as before, $\lambda'_{p_1p_2}=\frac{4T-2}{T\,(T+1)}$, $\lambda'_{p_1p_2p_3}=-\frac{6}{T\,(T+1)}$ for $p_1,p_2,p_3<\xi^{1/3}$, while all other

[37] This is rather cumbersome to carry out. As we anyway shall improve on (14.33) by other means, we do not go into the details here.

$\lambda'_d = 0$. That this is a lower bound sieve is seen from the fact that if d has r_1 prime factors $< \xi^{1/3}$ and r_2 prime factors with $\xi^{1/3} \le p \le \xi$, then

$$\theta_d = \sum_{\delta | d} \lambda'_\delta = (1 - r_1)\left(1 - \frac{r_1}{T}\right)\left(1 - \frac{r_1}{T+1}\right) - r_2 ,$$

and so $\theta_d \le 0$ for $d > 1$. Our T here is a positive integer which we will choose later in relation to k.

Writing again $y_p = \sigma\left(\frac{\log p}{\log z}\right)$, and proceeding as we did earlier to obtain the expression (10.15) and (10.15′), we get in the limit as $x \to \infty$, that the denominator of (14.34) leads to (10.15′) as before, but the numerator becomes (14.35)

$$\int_0^u \sigma^2\left(\frac{v}{u}\right) h'_k(v)dv - k\int_0^u h'_k(v)dv \int_0^1 \left(\sigma\left(\frac{v}{u}\right) - \sigma\left(\frac{v+t}{u}\right)\right)^2 \frac{dt}{t}$$
$$+ \frac{(4T-2)k^2}{T(T+1)} \int_0^u h'_k(v)dv \iint\limits_{0<t_1<t_2<1/3} \left(\sigma\left(\frac{v}{u}\right)\right.$$
$$\left. - \sum_{i=1}^2 \sigma\left(\frac{v+t_i}{u}\right) + \sigma\left(\frac{v+t_1+t_2}{u}\right)\right)^2 \frac{dt_1 dt_2}{t_1 t_2}$$
$$- \frac{6k^3}{T(T+1)} \int_0^u h'_k(v)dv \int\cdots\int\limits_{0<t_1<t_2<t_3<1/3} \left(\sigma\left(\frac{v}{u}\right) - \sum_{i=1}^3 \sigma\left(\frac{v+t_i}{u}\right)\right.$$
$$\left. + \sum_{1\le i<j\le 3} \sigma\left(\frac{v+t_i+t_j}{u}\right) - \sigma\left(\frac{v+t_1+t_2+t_3}{u}\right)\right)^2 \frac{dt_1 dt_2 dt_3}{t_1 t_2 t_3} .$$

The proof is patterned on that given earlier in this Section for (10.15), we handle the triple and quadruple sums that come from the terms when d has two or three prime factors and $\lambda'_d \ne 0$, in a way quite similar to that used for the double sum which gave the second term of (10.15) or (14.35). As before we have $\sigma = 0$ if the argument is > 1.

We now choose our σ as before, put $u = k'$ and $\sigma(v/u) = k' - v + \mu$ for $0 \le v \le k'$. The two first terms in (14.35) give of course the same contribution as before. We choose $T = [k/\vartheta]$, where ϑ is a positive constant to be chosen later. Then

$$\frac{4T-2}{T(T+1)} k^2 = 4\vartheta k + O(1) , \qquad \frac{6k^3}{T(T+1)} = 6\vartheta^2 k + O(1) ,$$

as $k \to \infty$. We also observe that since $\sigma(v/u)$ is linear in v for $v \le u = k'$, the second difference which occurs in the squared parenthesis in the third term of (14.35) is zero unless $v + t_1 + t_2 > k'$, thus certainly for $v \le k' - 2/3$. Similarly in the fourth term, the squared parenthesis is zero unless $v + t_1 + t_2 + t_3 > k'$, thus certainly for $v \le k' - 1$. For $k' - 1 \le v \le k'$ we also have from (14.23)

$$h'_k(v) = \frac{1}{\sqrt{\pi k}}\left(1 + O\left(\frac{1}{k}\right)\right) .$$

Thus the two new terms in (14.35) contribute

$$\frac{4\vartheta k}{\sqrt{\pi k}}\,\mathfrak{I}_5-\frac{6\vartheta^2 k}{\sqrt{\pi k}}\,\mathfrak{I}_6+O\left(\frac{1}{\sqrt{k}}\right), \tag{14.36}$$

where,

$$\mathfrak{I}_5=\iiint\limits_{\substack{0<t_1<t_2<1/3\\0<w<t_1+t_2}}\Big(\sigma^*(w)-\sum_{i=1}^{2}\sigma^*(w-t_i)+\sigma^*(w-t_1-t_2)\Big)^2\,\frac{dw\,dt_1\,dt_2}{t_1t_2},$$

and

$$\mathfrak{I}_6=\iiiint\limits_{\substack{0<t_1<t_2<t_3<1/3\\0<w<t_1+t_2+t_3}}\Big(\sigma^*(w)-\sum_{i=1}^{3}\sigma^*(w-t_i)+\sum_{1\le i<j\le 3}\sigma^*(w-t_i-t_j)-\sigma^*(w-t_1-t_2-t_3)\Big)^2\,\frac{dw\,dt_1\,dt_2\,dt_3}{t_1t_2t_3}.$$

Here we have written $\sigma^*(w)$ for $\sigma((k'-w)/k')$, so that $\sigma^*(w)=w+\mu$ for $w\ge 0$ and $\sigma^*(w)=0$ for $w\le 0$.

These integrals can be evaluated explicitly (the computation is rather tedious, so the details will not be given here, let me just remark that in $\mathfrak{I}_5$ we first perform the integration over w, dividing the interval $(0,t_1+t_2)$ into the three parts $(0,t_1)$, (t_1,t_2) and (t_2,t_1+t_2), in $\mathfrak{I}_6$ more parts are needed and one needs to distinguish the cases $t_3<t_1+t_2$ and $t_1+t_2<t_3$, the result of this computation is

$$\mathfrak{I}_5=\frac{2}{3}\mu^2+\frac{7}{2}3^{-6}, \tag{14.37}$$

and

$$\mathfrak{I}_6=\frac{4}{3}\mu^2+\frac{\pi^2-1}{4}3^{-6}. \tag{14.37$'$}$$

Combining the expression (14.36) with (14.31) we get for (14.35)

$$\sqrt{\frac{k}{\pi}}\left(-\frac{c}{2}+\frac{\mu}{2}-\mu^2-\frac{1}{9}+4\vartheta\,\mathfrak{I}_5-6\vartheta^2\mathfrak{I}_6\right)+O(1). \tag{14.38}$$

The optimal choice of ϑ is seen to be

$$\vartheta=\frac{1}{3}\frac{\mathfrak{I}_5}{\mathfrak{I}_6},$$

which gives

$$\sqrt{\frac{k}{\pi}}\left(-\frac{c}{2}+\frac{\mu}{2}-\mu^2-\frac{1}{9}+\frac{2}{3}\frac{\mathfrak{I}_5^2}{\mathfrak{I}_6}\right)+O(1). \tag{14.38$'$}$$

This is positive if

$$c < \mu - 2\mu^2 - \frac{2}{9} + \frac{4}{3}\,\frac{\mathfrak{I}_5^2}{\mathfrak{I}_6}\,. \tag{14.39}$$

The optimal choice of μ can be found by inserting the expressions for $\mathfrak{I}_5$ and $\mathfrak{I}_6$ and determining the maximum of the right hand side of (14.39), this leads to a fifth degree equation for μ. We shall instead proceed as follows: Noting that $\mathfrak{I}_5 \geq \frac{1}{2}\mathfrak{I}_6$ and that unless μ is quite small the ratio between $\mathfrak{I}_5$ and $\mathfrak{I}_6$ is close to $1/2$, we replace the last term in (14.39) by $\frac{2}{3}\mathfrak{I}_5$. In the resulting expression μ enters only in the form

$$\mu - 2\,\mu^2 + \frac{4}{9}\,\mu^2 = \mu - \frac{14}{9}\,\mu^2\,.$$

This expression takes its maximum value for $\mu = 9/28$, which is the value we shall use in (14.37), (14.37′) and (14.39). This gives that (14.35) is positive for

$$c \leq -0.05603, \qquad \text{and } k > k_0\,.$$

Since we have for $k > k_0$,

$$\alpha_k > \frac{1}{2k'+1} = \frac{1}{2k + 1/3 - 2c}\,,$$

we get finally

$$\alpha_k > \frac{1}{2k + 0.4454} \tag{14.40}$$

for $k > k_0$. We shall not try to push this further by modifying again the choice of our Λ^-. Undoubtedly improvements can be made by allowing say $\lambda'_d \neq 0$ for d of the form $p_1p_2p_3p_4$ and $p_1p_2p_3p_4p_5$ if $p_1, p_2, \ldots, p_5 \leq \xi^{1/5}$. But they are not likely to be significant. Undoubtedly we could replace the constant 0.4454 by 4/9 and perhaps even with 3/7 but it seems most unlikely that we could make it negative or even zero, probably already 2/5 may be unreachable by these means.

However, we have no reason to think that our result (14.40) is so close to being optimal that

$$\alpha_k \leq \frac{1}{2k}$$

for large k. While I would conjecture that

$$\lim_{x\to\infty} 2k\alpha_k = 1\,, \tag{14.41}$$

I am inclined to believe that for large enough k we have $2k\alpha_k > 1$. It is even conceivable that we may have

$$\alpha_k \geq \frac{1}{2k}$$

for all $k \geq 1/2$. As we shall see the equality sign cannot be excluded here. For $k = 1/2$ our results on the B^{2}R sieve showed that $\alpha_{1/2} = 1$. Later we shall show that $\alpha_1 = 1/2$. From the results on the B^{2}R sieve, we already know that $\alpha_1 \geq 1/2$.

Before we go on to this and also to establish some other upper bounds on α_k, as well as some other limitations on what can be obtained by sieve methods, we need to turn our attention briefly to sifting in situations where we no longer impose the stringent condition $|R_d| \leq u(d)$.

15. The effects of relaxing the conditions on the R_d in our sifting problem

We have so far only considered situations where $|R_d| \leq u(d)$, a restriction that is rather severe, but rather natural since it holds for the original setting, sifting an interval $\mathcal{I}_x$. We shall now look at the consequences of allowing rather larger error terms R_d, and see that we may relax this restriction very much.

Theorem 10. *The statements of Theorem 1 continue to hold if we replace the assumption that $|R_d| \leq u(d)$ with the assumption:*

For every positive δ, there exists an $\eta = \eta_\delta > 0$ and an x_δ such that

$$\sum_{d \leq x^{1-\delta}} |R_d| < x^{1-\eta}, \tag{15.1}$$

for all $x > x_\delta$.

To prove this, we apply the sieves that were constructed in connection with the proof of Lemma 3. For these sieves the $|\lambda_d|$ are bounded by $M\,4^r$ where M is a constant and r the number of prime factors of d.[38] Thus clearly $|\lambda_d| < A\,x^{\eta/2}$, and so

$$R_u(\Lambda) = \sum_{d \leq x^{1-\delta}} |\lambda_d|\,|R_d| < A\,x^{1-\eta/2},$$

this gives

$$\frac{R_u(\Lambda)}{E_u(W_x, \mathcal{P}(x^\alpha))} \to 0, \tag{15.2}$$

which completes the proof.

[38] More precisely, we could put $M\,3^r$ for the upper bound sieve and $M\,3^r\,(1+r)$ for the lower bound sieve. This comes in via the Λ^2 and $\Lambda^2\Lambda^-$ sieves that are combined with the sieves of Lemma 2. With a bit more effort in estimating the λ's of the Λ^2, we may even replace the 3^r in both expressions by 2^r.

The assumption of Theorem 10 is seen to be satisfied if for every positive δ there exists a positive $\eta = \eta_\delta$ and an $A = A_\delta$, such that

$$|R_d| < A\,u(d)\left(\frac{x}{d}\right)^{1-\eta}, \tag{15.3}$$

for $d \le x^{1-\delta}$. We get from (15.3) that

$$\sum_{d \le x^{1-\delta}} |R_d| < A\,x^{1-\eta}\,x^{(1-\delta)\eta} \sum_d \frac{u(d)}{d} = A\,x^{1-\delta\eta} \prod_{p\in\mathcal{P}(x^\alpha)} \left(1 + \frac{u(p)}{p}\right) < x^{1-\delta\eta/2},$$

for x large enough.

If we make stronger assumptions about $e_u(x^\alpha)$ than (4.4), say that for $\varepsilon > 0$

$$e_u(x^\alpha) > (\log x)^{-k-\varepsilon} \tag{15.4}$$

for $x > x(\varepsilon)$, where k is some positive constant, and weaken the assumption (15.3) to

$$|R_d| < A\,u(d)\,\frac{x}{d}\left(\log\frac{x}{d}\right)^{-2k-\eta}, \tag{15.5}$$

for $d \le x^{1-\delta}$, we get

$$R_u(\Lambda) = \sum_{d \le x^{1-\delta}} |\lambda_d|\,|R_d| < A\,x(\delta\,\log x)^{-2k-\eta} \sum_d \frac{u(d)\,|\lambda_d|}{d},$$

here, from (5.3) and (5.5)

$$\sum_d \frac{u(d)}{d}\,|\lambda_d| < e_u^{-1}(x^\alpha)\,(2 + |T(\Lambda)|) < A'\,(\log x)^{k+\varepsilon},$$

by (15.4) for x large enough. Thus we get

$$R_u(\Lambda) < A''_\delta x(\log x)^{-k-\eta+\varepsilon} \tag{15.6}$$

for x large enough. We also have

$$E_u(W_x, \mathcal{P}(x^\alpha)) = x\,e_u(x^\alpha) > x\,(\log x)^{-k-\varepsilon}, \tag{15.7}$$

and thus, by choosing $\varepsilon = \frac{1}{3}\eta$,

$$\frac{R_u(\Lambda)}{E_u(W_x, \mathcal{P}(x^\alpha))} \to 0\,. \tag{15.8}$$

This gives

Theorem 11. *The statements of Theorem 1 hold if we replace the assumption* $|R_d| \le u(d)$ *with (15.5), and (4.4) with (15.4).*

We may also consider situations where information about the R_d is available only for a considerably shorter range of d than in the previous theorems. We have for instance

Theorem 12. *Let* μ *be fixed with* $0 < \mu < 1$ *and suppose that for every* $\delta > 0$, *there exists an* $\eta = \eta_\delta > 0$ *and an* x_δ, *such that*

$$\sum_{d \le x^{\mu-\delta}} |R_d| < x^{1-\eta}, \tag{15.9}$$

for all $x > x_\delta$. *Let otherwise our earlier assumptions about* $u(p)$, $e_u(x^\alpha)$ *and* $g(t)$[39] *be satisfied. Write also*

$$g_\mu(t) = \frac{1}{\mu} g(\mu t). \tag{15.10}$$

Then

$$\lim_{x\to\infty} \frac{M_u^+(W_x, \mathcal{P}(x^\alpha))}{E_u(W_x, \mathcal{P}(x^\alpha))} = T_{g_\mu}^+\left(\frac{\alpha}{\mu}\right), \tag{15.11}$$

and, for $\alpha < \mu$,

$$\lim_{x\to\infty} \frac{M_u^=(W_x, \mathcal{P}(x^\alpha))}{E_u(W_x, \mathcal{P}(x^\alpha))} = T_{g_\mu}^-\left(\frac{\alpha}{\mu}\right), \tag{15.11$'$}$$

also for $\alpha < \mu$,

$$\lim_{x\to\infty} \frac{M_u^-(W_x, \mathcal{P}(x^\alpha))}{E_u(W_x, \mathcal{P}(x^\alpha))} = \max\left(0, T_{g_\mu}^-\left(\frac{\alpha}{\mu}\right)\right). \tag{15.11$''$}$$

The proof is quite similar to that of Theorem 10, only $x^* = x^\mu$ will play the role of x in the arguments. This explains the shift from g to g_μ and from α to α/μ in the expressions on the right hand sides of (15.11), (15.11′) and (15.11″). In the special case of constant sifting density k, we have $g_\mu(t) = g(t) = kt$.

It is obvious that we may in a similar way extend the results of Section 9 concerning sifting with weights to situations where we relax the demands made concerning the R_d in the same way as above, and may, for instance, prove theorems that relate to Theorem 3 in the same way as the Theorems 10, 11 and 12 relate to Theorem 1. We shall not go into the details of this.

[39] (4.2), (4.4), and (8.26).

16. Two examples in connection with the B^2R method

We shall in this Section construct some particular examples or models in connection with the sifting problem for constant density when $k = 1/2$ and when $k = 1$. Before we go on to this, we shall need some lemmas.

Lemma 10. *Let $\pi(x)$ denote the number of primes $< x$, $\pi_{3,4}(x)$ the number of primes $p \equiv 3\ (4)$ and $\le x$, and $\psi(x)$ the number of odd integers n with $1 \le n \le x$, which do not have any prime factors of the form $4n+3$, then as $x \to \infty$, we have*

$$\pi(x) \sim \frac{x}{\log x}, \tag{16.1}$$

$$\pi_{3,4}(x) \sim \frac{x}{2\log x}, \tag{16.1'}$$

and

$$\psi(x) \sim \frac{c\,x}{\sqrt{\log x}} \tag{16.2}$$

where c is a constant.

(16.1) of course is the well known prime number theorem and (16.1′) the analogue for a particular arithmetic progression, while (16.2) is essentially a (less known) result due to E. Landau. Since an elementary and simple proof of (16.2) is easily sketched, we include it here.

Let P_1 denote the primes $\equiv 1\ (4)$, P_3 those of the form $p \equiv 3\ (4)$, while $P_1(x)$ and $P_3(x)$ means the restriction to those p that are $< x$. Also (P_1), (P_3) or $(P_1(x))$ and $(P_3(x))$ will have their usual meaning here.

We consider the expression

$$\sum_{\substack{1\le n\le x\\ n\equiv 1\ (4)}} \sum_{\substack{d\mid n\\ d\in(P_3)}} \mu(d)\log\frac{x}{d} = \sum_{\substack{d\in(P_3)\\ d<x}} \mu(d)\log\frac{x}{d} \sum_{\substack{d\mid n\\ 1\le n\le x\\ n\equiv 1\ (4)}} 1\,. \tag{16.3}$$

Since, as one easily sees, when $n \equiv 1\ (4)$.

(16.4)

$$\sum_{\substack{d\mid n\\ d\in(P_3)}} \mu(d)\log\frac{x}{d} = \begin{cases} \log x\,, & \text{if } n \in (P_1),\\ \log p\,, & \text{if } n \text{ is divisible by exactly one of } p \text{ from } P_3,\\ 0, & \text{if } n \text{ has two or more distinct prime factors from } P_3, \end{cases}$$

we see that the left hand side of (16.3) equals

$$\log x\psi(x) + \sum_{\substack{p\in P_3\\ r\ge 1}} \log p\,\psi\left(\frac{x}{p^{2r}}\right) = \log x\,\psi(x) + O(x)\,. \tag{16.5}$$

For the right hand side of (16.3) we get
(16.6)

$$\frac{x}{4}\sum_{\substack{d\in(P_3)\\ d\le x}}\frac{\mu(d)\log(x/d)}{d}+O\left(\sum_{d\le x}\log\frac{x}{d}\right)=\frac{x}{4}\sum_{\substack{d\in(P_3)\\ d\le c'x}}\frac{\mu(d)\log(x/d)}{d}+O(x)\,,$$

where c' is a constant chosen so that we have

$$\frac{1}{2}\log z=\sum_{\substack{n\le c'z\\ n\equiv 1\ (2)}}\frac{1}{n}+O\left(\frac{1}{z}\right).$$

We may thus insert

$$\frac{1}{2}\log\frac{x}{d}=\sum_{\substack{n\le c'\,x/d\\ n\equiv 1\ (2)}}\frac{1}{n}+O\left(\frac{d}{x}\right),$$

in the first term on the right hand side of (16.6), this gives

$$\begin{aligned}\frac{1}{2}\sum_{\substack{d\in(P_3)\\ d\le c'x}}\frac{\mu(d)}{d}\log\frac{x}{d}&=\sum_{\substack{dn\le c'x\\ d\in(P_3)\\ n\equiv 1\ (2)}}\frac{\mu(d)}{dn}+O\left(\frac{1}{x}\sum_{\substack{d\in(P_3)\\ d\le c'x}}1\right)\\ &=\sum_{\substack{m\le c'x\\ m\in(P_1)}}\frac{1}{m}+O(1)=\sum_{\substack{m\le x\\ m\in(P_1)}}\frac{1}{m}+O(1)\\ &=\int_1^x\frac{d\psi(t)}{t}+O(1)=\int_1^x\frac{\psi(t)}{t^2}dt+O(1)\,.\end{aligned}$$

Inserting this on the left hand side of (16.6) and equating it with the right hand side of (16.5) we get

$$\log x\psi(x)-\frac{x}{2}\int_1^x\frac{\psi(t)}{t^2}dt=O(x)\,. \tag{16.7}$$

If we here write

$$A(x)=\int_1^x\frac{\psi(t)}{t^2}\,dt\,,$$

this takes the form

$$x^2\log x\,A'(x)-\frac{x}{2}A(x)=O(x)$$

from which we get

$$\left(\frac{A(x)}{\sqrt{\log x}}\right)'=O\left(\frac{1}{x\,(\log x)^{3/2}}\right),$$

which immediately gives

$$\frac{A(x)}{\sqrt{\log x}} = c'' + O\left(\frac{1}{\sqrt{\log x}}\right).$$

Inserting now the estimation for $A(x) = \int_1^x \frac{\psi(t)}{t^2}dt$ in (16.7) gives

$$\psi(x) = \frac{cx}{\sqrt{\log x}} + O\left(\frac{x}{\log x}\right)^{40} \tag{16.8}$$

which proves (16.2).

The constant c can of course easily be found, but we have no particular need for it here, we shall however need that

$$e(P_3(x)) \sim \frac{c'''}{\sqrt{\log x}}, \tag{16.9}$$

where $u(p) = 1$ for each p in P_3. (16.9) is of course an immediate consequence of

$$\prod_{p\le x}\left(1-\frac{1}{p}\right) \sim \frac{e^{-\gamma}}{\log x},$$

and

$$\prod_{p\in P_1(x)}\left(1-\frac{1}{p}\right)\prod_{p\in P_3(x)}\left(1+\frac{1}{p}\right) \sim \frac{4}{\pi}.$$

Lemma 11. *If $\lambda(n)$ denotes Liouville's function*

$$\lambda(n) = (-1)^{\nu(n)}$$

where ν denotes the number of prime factors in n (counted with multiplicities), then

$$\sum_{n\le x}\lambda(n) = O\left(\frac{x}{(\log x)^r}\right), \tag{16.10}$$

for any positive r.

This is a well known estimation, easiest derived from the similar estimation for $M(x) = \sum_{n\le x}\mu(n)$, where μ is the Möbius function, since we clearly have

$$\sum_{n\le x}\lambda(n) = \sum_{m^2n\le x}\mu(n) = \sum_{m\le\sqrt{x}} M\left(\frac{x}{m^2}\right).$$

Of course much sharper estimations than (16.10) are known, we may for instance for instance replace the right hand side of (16.10) by

$$O(x\,e^{-(\log x)^{\mu}}), \tag{16.10'}$$

[40] By estimating remainder terms more accurately in our argument, one easily gets $O(x/(\log x)^{3/2})$ instead of $O(x/\log x)$.

with $\mu < 2/3$, and if we assume the Riemann hypothesis for the Riemann zeta–function, we have

$$\sum_{n\le x} \lambda(n) = O(x^{1/2+\varepsilon}), \tag{16.10''}$$

for any positive ε.

Let now $\chi(n)$ denote the quadratic character modulo 4, and let us define two weighted sets by weighting the integers n with $1 \le n \le x$ with the weights

$$w_1(n) = (1-(-1)^n)(1+\chi(n)), \tag{16.11}$$

and

$$w_3(n) = (1-(-1)^n)(1-\chi(n)), \tag{16.11'}$$

respectively while the integers outside the interval are given weight zero. Let $W_1 = W_1(x)$ be the set with weights $w_1(n)$ and $W_3 = W_3(x)$ the set with weights $w_3(n)$. We sift these sets with respect to the primes in $P_3(x^\alpha)$ by sifting out the n that are divisible by a p from $P_3(x^\alpha)$. For brevity let us write

$$M(W_1(x), P_3(x^\alpha)) = \psi_1(x, x^\alpha), \tag{16.12}$$

and

$$M(W_3(x), P_3(x^\alpha)) = \psi_3(x, x^\alpha). \tag{16.12'}$$

It is now obvious that, for $\alpha' > \alpha$, we have

$$\psi_1(x, x^{\alpha'}) = \psi_1(x, x^\alpha) - \sum_{\substack{p\equiv 3\ (4)\\ x^\alpha\le p<x^{\alpha'}}} \psi_3\left(\frac{x}{p}, p\right), \tag{16.13}$$

and similarly

$$\psi_3(x, x^{\alpha'}) = \psi_3(x, x^\alpha) - \sum_{\substack{p\equiv 3\ (4)\\ x^\alpha\le p<x^{\alpha'}}} \psi_1\left(\frac{x}{p}, p\right). \tag{16.13'}$$

From (16.2), we see that for $\alpha > 1/2$,

$$\psi_1(x, x^\alpha) = 4\,\psi(x) \sim \frac{4c\,x}{\sqrt{\log x}}. \tag{16.14}$$

Also since $\psi_3(x,x) = 0$, we get from (16.13′) for $\alpha > 1/3$

$$\psi_3(x, x^\alpha) = \sum_{\substack{p\equiv 3\ (4)\\ x^\alpha\le p<x}} \psi_1\left(\frac{x}{p}, p\right) = \sum_{\substack{p\equiv 3\ (4)\\ x^\alpha\le p<x}} 4\psi\left(\frac{x}{p}\right).$$

Using (16.1′) and (16.2) one easily find that for $1/3 < \alpha < 1$, we have

$$(16.14') \qquad \psi_3(x, x^\alpha) = \frac{2c\,x}{\sqrt{\log x}} \int_\alpha^1 \frac{du}{u\sqrt{1-u}} + o\left(\frac{x}{\sqrt{\log x}}\right)$$
$$= 2c \log \frac{1+\sqrt{1-\alpha}}{1-\sqrt{1-\alpha}} \frac{x}{\sqrt{\log x}} + o\left(\frac{x}{\sqrt{\log x}}\right) .$$

In general we see from (16.13) and (16.13′) that

$$(16.15) \qquad \psi_1(x, x^\alpha) \sim a_1(\alpha) \frac{x}{\sqrt{\log x}} ,$$

and

$$(16.15') \qquad \psi_3(x, x^\alpha) \sim a_3(\alpha) \frac{x}{\sqrt{\log x}} ,$$

where $a_1(\alpha)$ and $a_3(\alpha)$ satisfy

$$(16.16) \qquad a_1(\alpha') = a_1(\alpha) - \frac{1}{2} \int_\alpha^{\alpha'} a_3\left(\frac{u}{1-u}\right) \frac{du}{u\sqrt{1-u}} ,$$

and

$$(16.16') \qquad a_3(\alpha') = a_3(\alpha) - \frac{1}{2} \int_\alpha^{\alpha'} a_1\left(\frac{u}{1-u}\right) \frac{du}{u\sqrt{1-u}} ,$$

for $1 > \alpha' > \alpha$. Here $a_3(\alpha)$ is understood to be zero for $\alpha \geq 1$, and $a_1(\alpha) = 4c$ for $\alpha > 1/2$.

Using (16.9), and rewriting (16.15) and (16.15′) as

$$(16.17) \qquad \psi_1(x, x^\alpha) \sim a_1^*(\alpha) x e(P_3(x^\alpha)) ,$$

and

$$(16.17') \qquad \psi_3(x, x^\alpha) \sim a_3^*(\alpha) x e(P_3(x^\alpha)) ,$$

so that

$$a_1^*(\alpha) = \frac{1}{c'''} \sqrt{\alpha}\, a_1(\alpha) ,$$

and

$$a_3^*(\alpha) = \frac{1}{c'''} \sqrt{\alpha}\, a_3(\alpha) ,$$

we may rewrite (16.16) and (16.16′) and differentiate. We get then

$$(16.18) \qquad \frac{d\,\alpha^{-\frac{1}{2}}\, a_1^*(\alpha)}{d\alpha} = \begin{cases} 0 & \text{for } \alpha \geq 1/2 \\ -\frac{1}{2}\,\alpha^{-3/2}\, a_3^*\left(\frac{\alpha}{1-\alpha}\right) , & \text{for } \alpha \leq 1/2, \end{cases}$$

and

$$(16.18') \qquad \frac{d\,\alpha^{-\frac{1}{2}}\,a_3^*(\alpha)}{d\alpha} = -\frac{1}{2}\,\alpha^{-3/2}\,a_1^*\left(\frac{\alpha}{1-\alpha}\right),$$

for $\alpha < 1$.

Going back to Lemma 9 in Section 11, we see that $a_1^*(\alpha)$ and $a_3^*(\alpha)$ satisfy precisely the relations given there for $t_{1/2}^+(\alpha)$ and $t_{1/2}^-(\alpha)$ for $\beta = 1$. These relations, together with the fact that $t_{1/2}^+(\alpha)$ and $t_{1/2}^-(\alpha)$ tended very strongly to 1 as $\alpha \to 0$, were enough to determine these two functions uniquely. But it is obvious that both a_1^* and a_3^* satisfy the inequality

$$t_{1/2}^-(\alpha) \le a^*(\alpha) \le t_{1/2}^+(\alpha),$$

since we may apply the B^2R sieve both to the weighted set $W_1(x)$ and to $W_3(x)$ with $\beta = 1$. Thus a_1^* and a_3^* tend at least as strongly towards 1 as $\alpha \to 0$ as $t_{1/2}^-$ and $t_{1/2}^+$ and the relations (16.18) and (16.18′) thus determine them uniquely so we must have

$$(16.19) \qquad a_1^*(\alpha) = t_{1/2}^+(\alpha),$$

and

$$(16.19') \qquad a_3^*(\alpha) = t_{1/2}^-(\alpha).$$

Since from (11.49) for $\beta = 1$ and $k = 1/2$ we have

$$\sqrt{2}t_{1/2}^+(1/2) = A_{1/2} = \frac{2\,e^{\gamma/2}}{\Gamma(1/2)} = \frac{2}{\sqrt{\pi}}\,e^{\gamma/2}$$

where γ is Euler's constant, we see that

$$c = \frac{e^{\gamma/2}}{2\sqrt{\pi}}c'''.$$

Actually, both c and c''' are easy to determine independently once one knows that asymptotic formulas (16.2) and (16.9) exist. One finds

$$c = \frac{1}{2\sqrt{2}}\prod_{p\in P_3}\left(1-\frac{1}{p^2}\right)^{1/2}, \qquad c''' = \sqrt{\frac{\pi}{2}}\,e^{-\gamma/2}\prod_{p\in P_3}\left(1-\frac{1}{p^2}\right)^{1/2}.$$

The examples $W_1(x)$ and $W_3(x)$ show that the B^2R sieve for $k = 1/2$ for all α gives the optimal result, so that in this case

$$(16.20) \qquad T_{1/2}^+(\alpha) = t_{1/2}^+(\alpha),$$

and

$$(16.20') \qquad T_{1/2}^-(\alpha) = t_{1/2}^-(\alpha),$$

for $0 \le \alpha \le 1$.

We next look at two other weighted sets obtained by defining weights w_n for $1 \le n \le x$. We define, for n in this interval, $w_+(n) = 1 - \lambda(n)$ and $w_-(n) = 1 + \lambda(n)$, while $w_+(n) = w_-(n) = 0$ for n outside the interval, and call the weighted sets $W_+ = W_+(x)$ and $W_- = W_-(x)$ respectively. Again, we wish to look at what remains when we sift out all n that are divisible by a prime from $\mathcal{P}(x^\alpha)$, the set of all primes $p < x^\alpha$.

We write for brevity

$$M(W_+(x), \mathcal{P}(x^\alpha)) = \pi_+(x, x^\alpha)\,, \tag{16.21}$$

and

$$M(W_-(x), \mathcal{P}(x^\alpha)) = \pi_-(x, x^\alpha)\,. \tag{16.21'}$$

It is readily seen that for $\alpha \ge 1/2$, we have

$$\pi_-(x, x^\alpha) = 2\,, \tag{16.22}$$

unless $\alpha = 1/2$ and $x = p^2$ in which case $\pi_-(p^2, p) = 4$. For $1/3 \le \alpha < 1$,

$$\pi_+(x, x^\alpha) = 2\left(\pi(x) - \pi(x^\alpha)\right) \sim \frac{2x}{\log x}\,, \tag{16.22'}$$

unless x^α is a prime in which case we have to add 2 to the right hand side of the equation, the same is the case if $\alpha = 1/3$ and x is the cube of a prime.

We observe that we have the recursion formulas

$$\pi_+(x, x^{\alpha'}) = \pi_+(x, x^\alpha) - \sum_{x^\alpha \le p < x^{\alpha'}} \pi_-\left(\frac{x}{p}, p\right)\,, \tag{16.23}$$

and

$$\pi_-(x, x^{\alpha'}) = \pi_-(x, x^\alpha) - \sum_{x^\alpha \le p < x^{\alpha'}} \pi_+\left(\frac{x}{p}, p\right)\,. \tag{16.23'}$$

Here both formulas are valid for $0 < \alpha < \alpha' \le 1/2$. (16.23′) gives with $\alpha' = 1/2$ and $1/4 < \alpha < 1/2$, that

$$\begin{aligned}\pi_-(x, x^\alpha) &= \sum_{x^\alpha \le p < \sqrt{x}} \pi_+\left(\frac{x}{p}, p\right) + 2\\ &= 2\sum_{x^\alpha \le p < \sqrt{x}} \pi\left(\frac{x}{p}\right) - 2\sum_{x^\alpha \le p < \sqrt{x}} \pi(p) + O(1)\\ &= 2\sum_{x^\alpha \le p < \sqrt{x}} \pi\left(\frac{x}{p}\right) + O\left(\frac{x}{\log^2 x}\right)\,.\end{aligned}$$

Using (16.1), it is easy to show that the first term on the right hand side is asymptotic to

$$2\log\left(\frac{1-\alpha}{\alpha}\right)\frac{x}{\log x},$$

so that for $1/4<\alpha<1/2$ we have

$$\pi_-(x,x^\alpha)\sim 2\log\left(\frac{1-\alpha}{\alpha}\right)\frac{x}{\log x}. \tag{16.24}$$

By use of (16.23) and (16.23′) one now finds in general for $0<\alpha<1/2$ that

$$\pi_+(x,x^\alpha)\sim a_+(\alpha)\frac{x}{\log x}, \tag{16.25}$$

and

$$\pi_-(x,x^\alpha)\sim a_-(\alpha)\frac{x}{\log x}, \tag{16.25′}$$

where

$$a_+(\alpha')=a_+(\alpha)-\int_\alpha^{\alpha'}a_-\left(\frac{u}{1-u}\right)\frac{du}{u(1-u)}, \tag{16.26}$$

and

$$a_-(\alpha')=a_-(\alpha)-\int_\alpha^{\alpha'}a_+\left(\frac{u}{1-u}\right)\frac{du}{u(1-u)}, \tag{16.26′}$$

These formulas are valid for $0<\alpha\le\alpha'\le 1/2$, if we take $a_+(u)=2$ for $1/3\le u<1$, and $a_-(u)=0$ for $u\ge 1/2$.

Writing

$$e(\mathcal{P}(x^\alpha))=\prod_{p\le x^\alpha}\left(1-\frac{1}{p}\right)\sim\frac{e^{-\gamma}}{\alpha\log x},$$

we recast (16.25) and (16.25′) as

$$\pi_+(x,x^\alpha)\sim a_+^*(\alpha)xe(\mathcal{P}(x^\alpha)), \tag{16.27}$$

and

$$\pi_-(x,x^\alpha)\sim a_-^*(\alpha)xe(\mathcal{P}(x^\alpha)), \tag{16.27′}$$

where now

$$a_\pm^*(\alpha)=e^\gamma\alpha\,a_\pm(\alpha).$$

Rewriting (16.26) and (16.26′) in terms of the a_+^* and a_-^* and differentiating we get

$$\frac{d\alpha^{-1}a_+^*(\alpha)}{d\alpha}=-\alpha^{-2}a_-^*\left(\frac{\alpha}{1-\alpha}\right), \tag{16.28}$$

and

$$\frac{d\alpha^{-1}\,a_{-}^{*}(\alpha)}{d\alpha} = -\alpha^{-2}a_{+}^{*}\left(\frac{\alpha}{1-\alpha}\right). \tag{16.28$'$}$$

Here, for $1/3 \le \alpha < 1$, we interpret the right hand side of (16.28) as zero, while (16.28′) is valid for $\alpha < 1/2$. Thus we see that $a_{+}^{*}(\alpha)$ and $a_{-}^{*}(\alpha)$ satisfy the same relations as $t_{1}^{+}(\alpha)$ and $t_{1}^{-}(\alpha)$ with $\beta = 2$ according to Lemma 9 of Section 11.

Remembering the definition of $W_{+}(x)$ and $W_{-}(x)$ and (16.10) of Lemma 11, we see that for $d < x$

$$\begin{aligned}\sum_{\substack{1\le n\le x\\ d|n}} w_{\pm}(n) &= \frac{x}{d} + O(1) + O\left(\sum_{\substack{d|n\\ 1\le n\le x}} \lambda(n)\right)\\ &= \frac{x}{d} + O(1) + O\left(\sum_{1\le n\le x/d} \lambda(n)\right)\\ &= \frac{x}{d} + O\left(\frac{x}{d\,(\log(x/d))^{r}}\right),\end{aligned}$$

for any fixed positive r. This, combined with Theorem 11, shows that

$$T_{1}^{-}(\alpha) \le a_{\pm}^{*}(\alpha) \le T_{1}^{+}(\alpha),$$

thus also, with $\beta = 2$,

$$t_{1}^{-}(\alpha) \le a_{\pm}^{*}(\alpha) \le t_{1}^{+}(\alpha).$$

This shows that the $a_{\pm}^{*}(\alpha)$ tend to 1 at least as strongly as $t_{1}^{\pm}(\alpha)$ as $\alpha \to 0$. Since the relations in Lemma 9 together with the strong convergence to 1 as $\alpha \to 0$, determine $t_{1}^{\pm}(\alpha)$ uniquely, and the $a_{\pm}^{*}(\alpha)$ satisfy the same conditions, we must clearly have, with $\beta = 2$,

$$a_{\pm}^{*}(\alpha) = t_{1}^{\pm}(\alpha).$$

Thus for $k = 1$ and $\beta = 2$ the B^2R sieve is optimal and we have

$$t_{1}^{\pm}(\alpha) = T_{1}^{\pm}(\alpha) = a_{\pm}^{*}(\alpha), \tag{16.29}$$

for $0 < \alpha \le 1/2$.

With the upper sign it is clear that (16.29) also holds for $1/2 \le \alpha < 1$.

For $1/2 < \alpha < 1$ however, both $t_{1}^{-}(\alpha)$ and $T_{1}^{-}(\alpha)$ are negative,[41] while for our model sequence $W_{-}(x)$ we have $a_{-}^{*}(\alpha) = 0$ for $1/2 < \alpha < 1$. It is still an open question whether the equation

$$t_1^-(\alpha) = T_1^-(\alpha)\,,$$

is valid for $1/2 < \alpha < 1$.

We can formulate the main results of this Section as follows.

Theorem 13. *The B^2R sieve is optimal for $k = 1/2$ if we choose $\beta = 1$, and we have then*

$$t_{1/2}^{\pm}(\alpha) = T_{1/2}^{\pm}(\alpha)\,,$$

for $0 < \alpha \leq 1$.

For $k = 1$ the B^2R sieve is optimal with the choice $\beta = 2$, we then have

$$t_1^{\pm}(\alpha) = T_1^{\pm}(\alpha)$$

in the case that $0 < \alpha \leq 1/2$ for the lower sign, and when $0 < \alpha \leq 1$ for the upper sign.

For $k = 1$ the sifting limit $\alpha_1 = 1/2$.

17. Some upper bounds for sifting limits for constant sifting density

As we have seen, $\alpha_{1/2} = 1$ (and of course $\alpha_k = 1$ for $0 < k < 1/2$ also), while $\alpha_1 = 1/2$. For no other values of k do we know the value of α_k, it is also unknown whether the $\mathrm{B}^2\mathrm{R}$ method leads to optimal results in any case except $k = 1/2$ and $k = 1$. The special sieves constructed in earlier sections do of course lead to lower bounds for the sifting limit α_k also for other values of k as was discussed in Section 14. In this Section we wish to establish an upper bound for α_2 and a general upper bound for α_k.

We shall need some lemmas.

---

[41] This is fairly obvious since for any sieve $\Lambda^-(\mathcal{P}(x^\alpha)) = \Lambda^-[\mathcal{P}(x^\alpha), x]$ we may consider its restriction to $\mathcal{P}(x^{1/2})$, and we have then clearly

$$T(\Lambda^-(\mathcal{P}(x^\alpha))) \leq T(\Lambda^-(\mathcal{P}(x^{1/2}))) + \sum_{\sqrt{x}<p<p'<x^\alpha} \frac{\theta_{pp'}}{f'(pp')}\,,$$

here, as $x \to \infty$ the first term on the right hand side could at most approach zero or $T_1^-(1/2)$ while in the second term the $\theta_{pp'} = 1 + \lambda_p + \lambda_{p'} \leq -1$ since each $\lambda_p \leq -1$. Thus the second term could at most approach $-\frac{1}{2}\,(\log 2\alpha)^2$, so we get

$$T_1^-(\alpha) \leq -\frac{1}{2}\,(\log 2\alpha)^2\,.$$

Lemma 12. *If $\tau(n)$ denotes the divisor function, we have*

$$\sum_{1\le n\le x} \tau(n) = x(\log x + c) + O(\sqrt{x}), \tag{17.1}$$

where c is a constant. We also have

$$\sum_{1\le n\le x} \tau(n)\,\lambda(n) = O\left(\frac{x}{(\log x)^r}\right), \tag{17.2}$$

for any fixed r.

Here (17.1) is well known, so we give no proof. For (17.2) we write the left hand side as

$$\begin{aligned}\sum_{1\le mn\le x} \lambda(m)\lambda(n) &= 2 \sum_{1\le m<n\le x/m} \lambda(m)\lambda(n) + [\sqrt{x}] \\ &= 2 \sum_{1\le m<\sqrt{x}} \lambda(m) \sum_{m<n\le x/m} \lambda(n) + O(\sqrt{x}) \\ &= O\left(\sum_{1\le m<\sqrt{x}} \frac{x}{m\,(\log x)^{r+1}}\right) + O(\sqrt{x}) \\ &= O\left(\frac{x}{(\log x)^r}\right).\end{aligned}$$

Here we have made use of Lemma 11 with $r+1$ instead of r.

Lemma 13. *We have, if $\nu(n)$ denotes the number of prime factors of n, counting them with multiplicities, that*

$$\sum_{\substack{1\le n\le x\\ n\equiv 1\ (2)}} 2^{\nu(n)} = c_1\,x(\log x + c_2) + O(x^{2/3}), \tag{17.3}$$

where c_1 and c_2 are constants. We also have

$$\sum_{\substack{1\le n\le x\\ n\equiv 1\ (2)}} 2^{\nu(n)}\lambda(n) = O\left(\frac{x}{(\log x)^r}\right) \tag{17.4}$$

for any fixed r.

If we compare the two Dirichlet series

$$f(s) = \sum_{n\equiv 2\ (2)} \frac{2^{\nu(n)}}{n^s} = \prod_{p>2}\left(1-\frac{2}{p^s}\right)^{-1}, \tag{17.5}$$

and

$$\zeta^2(s) = \sum_n \frac{\tau(n)}{n^s} = \prod_p \left(1-\frac{1}{p^s}\right)^{-2},$$

we see that

$$f(s) = \zeta^2(s)g(s) \tag{17.6}$$

where

$$g(s) = \sum_n \frac{b_n}{n^s} = (1-2^{-s})^2 \prod_{p>2} \frac{\left(1-\frac{1}{p^s}\right)^2}{1-\frac{2}{p^s}}. \tag{17.7}$$

We see that the series

$$\sum \frac{|b_n|}{n^\sigma}$$

converges for $\sigma > \log 2/\log 3$, thus certainly

$$\sum_n \frac{|b_n|}{n^{2/3}} \quad \text{and} \quad \sum_n \frac{|b_n|}{n^{2/3}} \log n$$

both converge. Denote the left hand side of (17.3) by $s(x)$ and the left hand side of (17.1) by $t(x)$, (17.6) gives

$$s(x) = \sum_n b_n t\left(\frac{x}{n}\right). \tag{17.8}$$

Since we for all n have, by (17.1), that

$$t\left(\frac{x}{n}\right) = \frac{x}{n}\left(\log\frac{x}{n} + c\right) + O\left(\left(\frac{x}{n}\right)^{2/3}\right),$$

we can insert this in the formula above, and get, since all resulting series converge absolutely

$$s(x) = g(1)\, x(\log x + c) + g'(1)\, x + O(x^{2/3}) \tag{17.9}$$

which proves (17.3).

To prove (17.4), call the left hand side of (17.4) $s_1(x)$ and the left hand side of (17.2) $t_1(x)$, it is then easy to show that in analogy with (17.8) we have

$$s_1(x) = \sum_{n\le x} b_n \lambda(n) t_1\left(\frac{x}{n}\right),$$

here we divide the sum into two parts: $n \le \sqrt{x}$ and $n > \sqrt{x}$. For the first part we get, using (17.2)

$$O\left(\frac{x}{(\log x)^r}\sum_n \frac{|b_n|}{n}\right) = O\left(\frac{x}{(\log x)^r}\right),$$

while for the second part using just the bound $\tau_1(x/n) = O(x/n)$, we get

$$O\left(x\sum_{n>\sqrt{x}} \frac{|b_n|}{n}\right) = O\left(x^{5/6}\sum_n \frac{|b_n|}{n^{2/3}}\right) = O(x^{5/6}).$$

Together this gives (17.4).

Suppose now that we have for the standard sifting problem with $u(p) = 2$ for all primes $p > 2$ a lower bound sieve $\Lambda^-(\mathcal{P}(x^\alpha))$ which gives a non negative lower bound, it will then also do so if we omit $p = 2$ from the sifting range (or by defining $u(2) = 0$). That is to say, we have, since for d odd $u(d) = \tau(d)$ when d is squarefree,

$$(17.10) \qquad x \sum_d \frac{\tau(d)}{d} \lambda_d - \sum_d \tau(d)\, |\lambda_d| \geq 0$$

so certainly

$$(17.11) \qquad \sum_d \frac{\tau(d)}{d} \lambda_d = e_u(\mathcal{P}(x^\alpha)) \sum_d \frac{\theta_d}{f'(d)} > 0\,,$$

where

$$f'(p) = \frac{p}{2} - 1\,.$$

Also, since the first term in (17.10) by (17.11) is less than $x\, e_u(\mathcal{P}(x^\alpha)) = O(x/\log^2 x)$ for $\alpha \geq \alpha_0 > 0$, we must have

$$(17.12) \qquad \sum_d \tau(d)\, |\lambda_d| = O\left(\frac{x}{\log^2 x}\right).$$

We now wish to apply the same sieve to another set.

We define weights $w(n)$ as follows for $x_1/2 \leq n < x_1$, where x_1 will be chosen later somewhat larger than x, we put

$$(17.13) \qquad w(n) = \frac{1 - (-1)^n}{2} \left(2^{\nu(n)}\, (1 - \lambda(n)) + 16\, \lambda(n)\right)$$

for $x_1/2 \leq n < x_1$, and $w(n) = 0$ outside this interval, and observe that $w(n) = 0$ for all even n, while for odd n, we have

$$(17.14) \qquad w(n) = \begin{cases} -12, & \text{for } \nu(n) = 1, \\ 16, & \text{for } \nu(n) = 2, \\ 0, & \text{for } \nu(n) = 3, \\ > 0, & \text{for } \nu(n) > 3. \end{cases}$$

$\nu(n) = 0$ does not occur in the range $x_1/2 \leq n < x_1$ for $x_1 > 2$. We shall assume $1/4 < \alpha < 1/2$, and sift out from the interval the $w(n)$ for which n is divisible by a prime from the sifting range $\mathcal{P}(x^\alpha)$. That we can use our sieve and obtain a valid lower bound for what remains after the sifting in spite of the fact that some $w(n)$ are negative, is due to the fact that the negative $w(n)$ correspond to n's which are primes and since we assume $x_1 > 2x$, these primes lie outside the sifting range and so are counted with their true weights since $\lambda_1 = 1$. We shall assume $x_1 < x^{4\alpha}$, and see then (since any $n < x_1$ with

$\nu(n) > 3$ must have a prime factor $< x^\alpha$) that what remains after the sifting is

$$-12\left(\pi(x_1) - \pi\left(\frac{x_1}{2}\right) + 16\left(\pi_2(x_1, x^\alpha) - \pi_2\left(\frac{x_1}{2}, x^\alpha\right)\right)\right), \tag{17.15}$$

where we have used the symbol $\pi_2(x, x^\alpha)$ to denote the number of integers $< x$, which are products of two primes (distinct or not) both of which are $\geq x^\alpha$. The same argument which gave (16.24) gives

$$\pi_2(x, x^\alpha) \sim \frac{x}{\log x} \log \frac{1-\alpha}{\alpha} \qquad \text{for } \frac{1}{4} \leq \alpha < \frac{1}{2}.$$

Since we will choose $x_1 = x^{1+\varepsilon}$ where $\varepsilon = \varepsilon(x)$ is positive and tending to zero as x goes to infinity,[42] we get that the quantity (17.15) is

$$\left(-6 + 8\log\frac{1-\alpha}{\alpha}\right)\frac{x_1}{\log x} + O\left(\frac{\varepsilon x_1}{\log x}\right). \tag{17.16}$$

We now see what we get by applying the sieve $\Lambda^-(\mathcal{P}(x^\alpha))$ to our weighted set, first, for $d \in (\mathcal{P}(x^\alpha))$, we have

$$\begin{aligned}
\sum_{\substack{x_1/2 \leq n < x_1 \\ d|n}} w(n) &= \sum_{\substack{x_1/2 \leq n < x_1 \\ d|n \\ n\equiv 1\,(2)}} 2^{\nu(n)} - \sum_{\substack{x_1/2 \leq n < x_1 \\ d|n \\ n\equiv 1\,(2)}} 2^{\nu(n)}\lambda(n) + 16 \sum_{\substack{x_1/2 \leq n < x_1 \\ d|n \\ n\equiv 1\,(2)}} \lambda(n) \\
&= 2^{\nu(d)} \sum_{\substack{x_1/(2d) \leq n < x_1/d \\ n\equiv 1\,(2)}} 2^{\nu(n)} \\
&\quad - 2^{\nu(d)} \sum_{\substack{x_1/(2d) \leq n < x_1/d \\ n\equiv 1\,(2)}} 2^{\nu(n)}\lambda(n) + 16\lambda(d) \sum_{\substack{x_1/(2d) \leq n < x_1/d \\ n\equiv 1\,(2)}} \lambda(n).
\end{aligned}$$

The first term in the last expression is estimated by using (17.3), the second term by using (17.4) and the last term by using (16.10), in this way we get

$$\sum_{\substack{x_1/2 \leq n < x_1 \\ d|n}} w(n) = \frac{\tau(d)}{d} x_1 \left(c_1 \log\frac{x_1}{d} + c_2\right) + R_d, \tag{17.17}$$

where if $d \leq x_1$

$$R_d = O\left(\frac{\tau(d)}{d} x_1 \left(1 + \log\frac{x_1}{d}\right)^{-r}\right), \tag{17.18}$$

and for $d > x_1$

$$R_d = O\left(\frac{\tau(d)}{d} x_1 \left(1 + \left|\log\frac{x_1}{d}\right|\right)\right). \tag{17.18'}$$

We have here written $\tau(d)$ instead of $2^{\nu(d)}$ since d is squarefree.

[42] But slowly enough so that $\varepsilon(x)\log x$ goes to infinity with x.

Writing $c_2/c_1 = c_3$, we can now write the result of applying the sieve as

$$c_1 x_1 \sum_d \frac{\tau(d)}{d}\left(\log\frac{x_1}{d} + c_3\right)\lambda_d - \sum_d |\lambda_d|\,|R_d|. \tag{17.19}$$

From (3.3) we have

$$\sum_d \frac{\tau(d)}{d}\,(\log x_1 + c_3)\,\lambda_d = (\log x_1 + c_3)\, e_u(\mathcal{P}(x^\alpha)) \sum_d \frac{\theta_d}{f'(d)}. \tag{17.20}$$

Also from (3.3) we have if s is a variable

$$\sum_d \tau(d)\, d^{s-1}\lambda_d = \prod_{p\in\mathcal{P}(x^\alpha)} (1 - 2p^{s-1}) \sum_d \frac{\theta_d}{f_s'(d)},$$

where $f_s(p) = p^{1-s}/2$ and $f_s'(p) = p^{1-s}/2 - 1$. Differentiating with respect to s and putting $s = 0$, we get

$$\sum_d \frac{\tau(d)}{d}\lambda_d \log d = e_u(\mathcal{P}(x^\alpha)) \sum_d \frac{\theta_d}{f'(d)} \left\{ - \sum_{p\in\mathcal{P}(x^\alpha)} \frac{2\log p}{p-2} + \sum_{p|d} \frac{p}{p-2}\log p \right\}.$$

Combining this with (17.20) we get for the first term in (17.19), if we leave out the factors $c_1 x_1 e_u(\mathcal{P}(x^\alpha))$, the expression

$$\sum_d \frac{\theta_d}{f'(d)} \left\{ \log x_1 + c_3 + 2 \sum_{p\in\mathcal{P}(x^\alpha)} \frac{\log p}{p-2} - \sum_{p|d} \frac{p}{p-2}\log p \right\}. \tag{17.21}$$

Since the θ_d for $d > 1$ are non positive, we see that this is greater or equal to

$$\left(\log x_1 + c_3 + 2 \sum_{p\in\mathcal{P}(x^\alpha)} \frac{\log p}{p-2} \right) \sum_d \frac{\theta_d}{f'(d)}, \tag{17.22}$$

and thus certainly positive according to (17.11).

We now choose $\varepsilon = \varepsilon(x) = 2/\sqrt{\log x}$, and look at the remainder term in (17.19), we put $x_2 = \sqrt{x\, x_1} = x^{1+\varepsilon/2}$, and divide the sum

$$\sum_d |\lambda_d|\,|R_d|$$

into two parts, one with $d \le x_2$ and the second with $d > x_2$. For the first part we have that it is

$$\begin{aligned} O\left(\sum_{d\le x_2} \tau(d)\,|\lambda_2|\, \frac{x_1}{d(\log(x_1/d))^r} \right) &= O\left(\frac{x_1}{(\log(x_1/x_2))^r} \sum_d \frac{\tau(d)\,|\lambda_d|}{d} \right) \\ &= O\left(\frac{x_1}{(\log x)^{r/2}} \,(e_u(\mathcal{P}(x^\alpha))^{-1}) \right) \\ &= O\left(\frac{x_1}{(\log x)^{r/2-2}} \right), \end{aligned} \tag{17.23}$$

where we have made use of (17.18), (5.5) and the fact that we assumed $T_u(\Lambda^-(\mathcal{P}(x^\alpha))) \geq 0$. In the second part, where $d > x_2$, we estimate R_d by (17.18) or (17.18′) according to whether $x_2 < d \leq x_1$ or $d > x_1$, we see that in the first case

$$R_d = O\left(\tau(d)\,\frac{x_1}{x_2}\left(\log\frac{x_1}{x_2}\right)^{-r}\right),$$

and in the second case

$$R_d = O(\tau(d)) = O\left(\tau(d)\,\frac{x_1}{x_2}\left(\log\frac{x_1}{x_2}\right)^{-r}\right),$$

since r is fixed. Inserting this, and using (17.12) we get for the second part

$$O\left(\frac{x_1 x}{x_2}(\log x)^{-r/2-2}\right) = O\left(\frac{x_1}{(\log x)^r}\right).$$

Thus, combining this with (17.23) we get

$$\sum_d |R_d|\,|\lambda_d| = O\left(\frac{x_1}{(\log x)^r}\right), \tag{17.24}$$

for any fixed r.

From (17.11) it follows that (17.22) is positive and a fortiori (17.21) and so the main term of (17.19). But (17.19) is less than or equal to (17.16), and since the only negative term in (17.19) (namely (17.24)) is of much smaller order than $x_1/\log x$ it is clear that we must have

$$-6 + 8\,\log\frac{1-\alpha}{\alpha} \geq 0\,. \tag{17.25}$$

This gives

$$\alpha \leq \frac{1}{1+e^{3/4}} = \frac{1}{3.117\ldots}\,.$$

And so the result

$$\alpha_1 \leq \frac{1}{1+e^{3/4}}\,. \tag{17.26}$$

It is not hard to get rid of the equality sign in (17.25). When we replaced (17.21) by (17.22), we threw away something that is not only positive, but also significant in relation to the whole. The expression in the curly bracket in (17.21) is essentially

$$(1+2\alpha)\,\log x - \log d\,,$$

in (17.22) we replaced this by essentially $(1+2\alpha)\,\log x$, which would not matter much if in the sum $\sum_d \theta_d/f'(d)$ it would be only the "small" d where

$\log d = o(\log x)$ that counted. Obviously this is not the case since we otherwise, by restricting our sieve to the narrower range $\mathcal{P}(x^\eta, x^\alpha)$ with some small η, would get a sieve which because

$$T_u(\Lambda^-(\mathcal{P}(x^\eta, x^\alpha))) > 1 + \sum_{d > x^\eta} \frac{\theta_d}{f'(d)},$$

would have a T_u very close to 1. This is clearly absurd. We must have $\sum_{d>x^\eta} \theta_d / f'(d) < -\delta$ where δ is a small positive constant, and so the sum

$$\sum_d \frac{\theta_d}{f'(d)} \log d < -\eta\delta \log x .$$

This would lead us, instead of to (17.25) to the condition

$$-6 + 8 \log \frac{1-\alpha}{\alpha} \geq \delta',$$

where δ' is a small positive constant, and so to a somewhat smaller upper bound than (17.26). More precisely, we may choose the δ above as any number less than $1 - T_2^-(\eta, \alpha)$. If we do this, we arrive in the end at the inequality

$$-6 + 8 \log \frac{1-\alpha}{\alpha} \geq e^{-2\gamma} \alpha^{-2} \eta (1 - T_2^-(\eta, \alpha)), \tag{17.27}$$

instead of (17.25).

The improvements in the upper bound for α_2 one can get from (17.27) are rather insignificant. If we choose $\eta = 1/5$ and assume $\alpha \leq 1/(1 + e^{3/4})$, it is possible to show that

$$T_2^- \left(\frac{1}{5}, \alpha\right) = (5\alpha)^2 \left(1 - 2 \log 5\alpha + 2 (\log 5\alpha)^2 - \frac{4}{3} (\log 5\alpha)^3\right).$$ [43]

From this one can show

$$\alpha_2 < \frac{1}{3.128} < \frac{8}{25}. \tag{17.28}$$

While it is possible to refine this method by making a more accurate assessment of

$$\sum_d \frac{\theta_d \log d}{f'(d)},$$

it becomes too complicated compared to the extremely small further gains.

While it is possible to adapt the method used here to obtain upper bounds for α_k for larger integral values of k, it becomes already for $k = 3$ much more complicated and would not seem to offer any prospect of giving a bound for

[43] This is given by the combinatorial sieve with $\lambda_p = -1$, $\lambda_{pp'} = 1$ and $\lambda_{pp'p''} = -1$, which in this case is optimal.

general large k which would show that $k\,\alpha_k$ remains bounded as k goes to infinity.

To obtain such a result, and in a fairly simple way, we go back to the ideas in Section 13. We got there in particular the result $v_R < R$ for $R > 1$ contained in (13.22″). This result is in a sense the analogue for the problem of v_R to the result we are seeking for α_k, and we will try to make some use of this.

Let us assume, for $k > 1$, that we have $\alpha_k > 1/h$, when h is an integer > 1. We have then

$$T_k^-\left(\frac{1}{h},\alpha_k\right) > 1 - T_k^-\left(\frac{1}{h}\right) > 0\,.^{44} \tag{17.29}$$

This means that we can give a sequence of sieves $\Lambda_x^-[\mathcal{P}(x^{1/h},x^{\alpha_k}),x]$ as $x\to\infty$ so that with $u(p)=k$,

$$\lim_{x\to\infty} T_u(\Lambda_x^-) = T_k^-\left(\frac{1}{h},\alpha_k\right) > 0\,.$$

Denoting the range $\mathcal{P}(x^{1/h},x^{\alpha_k})$ for brevity with $\mathcal{P}_x$, we now form the quantities

$$\lambda_x(0) = 1\,, \tag{17.30}$$

and for $r \geq 1$,

$$\lambda_x(r) = \frac{\displaystyle\sum_{p_i\in\mathcal{P}_x}\frac{\lambda_{p_1\cdots p_r}}{p_1\cdots p_r}}{\displaystyle\sum_{p_i\in\mathcal{P}_x}\frac{1}{p_1\cdots p_r}}\,, \tag{17.30′}$$

where the $\lambda_{p_1\cdots p_r}$ belong to the Λ_x^-. As when we considered (13.3), we see we can take out a subsequence of $x\to\infty$ such that the quantities $\lambda_x(r)$ tend to limits $\lambda(r)$ as $x\to\infty$ through this subsequence, obviously we have $\lambda(r)=0$ for $r\geq h$.

We have then with these $\lambda(r)$, that

$$\sum_{r=1}^{\infty}\frac{\theta(r)}{r!}(k\log h\alpha_k)^r = T_k\left(\frac{1}{h},\alpha_k\right) > 0\,.$$

Therefore we must have

$$k\log h\alpha_k < v_{h-1} < h-1\,,$$

[44] This follows from the general inequality for $0<\eta<\alpha$, $T_k^-(\eta)+T_k^-(\eta,\alpha) > 1 + T_k^-(\alpha)$, which is essentially trivial.

from which we derive

$$\alpha_k < \frac{1}{h}\, e^{(h-1)/k}. \tag{17.31}$$

If k is an integer the optimal choice of h is $h = k$, so that in that case we get

$$\alpha_k < \frac{e^{1-1/k}}{k}. \tag{17.31$'$}$$

If k is not an integer the best choice of h is either the nearest integer $\le k$ or the nearest integer above k. In all cases we easily see we have

$$\alpha_k < \frac{e}{k}. \tag{17.31$''$}$$

For k an odd integer we get if we observe that $h - 1$ then is even so $v_{k-1} = v_{k-2} < k - 2$, that instead of (17.31′) we in that case get

$$\alpha_k < \frac{e^{1-2/k}}{k}$$

for k odd and > 1.

Actually (17.31) is easily replaced by

$$\alpha_k < \frac{1}{h} e^{(h-2)/k}, \tag{17.32}$$

where now $h \ge 3$ is chosen as an odd integer, when we just remember from Section 13 that the optimal polynomial $\theta(r)$ for the lower bound is always of odd degree. It is not hard to see that (17.32) implies that

$$\alpha_k < \frac{1}{k}\, e^{(1-1/k)^2}, \tag{17.32$'$}$$

for all $k > 1$ (and so of course, for all k).

These results can be improved somewhat further by: (a) improving the estimate $v_R < R$, (b) improving the inequality (17.29). This leads to a lot of complicated detail work, with (so far) only very meager results to show for it.

We can finally combine this result with our result (14.40) for large k into

Theorem 14. *We have for all* $k \ge 1$

$$\alpha_k < \frac{1}{k}\, e^{(1-1/k)^2},$$

and for all sufficiently large k

$$\alpha_k > \frac{1}{2k + 0.4454}.$$

18. A historical digression, the parity principle and a further example

Looking at the weighted sets constructed in Section 16 in order to prove Theorem 13 and the use of the Liouville function $\lambda(n)$ there we see we made use of two essential facts: (1) The integers $1 \le n \le x$ are for large x quite evenly distributed between the class of numbers with $\lambda(n) = 1$ (or $\nu(n)$ even) and the class of n with $\lambda(n) = -1$ (or with $\nu(n)$ odd), and the same property is inherited when we look at the subsets of n which are divisible by d as long as x/d is not small. (2) The sieve method is as we saw in Section 15 quite tolerant of cruder remainder terms R_d in the expression for N_d. To this comes of course that a sieve in itself contains no feature able to make a distinction between the two classes of integers. It follows from this, that if we had not made the distinction by selecting our weighted sets with a built in bias towards odd or even $\nu(n)$, the result of applying the sieve would have been about evenly divided between the contribution of the two classes of integers.[45]

I first ran into this phenomenon in 1946 when I was exploring the Λ^2 method, first in connection with upper bounds, as for instance in the result

$$\pi(x+y) - \pi(y) \le \frac{2y}{\log y} + O\left(\frac{y}{\log^2 y}\right), \tag{18.1}$$

which I obtained very early, and then trying in various ways to adapt it so as to give me lower bounds. I did not at the time take much notice of the factor 2 in (18.1) (except to note that it was an improvement on what others had obtained), the thing that made me take notice and changed the focus of my interest in the sieve as I tried to understand it, was the following:

In the fall of 1946 I was considering the two expressions

$$Q_1(\lambda) = \sum_{x \le n \le (1+c)\,x} \left(\sum_{d|n} \lambda_d\right)^2,$$

and

$$Q_2(\lambda) = \sum_{x \le n \le (1+c)\,x} \tau(n) \left(\sum_{d|n} \lambda_d\right)^2,$$

where c is a positive constant and $\tau(n)$ the divisor function. As usual I put $\lambda_1 = 1$ and let the other λ's to be disposed of in the most favorable way. My goal was to see how small I could make the ratio $Q_2(\lambda)/Q_1(\lambda)$.

[45] As one easily convinces oneself, the same is *not* true if we divide the integers in separate classes according to the residue of $\nu(n)$ (mod q) for $q > 2$.

In $Q_2(\lambda)$ the main term is essentially (I leave out some additive functions which are insignificant compared to the logarithm),

$$Q_2'(\lambda) = cx \sum_{d,d'} \tau\left(\frac{dd'}{\kappa}\right) \frac{\kappa}{dd'} \lambda_d \lambda_{d'} \log \frac{x\kappa}{dd'},$$

where $\kappa = (d, d')$. This cannot be brought entirely in diagonal form by our usual transformation, but if we take just $\log x$ instead of $\log \frac{x\kappa}{dd'}$, it is easily diagonalized and minimized. If we require $\lambda_d = 0$ for $d > \sqrt{x}$, say, in order to keep remainder terms small enough, we get

$$\lambda_d \approx \mu(d) \left(1 - 2\frac{\log d}{\log x}\right)^2$$

for $1 \leq d \leq \sqrt{x}$. Inserting these values in $Q_2(\lambda)$ and $Q(\lambda_1)$. I could show that with this choice

$$\frac{Q_2(\lambda)}{Q_1(\lambda)} = 4 + O\left(\frac{1}{\log x}\right). \tag{18.2}$$

This of course shows that there are a significant number of n in the interval with $\tau(n) \leq 4$, that is: primes or products of two primes, since the prime squares (with $\tau(n) = 3$) are too few to contribute.

Since the ratio (18.2) was attained, not by trying to minimize Q_2/Q_1 but rather by just minimizing the most significant part of Q_2, I felt rather sure that the real minimum would turn out to be < 4, and so the presence of primes in the interval $x \leq n \leq (1+c)\,x$ for any fixed $c > 0$ and for sufficiently large x could be established. This seemed to me an interesting result, if it could be obtained by elementary means, so I put in a considerable amount of work trying to get at the actual minimum of the ratio. Eventually, and to my surprise, I found I could not bring the ratio below 4.

I then started analyzing what contributions the various classes of integers: primes, products of two primes, of three primes etc. actually made to the expressions Q_1 and Q_2, and noticed that my initial choice of λ_d essentially eliminated the contributions of the n with $\nu(n)$ even except for those with $\nu(n) = 2$ ($\nu(n) = 0$ does not occur in the interval considered) to both Q_2 and Q_1. At the same time I found both Q_2 and Q_1 rather evenly divided between the contributions coming from n with $\nu(n)$ even and those coming from n with $\nu(n)$ odd. Thus, the ratio clearly had to come out about 4. Also if the λ's were changed in any way that made the contribution of the n with $\nu(n) > 2$ and even become non negligible, the ratio would become larger so that the number 4 in (18.2) would have to be replaced by a larger number.

I now also understood why the constant 2 appeared in (18.1), and that it could not be replaced by a smaller one.

This is what started my interest in investigating the limitations of sieve methods in general. All of this work was motivated by and guided by, what

I when I finally decided to give it a name (in the mid seventies, when I did lecture on some of this material) called "the principle of parity."

This principle can be formulated as follows:

Sets of integers tend to be very evenly distributed with respect to the parity of their number of prime factors unless they have been particularly produced, constructed or selected in a way that has a built in bias. The same goes for sets of r–tuples $(n_1, \ldots, n_r)$ *when we divide them into* 2^r *classes according to the parities of* $\nu(n_1), \ldots, \nu(n_r)$*, if there in addition is no linking between the parities of the members of the r–tuples.*

This principle can of course not be "proved" to be valid in general. But it provides a good guideline for assessing what a particular approach via a sieve method can give, as well as for constructing models which show that certain results cannot be reached by direct application of a sieve whether it be the classical sifting or sifting with weights.

Once one realizes these limitations, it is also possible to exploit them. It was my, yet rather vague, understanding of the parity principle which gave me the idea of devising a "local" sieve[46] that would between 1 and x count only the numbers with one or two prime factors and eliminate the others. This then led to the elementary proof of Dirichlet's theorem on arithmetic progressions, as well as to the elementary proof of the prime number theorem. While I experimented with several such local sieves (all of which lead to asymptotic results when applied to the interval $(1, x)$ or $(x, 2x)$ say), I only published what I considered the simplest. Later the idea was developed much further, first particularly by Diamond in connection with improving the remainder term in the elementary proof of the prime number theorem, and then in a much more general way by Bombieri, who in what he called the asymptotic sieve (because it leads to asymptotic formulas) developed the local sieve into a very versatile and powerful tool.

We shall not be dealing with local sieves in this exposition. Returning to the topic of the last few sections, we shall construct a model designed to illustrate limitations in a problem of sifting with weights. The problem is the following:

Suppose we have given a set of positive integers n_i with associated weights w_i such that all $n_i \leq x$ and such that if we put

$$y = \sum_i w_i\,,$$

we have

$$\sum_{d\,|\,n_i} w_i = \frac{u(d)}{d}\, y + R_d\,,$$

[46] By a local sieve, we mean a Λ system that works as a sieve only on sets of numbers with absolute values below some specific bound.

where the $u(d)$ correspond to sifting density 1 in a strong sense, say

$$\sum_{p\le x^\alpha} \frac{u(p)}{p} \log p = \alpha \log x + O(1),$$

and the R_d satisfy conditions such that sifting may be carried out with a sieve $\Lambda^-[\mathcal{P}(x^\alpha), \sigma, z]$ as long as z is somewhat smaller than $\sqrt{x}$. Is it then possible to choose the weights σ and the Λ^- in such a way that we can prove that there are n_i which do not contain more than 2 prime factors?

This relates to several classical problems, most obviously to the problem of trying to show that there are primes between x and $x+\sqrt{x}$ for x large, or that an irreducible second degree polynomial like n^2+1 represents infinitely many primes. Here one tries to attack the problem by showing that the statement is true if one replaces "prime" by "number with not more than r prime factors" and then tries to prove it with an r as small as possible.

Also the twin prime problem and the Goldbach problem lead to this type of question, if we formulate for instance the twin prime problem as the question whether the set of $p+2$, with $x/2 \le p < x$ for large x always contains primes.

That one for the two first named of these problems could answer them in the affirmative with $r = 3$, was realized fairly early.[47] For the two last named problems on the other hand, one could not attack them in this way without assuming the general Riemann hypothesis for the zeta–function and Dirichlet's L–functions, which were needed to show that the R_d did not behave too badly. Only after Bombieri's theorem[48] in 1965 could one attain $r = 3$ without any hypothesis.

The model we shall construct will be particularly designed to model the twin prime problem, but it could just as well be designed so as to mimic the conditions of N_d and R_d in the other problems mentioned.

Bombieri's result gives the following for the twin prime problem:

If we consider the set of numbers

$$p+2$$

where p runs through the primes $\le x$ and weight them with the weight $\log p$, and put, for d odd,

$$N_d = \sum_{\substack{d\,|\,p+2\\ p\le x}} \log p = \frac{x}{\varphi(d)} + R_d\,, \tag{18.3}$$

we have

$$\sum_{d\le z} |R_d| = O(x^{1/2} z\,(\log x)^5) \tag{18.4}$$

for $x^{1/2}(\log x)^{-A} \le z \le x^{1/2}$, where $A > 0$ is any fixed number.

---

[47] The result for n^2+1 with $r=3$ is first mentioned in a footnote in [Selberg 2, p. 17].

[48] [Bombieri 1]

In the construction of our model we shall for simplicity assume the Riemann hypothesis for the zeta-function. It is well known that then

$$\sum_{n\leq x} \lambda(n) = O(x^{1/2+\varepsilon}). \tag{18.5}$$

for any fixed $\varepsilon > 0$. The same estimate clearly holds if we sum only over the odd numbers n.

We now define a totally multiplicative function $g(n)$ defined as follows, $g(2) = 0$ and $g(p) = p/(p-1)$ for $p > 2$. We then have for n odd, that

$$g(n) = \sum_{\delta\,|\,n} g'(\delta),$$

where g' is a multiplicative function with $g'(p^r) = p^{r-1}/(p-1)^r$ for $r > 0$, $p > 2$. We then can easily show

$$\sum_{n\leq x} g(n) = cx + O(x^{1/2}),$$

where

$$c = \frac{1}{2}\prod_{p>2}\left(1 + \frac{1}{p(p-2)}\right),$$

and

$$\sum_{n\leq x} \lambda(n)g(n) = O(x^{1/2+\varepsilon}).$$

From this we derive, for d odd and squarefree,

$$\sum_{\substack{d\,|\,n\\ n\leq x}} (1+\lambda(n))\,g(n) = \frac{cx}{\varphi(d)} + O\left(\left(\frac{x}{d}\right)^{1/2+\varepsilon}\frac{d}{\varphi(d)}\right). \tag{18.6}$$

We now define a set of n as follows: $n = mp$, where $1 < m \leq \sqrt{x/2}$ and $\sqrt{x} < p \leq \sqrt{2x}$, we assign these n weights

$$w_n = (1+\lambda(m))\,g(m)\rho_p, \tag{18.7}$$

where the ρ_p for the primes p in $\sqrt{x} < p \leq \sqrt{2x}$ are > 0 and such that

$$\sum_{\sqrt{x}<p\leq\sqrt{2x}} \rho_p = \frac{1}{c}\sqrt{2x}.$$

We then have for $d \leq \sqrt{x}$

$$N_d = \sum_{d\,|\,n} w_n = \frac{x}{\varphi(d)} + R_d \tag{18.8}$$

with

$$R_d = O\left(\frac{x^{3/4+\varepsilon}\, d^{1/2-\varepsilon}}{\varphi(d)}\right) \tag{18.9}$$

so that for $z \le \sqrt{x}$

$$\sum_{d\le z} |R_d| = O(x^{3/4+\varepsilon}\, z^{1/2-\varepsilon}) . \tag{18.10}$$

While this is not quite identical to the form of (18.4), we see that for say z of the form $x^{1/2}(\log x)^{-A}$, (18.4) has the O–term

$$O\left(\frac{x}{(\log x)^{A-5}}\right),$$

while (18.10) gives

$$O\left(\frac{x}{(\log x)^{(1/2-\varepsilon)\,A}}\right).$$

Since $u(p) = p/(p-1)$ for $p > 2$ corresponds to $k = 1$ and $e_u(\mathcal{P}(x^\alpha)) \sim c'/(\alpha \log x)$, we see that we may actually carry the sifting slightly further with the condition (18.10) than with (18.4), or otherwise expressed we may apply sieves $\Lambda^-[\mathcal{P}(x^\alpha), \sigma, z]$ with a $z = x^{1/2}(\log x)^{-A}$ with a somewhat smaller A, and should therefore be able to do at least as well with the model weighted set as with the $p+2$ with the weight $\log p$. But the model set simply contains no number with less than 3 prime factors. *Thus it is impossible by application of a weighted sieve to prove that there are $p+2$ with less than 3 prime factors, based just on the information contained in (18.3) and (18.4).*

The same holds for the other problems mentioned above, we could have adapted our model to mimic one of these. Still it is a fact that for all of these problems an answer has been given in the affirmative. First by Jing–Run Chen who proved that there are infinitely many p for which $p+2$ has not more than two prime factors, and at the same time proved the corresponding result related to the Goldbach problem. He also settled the problem of establishing that there are for large x always numbers between x and $x + \sqrt{x}$ which have not more than two prime factors. Finally the problem of showing that n^2+1 represents infinitely many numbers with not more than two prime factors was settled by H. Iwaniec.

In general, the limitations we can establish for sieve methods are valid *only when no information beyond what is given about the N_d and $|R_d|$ is available.* In a specific "concrete" problem like those mentioned here, there is always the possibility of bringing some other information into the argument which may make it possible to transcend these limitations.

19. Sifting on an interval, the uses of Fourier analysis

The general setting in which we have developed the theory of the sieve, that of the general weighted set, while it permits us to make a nice theory, is not of course the setting in which sieves mostly are used. Mostly sieves are used in the original setting referred to at the beginning of Section 2, the interval of length x, where we for each p in our sifting range $\mathcal{P}$ eliminate the numbers belonging to $u(p)$ distinct residue classes mod p.[49]

As we remarked at the beginning of Section 2, it seems difficult to develop a very satisfactory theory for this narrower sifting problem. We may introduce the notation $M_u(\mathcal{I}_x, \mathcal{P})$ to denote the number of integers left in the interval $\mathcal{I}_x$, and

$$M_u^+(\mathcal{I}_x, \mathcal{P}) \qquad \text{and} \quad M_u^-(\mathcal{I}_x, \mathcal{P})$$

to denote respectively the maximum and the minimum of $M_u(\mathcal{I}_x, \mathcal{P})$ when $\mathcal{P}$ and the $u(p)$ are fixed as well as x, while for each p in $\mathcal{P}$ the $u(p)$ residue classes $\operatorname{mod} p$ can be chosen freely. Clearly

$$M_u^+(\mathcal{I}_x, \mathcal{P}) \le M_u^+(W_x, \mathcal{P}), \tag{19.1}$$

and

$$M_u^-(\mathcal{I}_x, \mathcal{P}) \ge M_u^-(W_x, \mathcal{P}). \tag{19.1'}$$

Beyond this it becomes very difficult to see what the relationships between the two sides of the above inequalities are, whether the $M_u^\pm$ associated with $\mathcal{I}_x$ behave essentially as those associated with W_x or perhaps very differently.

It is however, not difficult to show that the $M_u^\pm$ associated with the $\mathcal{I}_x$ obey the same rule of monotonicity that we established for those associated with the weighted sets in Section 3. Thus, if we use the notation $u \le u'$ to mean that $u(p) \le u'(p)$ for all $p \in \mathcal{P}$, we have

$$\frac{M_u^+(\mathcal{I}_x, \mathcal{P})}{x\, e_u(\mathcal{P})} \le \frac{M_{u'}^+(\mathcal{I}_x, \mathcal{P})}{x\, e_{u'}(\mathcal{P})}, \tag{19.2}$$

[49] Clearly $u(p)$ now has to be an integer ≥ 0. One may generalize the concept of sifting out residue classes so that $u(p)$ is not thus restricted. Let there for each $p \in \mathcal{P}$ and each $a \pmod p$ be assigned a factor $\omega_{p,a}$ with $0 \le \omega_{p,a} \le 1$, and count an integer n in the interval $\mathcal{I}_x$ of length x with the weight

$$\omega_n = \prod_{p \in \mathcal{P}} \omega_{p,n}.$$

The situation we consider above corresponds to the case when $\omega_{p,a}$ can only take the two values 0 and 1. In the more general situation we have

$$u(p) = \sum_{a \pmod p} (1 - \omega_{p,a})$$

which clearly does not need to be an integer. It is not difficult to adapt the theory of the sieve to this more general situation but we shall not go into this here.

and

$$\frac{M_u^-(\mathcal{I}_x,\mathcal{P})}{x\,e_u(\mathcal{P})} \geq \frac{M_{u'}^-(\mathcal{I}_x,\mathcal{P})}{x\,e_{u'}(\mathcal{P})}\,. \tag{19.2'}$$

To prove these inequalities it is clearly enough to consider the case that u and u' differ for just one p, say p_0, so that $u'(p_0) = u(p_0) + 1$, while $u'(p) = u(p)$ for all other $p \in \mathcal{P}$.

Let us consider the case of M_u^+ and assume we have the $\mathcal{I}_x$ and the choice of the $u(p)$ residue classes for each $p \in \mathcal{P}$ which result in the largest $M_u(\mathcal{I}_x,\mathcal{P})$. If we now for p_0 add a new residue class to be excluded, we have clearly $p_0 - u(p_0)$ choices open for this residue class. Any number n that remains in $\mathcal{I}_x$ after the sifting will be excluded by one and only one choice of the new residue class mod p_0. Therefore, if we average over all possible choices, on the average

$$M_u^+(\mathcal{I}_x,\mathcal{P})\,\frac{p_0 - u(p_0) - 1}{p_0 - u(p_0)}$$

numbers remain after excluding the new residue classes mod p. Thus we have in this case

$$M_{u'}^+(\mathcal{I}_x,\mathcal{P}) \geq M_u^+(\mathcal{I}_x,\mathcal{P})\,\frac{p_0 - u(p_0) - 1}{p_0 - u(p_0)} \tag{19.3}$$

which is equivalent to (19.2) in this situation. We also see that equality holds in (19.3) if, and only if, the $M_u^+(\mathcal{I}_x,\mathcal{P})$ numbers remaining after the initial sifting are equidistributed among the $p_0 - u(p_0)$ non excluded residue classes mod p_0. Since this possibility cannot be excluded, we cannot remove the equality sign in (19.2). Thus , while the ratio (3.8) was strictly increasing when any $u(p)$ was increased this may not always be true for the ratio if we replace W_x by $\mathcal{I}_x$.

The proof of (19.2′) can be carried out in the same manner, and the same remarks about the possibility of equality apply.

If we consider the situation where $\mathcal{P} = \mathcal{P}(x^\alpha)$ with some fixed α as $x \to \infty$, and we require that the $u(p)$ are such that (4.2) and (4.4) hold as $x \to \infty$, we are in general unable to show that the ratios

$$\frac{M_u^+(\mathcal{I}_x,\mathcal{P}(x^\alpha))}{x\,e_u(x^\alpha)} \qquad \text{and} \qquad \frac{M^-(\mathcal{I}_x,\mathcal{P}(x^\alpha))}{x\,e_u(x^\alpha)}$$

tend to limits as $x \to \infty$. This is the situation even if we restrict ourselves to the case of constant sifting density k.

Only in the case when $k = 1/2$ does it follow from our earlier results that the two ratios tend to limits and these are

$$T_{1/2}^+(\alpha) \qquad \text{and} \quad T_{1/2}^-(\alpha)$$

respectively. This follows readily if we look at the examples constructed in Section 16 where we considered the weighted sets with the weights (16.11) and (16.11′). (16.11) is equivalent to looking at the numbers n with $1 \le n \le x$ which are of the form $n = 4m + 1$ and (16.11′) at the numbers in the same interval and of the form $n = 4m + 3$. Thus in the first case we have $0 \le m \le (x-1)/4$ and in the second case $0 \le m \le (x-3)/4$.[50] We may just as well consider the sifting as done with respect to the m's as to the n's. In each case we are excluding exactly one residue class for every prime $p \le x^\alpha$ for which $p \equiv 3 \pmod 4$. That we have changed the length of our interval to $(x-1)/4$ or $(x-3)/4$ is clearly immaterial, our results from Section 16 show that we do indeed have

$$\lim_{x\to\infty} \frac{M_u^{\pm}(\mathcal{I}_x, \mathcal{P}(x^\alpha))}{x\, e_u(x^\alpha)} = T_{1/2}^{\pm}(\alpha)\,, \tag{19.4}$$

when the sifting density is constant and equal to 1/2. Thus in this case there is no real difference between the sifting problem for the interval and for the weighted set.

On the other hand it seems very plausible that when the sifting density is high, so that the $u(p)$ tend to be large, the excluded residue classes mod d may for most d be fairly evenly distributed mod d. This would result in a reduction of the remainder term R_d in N_d which would then tend to be much smaller than $u(d)$ in absolute value for most d. This would indicate that we could expect to do rather better for the interval than for the weighted set if the sifting density is large.

Later we shall indeed show that this is actually the case. Our results will however not tell us much about the behaviour of the two ratios

$$\frac{M_u^{\pm}(\mathcal{I}_x, \mathcal{P}(x^\alpha))}{x\, e_u(x^\alpha)}$$

as x goes to infinity. This behaviour remains an open question except in the case of constant sifting density 1/2. It seems not unreasonable to conjecture that if the sifting density is constant and smaller than some absolute k_0, there may not be much difference between the case of W_x and that of the $\mathcal{I}_x$, the ratios above might well converge to $T_k^{\pm}(\alpha)$ for $k \le k_0$. We have no basis on which to conjecture how large k_0 might be. The behaviour for $k = 1$ for instance is an intriguing question since it would have very interesting and far reaching consequences in number theory both if it could be shown that for $k = 1$ the ratios above behave like those with W_x instead of $\mathcal{I}_x$ and if it could be shown that they on the contrary behave very differently. We shall not elaborate more on these speculations here,[51] but instead turn to the

[50] If we in Section 16 had used the character mod 3 instead of mod 4, and the numbers $1 \le n \le x$ with the weights $\frac{1}{2}(1+\chi(n))$ or $\frac{1}{2}(1-\chi(n))$ respectively, this would have been *directly* equivalent to sifting the interval $0 \le n \le x$ with respect to divisibility for the set of primes $\mathcal{P}_2(x^\alpha)$ with $p \equiv 2 \pmod 3$ and in addition excluding the residue classes 0 and 2 mod 3 in the first case and 0 and 1 mod 3 in the second case. This way we would not have altered the length of the interval.

[51] Some such consequences were discussed in [Selberg 5].

question of improving the efficiency of the sieve when applied to an interval by improving the treatment of the remainder term without sacrificing too much on the main term.

We say of two weighted sets W and W' that $W \le W'$ if for every integer n we have $w_n \le w'_n$. Let us consider functions $F(u)$ of a real variable defined on $(-\infty, \infty)$ and which are bounded, non negative and in $L^1(-\infty,\infty)$. In addition we will require some condition which allows us to use Poisson's summation formula for $F(u)$ and its Fourier transform

$$\hat{F}(v) = \int_{-\infty}^{\infty} e^{2\pi iuv} F(uv)du\,,$$

we shall assume that $F(u)$ is of bounded total variation on $(-\infty,\infty)$. We shall consider weighted sets $W(F)$ arising by choosing the weights

$$w_n = F(n)\,.$$

We consider the problem of sifting $W(F)$ by excluding the n's that belong to one of $u(p)$ residue classes $\bmod p$ for some p in the sifting range $\mathcal{P}$. For a d in $(\mathcal{P})$, we write $d \times n$ if n lies in some excluded residue class $\bmod p$ for every $p|d$. We also write $\mathcal{P}\bar{\times} n$ if n does not lie in an excluded residue class modulo any prime $p \in \mathcal{P}$.

By Poisson's summation formula we have

$$(19.5)\qquad \sum_{n\equiv \ell\ (d)} F(n) = \frac{\hat{F}(0)}{d} + \frac{1}{d}\sum_{n}{}' e^{-2\pi i n\ell/d}\hat{F}\left(\frac{n}{d}\right) = \frac{1}{d}\hat{F}(0) + R_{\ell,d}\,,$$

where the dash $'$ indicates that the term $n=0$ is omitted in the summation. From this we see that

$$(19.6)\quad \begin{aligned} N_d(W) = \sum_{d\times n} F(n) &= \frac{u(d)}{d}\hat{F}(0) + \frac{1}{d}\sum_{n}{}'\hat{F}\left(\frac{n}{d}\right)\sum_{\ell} e^{-2\pi i n\ell/d} \\ &= \frac{u(d)}{d}\hat{F}(0) + R_d\,, \end{aligned}$$

here $\sum_\ell$ runs over the residue classes $\bmod d$ which are excluded by d.

The form of (19.6) indicates that we will succeed better with the sieve on the weighted set $W(F)$ if the Fourier transform $\hat{F}(v)$ has a narrow peak at the origin and then decreases fairly rapidly towards zero as $|v|$ goes to infinity.

Let now $F_x(u)$ be the characteristic function of an interval $\mathcal{I}_x$ of length x, we can assume that the interval is centered at $u=0$, since shifting the center only involves multiplying $\hat{F}_x(v)$ by the factor $e^{2\pi ibv}$ if we shift by the amount b to the right. We find

$$(19.7)\qquad \hat{F}_x(v) = \frac{\sin \pi xv}{\pi v}\,.$$

We shall also use the function

$$f_z(u) = \frac{1}{z}\, F_z(u)\,,$$

and have

$$\hat{f}_z(v) = \frac{\sin \pi z v}{\pi z v}\,. \tag{19.8}$$

If we assume that $x > rz$, where r is a positive integer, and form the functions

$$F^r_{x+rz,z}(u) = F_{x+rz} * f_z * f_z * \ldots * f_z\,, \tag{19.9}$$

and

$$F^r_{x-rz,z}(u) = F_{x-rz} * f_z * f_z * \ldots * f_z\,, \tag{19.9'}$$

where $*$ denotes convolution and f_z is repeated r times, it is easy to see that

$$0 \le F^r_{x-rz,z}(u) \le F_x(u) \le F^r_{x+rz,z}(u)\,, \tag{19.10}$$

for all u. Also we have

$$\hat{F}^r_{x\pm rz,z}(0) = x \pm rz\,, \tag{19.11}$$

and since

$$\hat{F}^r_{x\pm rz,z}(v) = \frac{\sin \pi\,(x \pm rz)\,v}{\pi v}\,\left(\frac{\sin \pi z v}{\pi z v}\right)^r,$$

we get

$$|\hat{F}^r_{x\pm rz,z}(v)| \le \frac{z}{(\pi z\,|v|)^{r+1}}\,. \tag{19.12}$$

We see from this that we for purposes of obtaining lower bounds for sifting the interval $\mathcal{I}_x$, may instead sift the weighted set given by $W(F)$ where

$$F = F^r_{x-rz,z},$$

similarly we may for the purpose of getting upper bounds for $\mathcal{I}_x$ use the weighted set $W(F)$ with

$$F = F^r_{x+rz,z}\,.$$

In each case we are clearly losing something in the main term, but if rz is small compared with x this will not be significant. On the other hand we are because of (19.12) gaining considerably in the remainder term. If we assume our sieve is a $\Lambda^{\pm}[\mathcal{P}, z_1]$ and assume $r > 1$, we get for the main term

$$(x \pm rz) \sum_d \frac{u(d)\lambda_d}{d} \tag{19.13}$$

and from the remainder term that it is less than

$$(19.14)\quad 2\pi^{-r-1}z^{-r}\zeta(r+1)\sum_d u(d)\,|\lambda_d|\,d^r < (\pi z)^{-r}\sum_d u(d)\,|\lambda_d|\,d^r < \left(\frac{z_1}{\pi z}\right)^r \sum_d u(d)\,|\lambda_d|\,.$$

We have thus reduced the remainder term by a factor $\left(\frac{z_1}{\pi z}\right)^r$ over what direct application of the sieve to the interval would give. For $z_1 = z$ and $r = \frac{\theta}{\log \pi}\log x$, this factor becomes $x^{-\theta}$ and if z is $o(x/\log x)$ the main term (19.13) has not been appreciably changed.

So far, this idea seems to work equally well for the lower and the upper bound sifting problem for the interval. There is however a significant difference rooted in the fact that a function F and its Fourier transform $\hat{F}$ cannot both have compact support.

For the upper bound problem we may seek to find a suitable majorant $F(u)$ of $F_x(u)$ which is such that $\hat{F}(v) = 0$ for $v \geq 1/z$. For the lower bound it is clearly impossible to find a minorant $F(u)$ with $0 \leq F(u) \leq F_x(u)$ for all u which has a $\hat{F}(v)$ with compact support since $F(u)$ has to have its support limited to the interval $\mathcal{I}_x$.

Let us assume that we have a majorant of $F_x(u)$ which is such that $\hat{F}(v)$ has its support on $|v| < 1/z$. Applying an upper bound sieve $\Lambda^+[\mathcal{P}, z]$ to $W(F)$ we get the upper bound

$$(19.15)\qquad \hat{F}(0)\sum_d \frac{u(d)\,\lambda_d}{f(d)}\,,$$

which is therefore also an upper bound for the interval $\mathcal{I}_x$. Obviously we have $\hat{F}(0) > x$. The question naturally arises: How small can we make $\hat{F}(0)$ given x and z? We shall try to answer this in the next Section and also make some applications of the answer.

20. An extremal problem with application to the upper bound sieve for the interval and a digression on the large sieve and Hilbert's inequality

Looking at the question raised at the end of the last Section, we see that we can reduce it to a problem containing just one parameter x instead of two (x and z). This is so since if $\hat{F}(v)$ is the Fourier transform of $F(u)$ then $a\,\hat{F}(va)$ is the Fourier transform of $F(u/a)$. Therefore, if then we put $z = 1$ and designate the minimum value of $\hat{F}(0)$ for a majorant of $F_x(u)$ with $\hat{F}(v) = 0$ for $|v| \geq 1$ by $x + \theta(x)$, the minimum for the general problem will be $x + \theta(x/z)\,z$.

For our purposes now, we wish to shift the interval $\mathcal{I}_x$ to $(0, x)$, and shall at first assume that x is an integer. Use of the Poisson summation formula now gives if $F(u) \geq F_x(u)$ that

$$\hat{F}(0) = \sum_n F(n) \geq x + 1 . \tag{20.1}$$

Thus $\theta(x)$ in this case could not be less than 1.

We wish now to see if there could exist an $F(u) \geq F_x(u)$ whose Fourier transform has its support restricted to $(-1, 1)$, with $\hat{F}(0) = x + 1$, and which form it in that case has to have. Clearly we must have

$$F(n) = 1 ,$$

for $0 \leq n \leq x$, and

$$F(n) = 0 ,$$

for n outside the interval $(0, x)$. We also must have

$$F'(n) = 0 ,$$

for all integers except 0 and x. Also since $\hat{F}(v)$ has its support on $(-1, 1)$ and vanishes at the end points, it follows easily that $F(u)$ is an integral function such that

$$F(u + iv) = o\left(\frac{1}{1 + |v|}\, e^{2\pi |v|}\right) .$$

From this we get that if F exists it must have the form

$$\frac{F(w)}{\sin^2 \pi w} = \frac{1}{\pi^2} \left\{ \sum_{n=0}^{x} \frac{1}{(w-n)^2} + \frac{\lambda}{w} + \frac{\lambda}{x - w} \right\} , \tag{20.2}$$

or

$$F(w) = \frac{\sin^2 \pi w}{\pi^2} \left\{ \sum_{n=0}^{x} \frac{1}{(w-n)^2} + \frac{\lambda x}{w(x - w)} \right\} , \tag{20.3}$$

here λ is some constant. We must now establish whether any choice of λ makes this F a majorant of F_x on the real line, if it does, the resulting F fulfills all the requirements and has $\hat{F}(0) = x + 1$. We first look at the case when w is real and $0 \leq u \leq x$, we must then have $F(u) \geq 1$. We make use of the well known formula

$$1 = \frac{\sin^2 \pi u}{\pi^2} \sum_{-\infty}^{\infty} \frac{1}{(u-n)^2} , \tag{20.4}$$

and subtract this from (20.3) writing u instead of w in the latter formula. We get

$$F(u) - 1 = \frac{\sin^2 \pi u}{\pi^2} \left\{ \frac{\lambda x}{u(x-u)} - \sum_{m=1}^{\infty} \left(\frac{1}{(m+u)^2} + \frac{1}{(m + x - u)^2} \right) \right\} .$$

Thus we must have for all $0 \le u \le x$

$$\lambda \ge \frac{u(x-u)}{x} \sum_{m=1}^{\infty} \left\{ \frac{1}{(m+u)^2} + \frac{1}{(m+x-u)^2} \right\}. \tag{20.5}$$

The maximum of the right hand side of (20.5) for u in $(0,x)$ thus gives us a lower bound for λ. We shall see that the maximum is attained for $u = x/2$. If we write $u = x/2+t$, the expression on the right hand side of (20.5) becomes

$$\frac{(x/2)^2 - t^2}{x} \sum_{m=1}^{\infty} \left\{ \frac{1}{(m+x/2+t)^2} + \frac{1}{(m+x/2-t)^2} \right\} = A(t).$$

Since $A(t)$ is an even function we need only consider the case $0 \le t \le x/2$. We write

(20.6)

$$\begin{aligned} &x\,(A(t) - A(0)) \\ &= \left(\left(\frac{x}{2}\right)^2 - t^2\right) \sum_{m=1}^{\infty} \left\{ \frac{1}{(m+x/2+t)^2} + \frac{1}{(m+x/2-t)^2} - \frac{2}{(m+x/2)^2} \right\} \\ &\quad - 2t^2 \sum_{n=1}^{\infty} \frac{1}{(m+x/2)^2}, \end{aligned}$$

and wish to show that this is negative for $t \ne 0$. Since the expression in the curly bracket on the right hand side of (20.6) as a function of m is decreasing, positive and with positive second derivative,[52] we see that

(20.7)

$$\begin{aligned} \sum_{m=1}^{\infty} \{\ \} &\le \int_{1/2}^{\infty} \left\{ \frac{1}{(m+x/2+t)^2} + \frac{1}{(m+x/2-t)^2} - \frac{2}{(m+x/2)^2} \right\} dm \\ &= \frac{1}{(1+x)/2+t} + \frac{1}{(1+x)/2-t} - \frac{4}{1+x} \\ &= \frac{4t^2}{(1+x)\left(((1+x)/2)^2 - t^2\right)}. \end{aligned}$$

Also

$$\sum_{m=1}^{\infty} \frac{1}{(m+x/2)^2} > \int_1^{\infty} \frac{du}{(u+x/2)^2} = \frac{1}{1+x/2}. \tag{20.8}$$

Inserting these in (20.6) we get an upper bound for (20.6) as follows

$$x\,(A(t) - A(0)) \le 4t^2 \left\{ \frac{(x/2)^2 - t^2}{(1+x)\left(((1+x)/2)^2 - t^2\right)} - \frac{1}{2+x} \right\}.$$

[52] It can be written as

$$\int_0^t \int_0^t \frac{6}{(m+x/2+u-v)^4} du dv.$$

It is here enough to show that the expression in the curly bracket is negative for $0 \le t \le x/2$. Since the first term takes its largest value for $t = 0$ and the second term does not contain t, we just have to verify that for $x \ge 0$

$$\frac{x^2}{(1+x)^3} - \frac{1}{2+x} < \frac{x^2}{3x^2+x^3} - \frac{1}{2+x} = \frac{1}{3+x} - \frac{1}{2+x} < 0\,.$$

(20.5) now gives

$$\lambda \ge \frac{x}{2} \sum_{m=1}^{\infty} \frac{1}{(m+x/2)^2}\,, \tag{20.9}$$

which is the condition λ must satisfy in order that we should have

$$F(u) \ge 1 \qquad \text{for} \quad 0 \le u \le x\,.$$

Next we see what condition λ must satisfy if we are to have $F(u) \ge 0$ for all u. This will be the case if the expression in the curly bracket on the right hand side of (20.3) is always non negative. If we multiply this expression by w^2 and let $w \to \infty$, the limit is

$$x + 1 - \lambda x\,,$$

so we must have

$$\lambda \le \frac{x+1}{x}\,. \tag{20.10}$$

Because of the symmetry around $x/2$, it is enough to look at the expression in question for $w < 0$ only. If we write $-w$ instead of w it becomes

$$\sum_{n=0}^{x} \frac{1}{(n+w)^2} - \frac{\lambda x}{w\,(w+x)}\,.$$

For $w > 0$, we have

$$\begin{aligned}\sum_{0}^{x} \frac{1}{(n+w)^2} &> \frac{1}{2w^2} + \frac{1}{2\,(x+w)^2} + \int_0^x \frac{du}{(u+w)^2}\\ &= \frac{1}{2w^2} + \frac{1}{2\,(x+w)^2} + \frac{x}{w\,(x+w)} > \frac{1}{w\,(x+w)} + \frac{x}{w\,(x+w)}\\ &= \frac{x+1}{w\,(x+w)}\,.\end{aligned}$$

Therefore the condition (20.10) is sufficient in addition to being necessary.

We now have the result that *the function F given by* (20.3) *fulfills all conditions required if*

$$\frac{x}{2}\sum_{m=1}^{\infty}\frac{1}{(m+x/2)^2}\le\lambda\le\frac{x+1}{x}.\text{[53]}$$

It is possible to prove in the same way, that if x is a positive integer and we seek a minorant $F(u)$ of $F_x(u)$ such that $F(u)\le F_x(u)$ for all u, and $\hat{F}(v)$ has its support restricted to $(-1,1)$ and which maximizes $\hat{F}(0)$ then the only solutions are the functions

$$F(u)=\frac{\sin^2\pi u}{\pi^2}\left\{\sum_{n=1}^{x-1}\frac{1}{(u-n)^2}+\frac{\lambda x}{u(x-u)}\right\}. \tag{20.12}$$

with

$$\frac{x-1}{x}\le\lambda\le\frac{x}{2}\sum_{m=0}^{\infty}\frac{1}{(m+x/2)^2}. \tag{20.13}$$

The maximal value of $\hat{F}(0)$ is of course

$$\hat{F}(0)=x-1. \tag{20.14}$$

In both cases we see that $\lambda=1$ is always a permissible choice of λ.

When x is not an integer we cannot make a similar guess of what the value of $\theta(x)$ might be, and what form the extremal function or functions might have. It is possible to get a bound <1 on $\theta(x)$ when x is not an integer by modifying the function (20.3) for an integer near x.[54] We shall not go into this but choose a different but simpler approach. If a is an integer >0 we denote the function (20.3) with a instead of x and with $\lambda=1$ by $f_a(u)$, and write it as

$$f_a(u)=\frac{\sin^2\pi u}{\pi^2}\left\{\frac{1}{u}+\sum_{n=0}^{a}\frac{1}{(u-n)^2}+\frac{1}{a-u}\right\}. \tag{20.15}$$

If we let $a\to\infty$ it tends to the limit function

$$f_\infty(u)=\frac{\sin^2\pi u}{\pi^2}\left\{\frac{1}{u}+\sum_{n=0}^{\infty}\frac{1}{(u-n)^2}\right\}, \tag{20.15'}$$

---

[53] That the lower bound for λ is smaller than the upper bound follows for instance from

$$\frac{x}{2}\sum_{m=1}^{\infty}\frac{1}{(m+x/2)^2}<\frac{x}{2}\int_{1/2}^{\infty}\frac{du}{(u+x/2)^2}=\frac{x}{x+1}.$$

The solution with $\lambda=(x+1)/x$ is distinguished by tending faster to zero as this $F(u)$ tends to zero as u^{-4} for large $|u|$ while the others only tend to zero as u^{-2}. Similar remarks hold for (20.13) and (20.12).

[54] We would use the $F(u)$'s corresponding to the two extremal values of λ, the largest λ if the nearest integer is below x, the smallest λ if it is above x. We can in this way even determine the right and left derivatives of $\theta(x)$ at the integers.

with the property that $f_\infty(u) \ge 0$ for all u and $f_\infty(u) \ge 1$ for $u \ge 0$, also it equals 0 for the negative integers and 1 for 0 and the positive integers. We see from this that

$$B(u) = 2\, f_\infty(u) - 1\,, \tag{20.16}$$

is a majorant of the function $\operatorname{sgn} u$. Using (20.4) we get from (20.16) and (20.15′) the expression

$$B(u) = \frac{\sin^2 \pi u}{\pi^2}\left\{\frac{2}{u} + \sum_{n=0}^{\infty} \frac{1}{(u-n)^2} - \sum_{n=1}^{\infty} \frac{1}{(u+n)^2}\right\}. \tag{20.17}$$

Since

$$B(u) - \operatorname{sgn} u + B(-u) - \operatorname{sgn}(-u) = B(u) + B(-u) = 2\,\frac{\sin^2 \pi u}{\pi^2 u^2}\,,$$

we get

$$\int_{-\infty}^{\infty} (B(u) - \operatorname{sgn} u)\, du = \int_{-\infty}^{\infty} \frac{\sin^2 \pi u}{\pi^2 u^2}\, du = 1\,. \tag{20.18}$$

Since the characteristic function of the interval $(0, x)$ can be written as

$$F_x(u) = \frac{1}{2} \operatorname{sgn} u + \frac{1}{2} \operatorname{sgn}(x-u)\,,$$

it is clear that

$$F(u) = \frac{1}{2}\, B(u) + \frac{1}{2}\, B(x-u) \tag{20.19}$$

is a majorant and we have from (20.18) that

$$\hat{F}(0) = \int_{-\infty}^{\infty} F(u) du = x + \int_{-\infty}^{\infty} (F(u) - F_x(u))\, du = x + 1\,. \tag{20.20}$$

$\hat{F}$ does (as follows by looking at the growth of $F(w)$ in the complex plane) have its support restricted to $(-1, 1)$, this shows that $\theta(x) \le 1$ for all x. However, if x is not an integer it is easy to see that the majorant given by (20.19) is not the best, since it is actually > 1 for $0 \le u \le x$ because for any u in this interval $B(u)$ and $B(x-u)$ are both ≥ 1, but they only are equal to 1 if the argument is a non negative integer and u and $x-u$ cannot be integers at the same time. Since we thus must have $\min_{0 \le u \le x} F(u) > 1$, it follows that we could divide $F(u)$ by this minimum in $(0, x)$ and get a better majorant, thus $\theta(x) < 1$ unless x is an integer. We can actually show

$$\theta(x) \le 1 - c\,\frac{|\sin \pi x|}{x+1}\,, \tag{20.21}$$

with a suitable constant $c > 0$.[55]

[55] This is obtained not by using the method given here, but by the method mentioned in the previous footnote.

It is possible in the same way to show in connection with the extremal problem for a minorant function which led to (20.12), (20.13) and (20.14) for integral $x > 0$, that if we denote the maximal value of $\hat{F}(0)$ by $x - \theta_1(x)$ for general $x > 0$, we have

$$\theta_1(x) \le 1 - c\,\frac{|\sin \pi x|}{x+1}\,, \tag{20.21'}$$

for all $x > 0$, actually for $0 < x \le 1$, it is easy to show that $\theta_1(x) = x$.

It follows from what we said at the beginning of this Section that *we can majorize the characteristic function of an interval of length x by a function $F(u)$ such that $\hat{F}(v)$ has its support restricted to $(-1/z, 1/z)$ and so that*

$$\hat{F}(0) = x + \theta\left(\frac{x}{z}\right) z \le x + z\,. \tag{20.22}$$

Similarly we can find a minorant $F(u)$ such that $\hat{F}(u)$ has its support restricted to $(-1/z, 1/z)$ and with

$$\hat{F}(0) = x - \theta_1\left(\frac{x}{z}\right) z \ge x - z\,. \tag{20.22'}$$

The graphs of the functions $\theta(x)$ and $\theta_1(x)$:

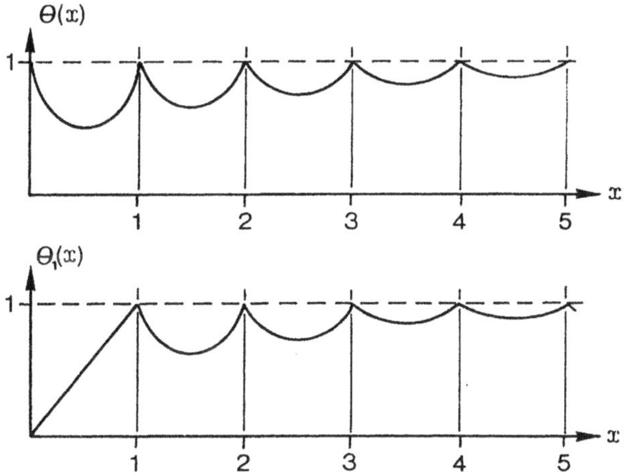

For the upper bound sifting problem we may now replace the interval of length x with a weighted set with the weights $F(n)$, *using a sieve $\Lambda^+[\mathcal{P}, z]$ we get from* (20.22) *the upper bound*

$$(x+z)\sum_d \frac{\lambda_d}{f(d)}\,, \tag{20.23}$$

in particular, we get for the Λ^2 sieve the upper bound

$$\frac{x+z^2}{\displaystyle\sum_{\substack{d<z\\ d\in(\mathcal{P})}} \frac{1}{f'(d)}}\,. \tag{20.23'}$$

If the sifting density is high this is not only more convenient, but also quite a bit better than the estimations given earlier.

We can of course *not* use the minorant of $F_x(u)$ in any similar way in connection with a lower bound sieve, it does however have some uses as we shall see later.

Though it lies somewhat outside the main theme of these lectures, it seems not unreasonable to give some further applications of the functions constructed above.

Let $x_{j+1} - x_j \geq \delta > 0$ for $j = 1, 2, \ldots, r-1$, and $1 + x_1 - x_r \geq \delta$. Let also $F(u)$ majorize the characteristic function of the interval $(M, M+N)$ and be such that $\hat{F}(v) = 0$ for $|v| \geq \delta$ and $\hat{F}(0)$ is minimal. This means

$$\hat{F}(0) = N + \theta(\delta N)\frac{1}{\delta} \leq N + \frac{1}{\delta}\,. \tag{20.24}$$

We define on the integers n, r functions as

$$\varphi_j(n) = \sqrt{\frac{F(n)}{\hat{F}(0)}}\, e^{-2\pi\, i n x_j}\,, \tag{20.25}$$

for $r = 1, 2, \ldots, r$. Then by Poisson's summation formula we have

$$\begin{aligned}\sum_{n=-\infty}^{\infty} \varphi_j(n)\,\overline{\varphi_k(n)} &= \frac{1}{\hat{F}(0)} \sum_{n=-\infty}^{\infty} F(n)\, e^{2\pi\, i n (x_j - x_k)} \\ &= \frac{1}{\hat{F}(0)} \sum_{n=-\infty}^{\infty} \hat{F}(n + x_j - x_k) \\ &= \delta_{j,k}\,,^{56}\end{aligned} \tag{20.26}$$

where $\delta_{j,k}$ is the Kronecker symbol. Let $\{a_n\}$ be a set of numbers defined for all $M \leq n \leq M+N$ and put $a_n = 0$ if n is outside this interval and define $g(n) = a_n/\sqrt{F(n)}$ for $M \leq n \leq M+N$, and $g(n) = 0$ for all other n. Then we have for any numbers ξ_j, $j = 1, 2, \ldots, r$ that

$$\sum_n \left| g(n) + \sum_{j=1}^{r} \xi_j\, \varphi_j(n) \right|^2 \geq 0\,,$$

which gives

$$\sum_{n=M}^{M+N} \frac{|a_n|^2}{F(n)} + 2\,\mathrm{Re} \sum_{j=1}^{r} \frac{\bar{\xi}_j}{\sqrt{\hat{F}(0)}} \sum_n a_n\, e^{2\pi\, i n x_j} + \sum_{j=1}^{r} |\xi_j|^2 \geq 0\,. \tag{20.27}$$

If we write[57]

$$S(x_j) = \sum_{n=M}^{M+N} a_n\, e^{2\pi i n x_j}\,, \tag{20.28}$$

[56] We here have used that if $j \neq k$, $\delta \leq |x_j - x_k| \leq 1 - \delta$.

[57] It is clear that one loses nothing by assuming M and N to be integers.

and put

$$\xi_j = \frac{-1}{\sqrt{\hat{F}(0)}} S(x_j)$$

in (20.27), this inequality becomes

$$\sum_{n=M}^{M+N} \frac{|a_n|^2}{F(n)} - \frac{1}{\hat{F}(0)} \sum_{j=1}^{r} |S(x_j)|^2 \geq 0$$

or

$$(20.29)\quad \sum_{j=1}^{r} |S(x_j)|^2 \leq \hat{F}(0) \sum_{n=M}^{M+N} \frac{|a_n|^2}{F(n)} = \left(N + \frac{1}{\delta}\,\theta\,(\delta N)\right) \sum_{n=M}^{M+N} \frac{|a_n|^2}{F(n)}$$
$$\leq \left(N + \frac{1}{\delta}\right) \sum_{n=M}^{M+N} |a_n|^2 .$$

This is a form of the so called "large sieve inequality." It is fairly easy to see that *we always have inequality except if the following conditions are fulfilled:*

We must have $\delta = 1/r$ *so that the points* $x_j = x_1 + (j-1)/r$ *are equidistributed,* r *must divide* N, a_n *can be* $\neq 0$ *only for* $n \equiv M \pmod r$, *and for these* n, $a_n\, e^{2\pi\, i n x_i}$ *must be constant.*

One can see this by remembering that in order to have equality, the original inequality must reduce to an equation with our choice of ξ_j, and that a_n can be $\neq 0$ only if $F(n) = 1$. Also, if n is outside $(M, M+N)$ we must have

$$\sum_{j=1}^{r} S(x_j) e^{-2\pi\, i n x_j} = 0 ,$$

except when $F(n) = 0$.

The inequality (20.29) is easily seen to be equivalent to the following inequality involving a bilinear form

$$(20.30)\qquad \left| \sum_{n=M}^{M+N} \sum_{j=1}^{r} a_n b_j\, e^{2\pi\, i n x_j} \right| \leq \left(N + \frac{1}{\delta}\right) \sum_{n=M}^{M+N} |a_n|^2 \sum_{j=1}^{r} |b_j|^2 .$$

This inequality, which in the exponential involves two sequences, the n which are perfectly regularly distributed in an interval, and the x_r which may be irregularly distributed, but still are well spaced by the condition $x_{j+1} - x_j \geq \delta$, suggests that there may be a more general form of the inequality involving two sequences each of which is distributed in some interval and with some condition of well spacing.

In order to prove this more general inequality we shall need to look at a third extremal problem that is closely related to our problem of majorizing $F_x(u)$ with an $F(u)$ with $\hat{F}(v)$ having its support on $(-\delta, \delta)$ and with minimal $\hat{F}(0)$. We ask now for a complex valued function $f(u)$ such that $|f(u)| \geq 1$

for u in $\mathcal{I}_x$, with a Fourier transform that has its support on $(-\delta/2, \delta/2)$ and which minimizes

$$\int_{-\infty}^{\infty} |f(u)|^2 du .$$

It is not difficult to see that if we have a solution to this problem $f(u)$ and for general complex w form the function

$$F(w) = f(w)\,\overline{f(\bar{w})},$$

then $F(u)$ will be a majorant of $F_x(u)$, its Fourier transform $\hat{F}(v)$ is the convolution of $\hat{f}(v)$ and $\hat{f}(-v)$ and will thus have its support in $(-\delta, \delta)$ and finally

$$\hat{F}(0) = \int_{-\infty}^{\infty} |f(u)|^2\, du .$$

Conversely, if we have a solution to the first problem, we may assume the interval $\mathcal{I}_x$ centered at the origin and that $F(u)$ is even. The zeroes of $F(w)$ are then symmetrical with respect to both the real and the imaginary axis. The real zeroes have to be of even order, and the zeroes on the imaginary axis are clearly finite in number. Zeroes that are neither real nor purely imaginary occur in quadruples $\rho, \bar{\rho}, -\rho, -\bar{\rho}$. We now form a canonical product

$$f(w) = \prod\nolimits_1 \left(1 - \frac{w}{\rho}\right) \prod\nolimits_2 \left(1 - \frac{w^2}{\rho^2}\right),$$

where in $\prod_1$ the ρ run over the finite number of zeroes on the upper imaginary axis, while in $\prod_2$ the ρ run over the real positive zeroes of $F(w)$ with half multiplicity and of the complex quadruples takes only those in the first quadrant. Then

$$F(w) = f(w)\,\overline{f(\bar{w})},$$

and it can be established that the growth of $f(w)$ in the imaginary direction is at most $f(u+iv) = o(e^{\pi\delta|v|})$ which gives that $\hat{f}$ has its support in $(-\delta/2, \delta/2)$, and we have

$$\int_{-\infty}^{\infty} |f(u)|^2 du = \hat{F}(0) = x + \theta(\delta x)\,\frac{1}{\delta}, \tag{20.31}$$

which thus gives the minimal value of

$$\int_{-\infty}^{\infty} |f(u)|^2 du .$$

We now assume we have a bilinear form

$$B(x, y) = \sum_m \sum_n x_m y_n e^{2\pi i \lambda_m \mu_n}, \tag{20.32}$$

where the λ_m and the μ_n are two strictly increasing sequences. We assume that the λ_m lie in an interval of length L and that $\eta > 0$ is such that there is no other λ inside the interval $(\lambda_m - \eta, \lambda_m + \eta)$ for all λ_m in the sequence. Similarly we assume the μ_n lie in an interval of length M and that there is a $\delta > 0$ such that there is no other μ inside the interval $(\mu_n - \delta, \mu_n + \delta)$ for all μ_n in the sequence. We write

$$S(\mu) = \sum_m x_m e^{2\pi i \lambda_m \mu}, \tag{20.33}$$

and

$$S^*(\mu) = \sum_m \frac{x_m}{f(\lambda_m)} e^{2\pi i \lambda_m \mu}, \tag{20.33'}$$

where f is a solution of the extremal problem mentioned above for the interval $\mathcal{I}_L$ in which the λ_m lie, so we have $|f(\lambda_m)| \geq 1$ for all λ_m. Then

$$S(\mu_m) = \int_{\mu_n - \delta/2}^{\mu_n + \delta/2} S^*(\mu)\, \hat{f}(\mu - \mu_n)\, du,$$

so

$$\begin{aligned} |S(\mu_m)|^2 &\leq \int_{\mu_n - \delta/2}^{\mu_n + \delta/2} |S^*(\mu)|^2 du \int_{-\infty}^{\infty} \hat{f}(\mu)\, d\mu \\ &= \left(L + \theta(\delta L)\frac{1}{\delta}\right) \int_{\mu_n - \delta/2}^{\mu_n + \delta/2} |S^*(\mu)|^2 du. \end{aligned} \tag{20.34}$$

Now

$$B(x, y) = \sum_n y_n S(\mu_n),$$

so

$$|B(x, y)|^2 \leq \sum_n |S(\mu_n)|^2 \sum_n |y_n|^2. \tag{20.35}$$

Now, since the intervals $(\mu_n - \delta/2, \mu_n + \delta/2)$ do not overlap but all lie in an interval of length $M + \delta$, we get from (20.34) that

$$\sum_n |S(\mu_n)|^2 \leq \left(L + \theta(\delta L)\frac{1}{\delta}\right) \int_{\mathcal{I}_{M+\delta}} |S^*(\mu)|^2 d\mu. \tag{20.36}$$

The integral on the right hand side we can estimate by putting in a factor $F(\mu)$ where $F(\mu)$ is a majorant of the characteristic function of the interval $\mathcal{I}_{M+\delta}$ which has a Fourier transform $\hat{F}$ which has its support in $(-\eta, \eta)$ and $\hat{F}(0)$ minimal, that is to say

$$\hat{F}(0) = \left(M + \delta + \theta(\eta\,(M + \delta))\frac{1}{\eta}\right).$$

Then

$$\int_{\mathcal{I}_{M+\delta}} |S^*(\mu)|^2 d\mu < \int_{-\infty}^{\infty} |S^*(\mu)|^2\, F(\mu) d\mu = \hat{F}(0) \sum_n |x_n|^2.$$

Inserting this with the value of $\hat{F}(0)$ in (20.36) we get

$$\sum_n |S(\mu_n)|^2 < \left(L + \theta(\delta L)\frac{1}{\delta}\right)\left(M + \delta + \theta(\eta(M+\delta))\frac{1}{\eta}\right)\sum_n |x_n|^2,$$

(20.35) *now gives*
(20.37)

$$|B(x,y)|^2 < \left(L + \theta(\delta L)\frac{1}{\delta}\right)\left(M + \delta + \theta(\eta(M+\delta))\frac{1}{\eta}\right)\sum_n |x_n|^2 \sum_m |y_m|^2,$$

or if we replace the θ *by the upper bound 1,*

$$(20.37') \qquad |B(x,y)|^2 \le \left(L + \frac{1}{\delta}\right)\left(M + \delta + \frac{1}{\eta}\right)\sum_n |x_n|^2 \sum_m |y_m|^2 .$$

The result is not quite symmetric, though the problem clearly is, *so we may rewrite it as*
(20.37″)

$$|B(x,y)| \le \left(\left(L + \frac{1}{\delta}\right)\left(M + \frac{1}{\eta}\right) + 1 + \min(\delta L, \eta M)\right)\sum_n |x_n|^2 \sum_m |y_m|^2 .$$

My belief is that the term $\min(\delta L, \eta M)$ *here can be dropped and* (20.37″) *would still be true. I cannot prove this, the term 1 however can be shown to be necessary.*

The inequalities (20.37) and (20.37″) are, as one can easily give examples of, in many cases quite trivial, on the other hand it is easy from these inequalities to recover (20.29)[58] in the sharp form if we assume one of the sequences to be completely regular. If the $\mu_n = n$ in the interval $M \le n \le M+N$, and the λ_n are the x_j in the interval of length < 1, we may just increase the x_j sequence by adding multiples of 1 to these numbers so they fill an interval of length k, writing down (20.37″) or (20.37) for these sequences, dividing the resulting inequality by k^2 and letting $k \to \infty$, one gets back to the sharp form of (20.30) or (20.29).

Above we saw the use made of the majorant function $F(u)$ to find an upper bound for the integral of $|S^*(\mu)|^2$ over an interval. This, of course, is something that can be done in general and it is also clear that we may use the minorant function referred to in (20.22′) in a similar way to get lower bounds.

If

$$S(u) = \sum_n a_n\, e^{2\pi i \lambda_n u},$$

where $\lambda_{n+1} - \lambda_n \ge \delta > 0$, we get in this way that[59]

$$(20.38) \quad \left(x - \theta_1(\delta x)\frac{1}{\delta}\right)\sum_n |a_n|^2 < \int_{I_x} |S(u)|^2 du < \left(x + \theta(\delta x)\frac{1}{\delta}\right)\sum_n |a_n|^2,$$

58 In the form (20.30) of course.

59 We assume that not all $a_n = 0$, so equality is excluded.

or replacing the θ's by their upper bound 1,

$$\left(x-\frac{1}{\delta}\right)\sum_n |a_n|^2 < \int_{I_x} |S(u)|^2 du < \left(x+\frac{1}{\delta}\right)\sum_n |a_n|^2 . \tag{20.38'}$$

If we denote the end points of the interval by α and β so $\beta-\alpha=x$, we have

$$\int_\alpha^\beta |S(u)|^2 du = x\sum_n |a_n|^2 + \sum_{m\neq n} \frac{a_m\bar{a}_n}{\lambda_m-\lambda_n}\left(\frac{e^{2\pi i\beta(\lambda_m-\lambda_n)}-e^{2\pi i\alpha(\lambda_m-\lambda_n)}}{2\pi i}\right),$$

so that

$$\left|\sum_{m\neq n} \frac{a_m\bar{a}_n}{\lambda_m-\lambda_n}\left(e^{2\pi i\beta(\lambda_m-\lambda_n)}-e^{2\pi i\alpha(\lambda_m-\lambda_n)}\right)\right| < \frac{2\pi}{\delta}\sum_n |a_n|^2 .$$

If the numbers λ_n are linearly independent we may choose $\beta=0$ and choose (by Kronecker's theorem) a sequence of α_j such that

$$\frac{a_m}{\bar{a}_m} e^{2\pi i\alpha_j\lambda_m} \longrightarrow 1$$

for all m as $j\to\infty$. In the limit the above inequality becomes

$$\left|\sum_{m\neq n} \frac{a_m\bar{a}_n-\bar{a}_m a_n}{\lambda_m-\lambda_n}\right| \le \frac{2\pi}{\delta}\sum_n |a_n|^2$$

or

$$\left|\sum_{m\neq n} \frac{a_m\bar{a}_n}{\lambda_m-\lambda_n}\right| \le \frac{\pi}{\delta}\sum_n |a_n|^2 .$$

If the numbers λ_n are not linearly independent we may perturb them slightly by considering

$$\lambda_n' = \lambda_n + \frac{\varepsilon_n}{N}$$

where the ε_n are linearly independent of each other and of the λ_m and N is a large positive integer. Then the λ_n' are always linear independent, writing down the above inequality for the λ_n' and letting $N\to\infty$, we then get in general Hilbert's inequality

$$\left|\sum_{m\neq n} \frac{a_m\bar{a}_n}{\lambda_m-\lambda_n}\right| \le \frac{\pi}{\delta}\sum_n |a_n|^2 . \tag{20.39}$$

21. Some historical comments on the last section

While I had made some use of Fourier analysis and the use of smoother functions that were majorants or minorants of the characteristic function of the interval $\mathcal{I}_x$ already in the mid sixties, it was essentially only the kind of majorants and minorants discussed in Section 19 connected with the expressions (19.9) and (19.9′). The idea to see what could be done with functions whose Fourier transform has compact support occurred to me rather later in the early seventies. When I lectured first on this at the meeting in Boulder in the summer of 1972, I had not yet solved the extremal problem in the case of x being an integer and $\hat{F}(v)$ having its support restricted to $(-1, 1)$. Only after I had sent in my manuscript for the conference report, did I manage to solve this problem. Somewhat later I mentioned the problem and my result to my colleague Professor Beurling. He then told me he had, I believe around 1940 in connection with some question about analytic functions that I can no longer recall, run across the problem of majorizing the function $\operatorname{sgn} u$ with a function whose (generalized) Fourier transform had its support in $(-1, 1)$ and such that the integral over $(-\infty, \infty)$ of the difference between this function and $\operatorname{sgn} u$ attained its minimal value. He had solved this with the function (20.17) (which I have always later referred to as the Beurling function $B(u)$), but never published anything about it. Neither have I before, though I have over the years frequently lectured on the material in the last Section. The inequality (20.37) and (20.37″) for instance was obtained in 1976 and I lectured on it for the first time at the Mittag–Leffler Institute in the fall of 1977.[60]

I never did determine $\theta(x)$ (nor $\theta_1(x)$) exactly, except in the case when x is an integer. However, I have been told that the problem for $\theta(x)$ has been solved by B.F. Logan. As I have not seen any paper or review on this, his solution may still be unpublished.

In this account we shall not go into applications of these ideas to the large sieve as it is used in number theory, some such applications are indicated in [Selberg 5].

Section 20 brings essentially to a close the general theoretical material in connection with sieves that I have lectured on over the years. As mentioned in the Introduction, I was, except for the earliest years, not much interested in specific applications to problems in number theory. It seems, however, not out of place to end this account with two specific applications to number theory, and this will be the subject of the two next, and last, Sections.

22. Remarks on the Brun–Titchmarsh theorem

We have throughout this account always used the natural ordering of the integers. There are cases where it may be advantageous to order the d in $(\mathcal{P})$ in a different way, not according to the size of d, but the size of some other multiplicative function of d. The problem we shall consider in this Section is

[60] In a survey article, [Montgomery 1], some of the work touched on in Section 19 is referred to, however he gives an inferior version of (20.37) or (20.37″) which I had obtained a few years prior to the result given here.

chosen partly for this reason, and partly because it concerns sharpening the estimation in one of the oldest results in sieve theory.

Before proceeding to the actual problem we shall prove some lemmas.

Lemma 14. *Let q be a prime, denote by Q the product of all $p \le q$, and define*

$$\theta_q = \max_{a,x} \left| x \prod_{p \le q} \left(1 - \frac{1}{p}\right) - \sum_{\substack{a \le n \le a+x \\ (n,Q)=1}} 1 \right| . \tag{22.1}$$

Then θ_q grows faster than any power of q. Also $\theta_1 = 1$, $\theta_2 = 1$, $\theta_3 = 4/3$, $\theta_5 = 28/15$ and $\theta_7 = 106/35$.

As one to determine θ_q may assume $0 \le a \le Q$, and $0 \le x \le Q$, because of the periodicity $\bmod Q$, θ_q can always be found by inspecting a finite number of cases. This is the way the values of θ_q for $q \le 7$ listed in the Lemma were obtained.[61]

As to the first part of the Lemma: Let us assume that $\theta_q = O(q^\alpha)$ for some positive fixed α. We see then from (22.1) that for $u > \alpha$ we would have

$$\pi(q^u, q) = q^u \prod_{p \le q} \left(1 - \frac{1}{p}\right) + O(q^\alpha),$$

where we use the notation $\pi(x, y)$ to designate the number of integers n with $0 < n \le x$ and no prime factors $\le y$. Thus as $q \to \infty$

$$\pi(q^u, q) \sim q^u \frac{e^{-\gamma}}{\log q},$$

where γ is Euler's constant, or writing x instead of q^u,

$$\pi(x, x^{1/u}) \sim u\, e^{-\gamma} \frac{x}{\log x}, \tag{22.2}$$

for $u > \alpha$. But we have for $u \ge 2$ that

$$\pi(x, x^{1/u}) \sim \frac{1}{2} \left(a_+ \left(\frac{1}{u}\right) + a_- \left(\frac{1}{u}\right)\right) \frac{x}{\log x}, \tag{22.3}$$

where a_+ and a_- are the quantities defined by (16.25) and (16.25′). Thus for $u > \alpha$, we would have

$$u\, e^{-\gamma} = \frac{1}{2} \left(a_+ \left(\frac{1}{u}\right) + a_- \left(\frac{1}{u}\right)\right). \tag{22.4}$$

[61] Presumably use of a computer would make it feasible to extend this list quite a bit.

Denoting the right hand side by $a(1/u)$, we see by adding the equations (16.26) and (16.26′) that a satisfies the recursion formula

$$a(w) = a(v) - \int_v^w a\left(\frac{t}{1-t}\right) \frac{dt}{t(1-t)}\,.$$

The same recursion formula is satisfied by the function $u\,e^{-\gamma}$ (considered as a function of $1/u$). If by (22.4) the two functions are equal for $u > \alpha$, they must by the recursion formula be equal for $u \geq 2$. But for $u = 2$ the right hand side of (22.4) equals 1 while the left hand side is $2\,e^{-\gamma} > 1$, so our original assumption must be false, which proves the first part of the Lemma.

Lemma 15. *Let $\sigma(\rho)$ denote the sum of the divisors of ρ, $\varphi(\rho)$ Euler's function, Q be defined as in Lemma 14 and write for $z > 1$,*

$$\sum\nolimits_1 = \sum_{\substack{\sigma(\rho)\leq z\\ (\rho,Q)=1}} \mu^2(\rho)\,\frac{\sigma(\rho)}{\varphi(\rho)}\left(1 - \frac{\sigma(\rho)}{z}\right),$$

$$\sum\nolimits_2 = \sum_{\substack{\sigma(\rho)\leq z\\ (\rho,Q)=1}} \mu^2(\rho)\,\frac{1}{\varphi(\rho)}\left(1 - \frac{\sigma(\rho)}{z}\right),$$

and

$$\sum\nolimits_3 = \sum_{\substack{\sigma(\rho)\leq z\\ (\rho,Q)=1}} \mu^2(\rho)\,\frac{1}{\varphi(\rho)}\left(1 - \frac{\sigma(\rho)}{z}\right)^2.$$

Then

(22.5) $$\sum\nolimits_1 = \frac{z}{2}\prod_{p\leq q}\left(1 - \frac{1}{p}\right) + O(z\,e^{-\sqrt{\log z}}),$$

(22.5′) $$\sum\nolimits_2 = \prod_{p\leq q}\left(1 - \frac{1}{p}\right)\{\log z + \gamma + \kappa_1 + \kappa'(q) - 1\} + O(e^{-\sqrt{\log z}}).$$

and

$$\sum\nolimits_3 = \prod_{p\leq q}\left(1 - \frac{1}{p}\right)\left\{\log z + \gamma + \kappa_1 + \kappa'(q) - \frac{3}{2}\right\} + O(e^{-\sqrt{\log z}}).$$

Here γ is Euler's constant, while

(22.6) $$\kappa_1 = \sum_p \left(\frac{\log p}{p-1} - \frac{\log(p+1)}{p}\right),$$

and

(22.6′) $$\kappa'(q) = \sum_{p\leq q} \frac{\log(p+1)}{p}\,.$$

It is possible to give a quite elementary proof of this with the factor $e^{-\sqrt{\log z}}$ replaced by $1/\log z$ in the remainder terms. Since this however is more cumbersome and would take more space, we shall sketch an analytic proof along very standard lines.

We write $s = \sigma + it$, and define for $\sigma > 1$ the function

$$f(s) = \sum_{(\rho,Q)=1} \mu^2(\rho) \frac{1}{\varphi(\rho)} (\sigma(\rho))^{1-s} = \prod_{p>q} \left(1 + \frac{1}{p-1} (p+1)^{1-s}\right).$$

We also define

$$g(s) = f(s)(\zeta(s))^{-1} = \prod_{p \le q} (1 - p^{-s}) \prod_{p>q} \left(1 + \frac{1}{p-1} (p+1)^{1-s}\right) (1 - p^{-s}),$$

and see that $g(s)$ is regular analytic for $\sigma > 1/2$, since the infinite product converges absolutely there. We also see that for $\sigma \ge \alpha > 1/2$, the logarithm of $g(s)$ is equal to

$$\sum_{p>q} \frac{1}{p} \left((p+1)^{1-s} - p^{1-s}\right)$$

plus a bounded function. Now, for $p \le |t|$, we have

$$|(p+1)^{1-s} - p^{1-s}| \le 3\, p^{1-\sigma},$$

and for $p > |t|$ we have

$$|(p+1)^{1-s} - p^{1-s}| \le |s|\, p^{-\sigma}.$$

Using these inequalities it is easily established that for $\sigma \ge \alpha$ and

$$\sigma \ge 1 - \frac{a}{\log(|t| + 2)},$$

where a is a positive constant, we have

$$g(s) = O((\log(|t| + 2))^c),$$

where c is a constant depending on a. Since for $|t| \ge 1$, we have

$$\zeta(s) = O(\log(|t| + 2))$$

in the same region, we get

$$f(s) = O((\log(|t| + 2))^{c+1}), \tag{22.7}$$

in the region $|t| \ge 1$; $\sigma \ge \alpha$ and

$$\sigma \ge 1 - \frac{a}{\log(|t| + 2)}.$$

Also, for $\sigma > 1/2$, $f(s)$ is regular except for a pole of first order at $s = 1$. Since $g(1) = \prod_{p \le q}(1 - 1/p)$, and

$$\frac{g'}{g}(1) = \sum_p \left(\frac{\log p}{p-1} - \frac{\log(p+1)}{p} \right) + \sum_{p \le q} \frac{\log(p+1)}{p} = \kappa_1 + \kappa'(q) \,,$$

we get that $f(s)$ around the pole at $s = 1$ has a Laurent expansion that begins with

$$g(1) \left\{ \frac{1}{s-1} + \gamma + \kappa_1 + \kappa'(q) + \ldots \right\} . \tag{22.8}$$

By standard formulas we now have

$$\sum\nolimits_1 = \frac{1}{2\pi i} \int_{2-i\infty}^{2+i\infty} \frac{z^s f(s)}{s(s+1)} \, ds \,,$$

$$\sum\nolimits_2 = \frac{1}{2\pi i} \int_{2-i\infty}^{2+i\infty} \frac{z^s f(s+1)}{s(s+1)} \, ds \,,$$

and

$$\sum\nolimits_3 = \frac{2}{2\pi i} \int_{2-i\infty}^{2+i\infty} \frac{z^s f(s+1)}{s(s+1)(s+2)} \, ds \,.$$

Shifting now in all these integrals the path of integration to the left, to the boundary of the region for which we proved (22.7), taking into account the residue at the pole of the integrand at $s = 1$ in the first integral and the residue at the second order pole in the two other integrals at $s = 0$, these residues give the main terms in the asymptotic expressions for $\sum_1, \sum_2$ and $\sum_3$ (22.5), (22.5′) and (22.5″), while the integrals over the remaining contour are easily estimated by taking absolute values and utilizing (22.7) and give the remainder terms in (22.5), (22.5′) and (22.5″).

The problem we shall now consider is to find an upper bound for the number of integers left in an interval $\mathcal{I}_x$; $a \le n \le a + x$, when we exclude one residue class for each prime $< z$, and we choose z just large enough to get the smallest possible upper bound that the method will allow. Since there is no loss in generality if we assume the residue class is always that represented by zero, we look at the case where we are excluding the n divisible by p for $p < z$. We shall do this by assuming that we already have excluded the n divisible by a $p \le q$, and then apply a Λ^2 sieve with the range $\mathcal{P}(q+1, z)$. Using Lemma 14 we get the upper bound

$$\frac{\varphi(Q)}{Q} x \sum_{d,d'} \frac{(d,d')}{dd'} \lambda_d \lambda_{d'} + \theta_q \left(\sum_d |\lambda_d| \right)^2 ,$$

or if we use the formulas (7.3) and (7.3′) to introduce the new variables y_ρ (in this case $f(d) = d$ and $f'(\rho) = \varphi(\rho)$), the first term becomes

$$\frac{\varphi(Q)}{Q} x \sum_{(\rho,Q)=1} \frac{y_\rho^2}{\varphi(\rho)},$$

we also find

$$\sum_d |\lambda_d| \le \sum_{(\rho,Q)=1} \frac{\sigma(\rho)}{\varphi(\rho)} |y_\rho|$$

and have the side condition

$$\sum_{(\rho,Q)=1} \frac{y_\rho}{\varphi(\rho)} = 1. \tag{22.9}$$

Since it is evident that we lose nothing by assuming the $y_\rho \ge 0$, we may write the upper bound in the homogeneous form

$$\frac{\dfrac{\varphi(Q)}{Q} x \displaystyle\sum_{(\rho,Q)=1} \frac{y_\rho^2}{\varphi(\rho)} + \theta_q \left(\sum_{(\rho,Q)=1} \frac{\sigma(\rho)}{\varphi(\rho)} y_\rho \right)^2}{\left(\displaystyle\sum_{(\rho,Q)=1} \frac{y_\rho}{\varphi(\rho)} \right)^2}, \tag{22.10}$$

and drop the condition (22.9).

We now seek to choose the $y_\rho \ge 0$ and such that they minimize the expression (22.10) with z chosen as large as possible. Our choice is

$$y_\rho = 1 - \frac{\sigma(\rho)}{z}$$

for $\sigma(\rho) < z$ and $y_\rho = 0$ for $\sigma(\rho) \ge z$.[62] (22.10) then becomes

$$\frac{\dfrac{\varphi(Q)}{Q} x \sum_3 + \theta_q \left(\sum_1\right)^2}{\left(\sum_2\right)^2}, \tag{22.11}$$

where $\sum_1, \sum_2$ and $\sum_3$ are defined in Lemma 15. Using now the formulas (22.5), (22.5′) and (22.5″) and writing $z = e^u \sqrt{x}$, where the parameter u (independent of x) will be chosen later, we get after a little manipulation for the upper bound

$$\frac{2x}{\log x + 2\gamma + 2\kappa_1 + 2\kappa'(q) - 1 + 2u - \dfrac{\theta_q}{2} e^{2u} + O(e^{-\frac{1}{2}\sqrt{\log x}})}. \tag{22.12}$$

[62] This choice is easy to motivate when we look at the condition for minimum of (21.10) for the $y_\rho > 0$.

The optimal choice of u is seen to be given by

$$e^{2u} = \frac{2}{\theta_q}\,.$$

This gives the upper bound

$$\frac{2x}{\log x + 2\gamma + 2\kappa_1 + 2\kappa'(q) - 2 + \log\dfrac{2}{\theta_q} + O(e^{-\frac{1}{2}\sqrt{\log x}})}\,. \tag{22.13}$$

We still can choose q so that

$$2\kappa'(q) - \log\frac{2}{\theta_q}$$

becomes as large as possible. That there is an optimal choice is clear since by Lemma 14 this expression tends to $-\infty$ as $q \to \infty$. Since we have not determined any value of θ_q beyond $q = 7$ we shall use $q = 7$, though it is quite possible that some larger q would do better. Using the value of θ_7 from Lemma 14, calculating $\kappa'(q)$ (easy, there are only 4 simple terms) and κ_1 (harder, the infinite series converges slowly, we sum it for the $p < 200$) we find

$$2\gamma + 2\kappa_1 + 2\kappa'(7) - 2 + \log\frac{2}{\theta_7} > 2.8\,,$$

and

$$z = \sqrt{\frac{35}{53}\,x} < \sqrt{x}\,.$$

Thus if we exclude from $\mathcal{I}_x$ one residue class for each prime $p < \sqrt{x}$, there remain at most

$$\frac{2x}{\log + 2.8} \tag{22.14}$$

numbers if $x > x_0$. To determine a x_0 beyond which this would be true we should have to estimate the error term $O(e^{-\frac{1}{2}\sqrt{\log x}})$ in the denominator of (22.13) more explicitly. We shall not go into that here. It is easy to see that the estimation (21.24) is false for small x, for instance for $x = 16$ when it would give an upper bound < 6 while it is easy to see that 6 numbers may remain even if we extend the sifting range to all primes < 16. It is probable that (22.14) becomes valid with an x_0 not very much larger than 16.

As a consequence of (22.14) we have

$$\pi(x+y) - \pi(x) < \frac{2y}{\log y + 2.8}\,, \tag{22.15}$$

for $y > x_0$ *and all* $x > 0$. *Similarly for* $(k, \ell) = 1$, *if* $\pi_{k,\ell}(x)$ *denotes the number of primes* $\leq x$ *which are* $\equiv \ell (\bmod k)$, *we have*

$$\pi_{k,\ell}(x+y) - \pi_{k,\ell}(x) < \frac{2y}{\varphi(k)\,(\log(y/k) + 2.8)}, \tag{22.15'}$$

for $y/k > x_0$. (22.15) is obvious, while (22.15′) follows by noting that sifting the section of the arithmetic progression that lies in an interval of length y is equivalent to sifting an interval of length y/k, and observing that the rule of monotonicity holds for sifting of an interval, and that the sifting range for the arithmetic progression does not include the primes dividing k. It should be noted that if k contains other primes than 2, particularly if it contains small primes, we could improve this result, by not relying on the monotonicity but instead going through the argument *ab ovo* for the case of the arithmetic progression. One then gets a larger number instead of 2.8. Inequalities of the form (22.15) or (22.15′) have since more than 50 years been known as the Brun–Titchmarsh theorem. Brun established it first in the form (22.15) (with a larger constant than 2 in the numerator), then Titchmarsh established it in the form (22.15′) for the arithmetic progression (also with a larger constant), at the time it was not realized that the second result could be derived by appealing to monotonicity. If one is willing to invest more effort it is clear that the constant 2.8 in the denominator may be improved by using the form

$$\sum_d \left| \sum_{[d_1,d_2]=d} \lambda_{d_1}\lambda_{d_2} \right|,$$

in the remainder term instead of the simpler

$$\left(\sum_d |\lambda_d| \right)^2,$$

it is possible to increase our constant somewhat in this way, but not significantly, and the cost in complication is very high.

23. An early approach to the twin prime and the Goldbach problem

We shall for simplicity confine ourselves to the twin prime problem, and so avoid the complication of dealing with a general even difference or even sum. This is a complication of detail only, and it is easily seen how one would modify the approach for the problem of a general even difference or even sum. The result we shall give is by now of historical interest only since after the appearance of E. Bombieri's theorem on primes in arithmetic progressions, one has been able to treat these problems as problems with sifting density one rather than two, and with the additional ingenious ideas brought in by Jing–Run Chen results have been obtained that go far beyond what the approach here described could be expected to yield even with further refinements.

As an illustration of the variety of ways in which the Λ^2 method may be adapted to a particular problem, it remains however of some interest. The reader may notice the affinity with the ideas in my attempt of 1946 to prove the existence of primes in the interval $(x,\ (1+c)x)$, which was briefly described in Section 18. The proof here dates from late 1950 or early 1951, the result was first proved about a year earlier using a set of weights similar to those used in [Ankeny 1] to prove the result.

The method could equally well have been applied to the Goldbach problem, but to avoid having to pay special attention to the prime factors of the number which we wish to show can be represented as a sum or difference of primes or at least "almost primes" (numbers with very few prime factors), we choose here the twin prime problem.

We shall need a series of lemmas.

Lemma 16. *Let k be a positive integer, $\tau(n)$ be the divisor function and*

$$D_k(x) = \sum_{\substack{(n,k)=1 \\ n\le x}} \tau(n)\,,$$

then we have

(23.1)

$$D_k(x) = \left(\frac{\varphi(k)}{k}\right)^2 x\left(\log x + c + 2\sum_{p\,|\,k}\frac{\log p}{p-1}\right) + O\left(\sqrt{x}\prod_{p\,|\,k}\left(1+\frac{1}{\sqrt{p}}\right)^2\right).$$

Here, if k is not squarefree the sum and the product taken over the $p\,|\,k$ on the right hand side are extended only over distinct prime factors of k.

It is of course enough to prove this for squarefree k. Lemma 12 gives

$$D(x) = D_1(x) = x\,(\log x + c) + O(\sqrt{x})\,. \tag{23.2}$$

It is easily seen that

$$D_k(x) = \sum_{\substack{d_1\,|\,k \\ d_2\,|\,k}} \mu(d_1)\mu(d_2)\,D\left(\frac{x}{d_1 d_2}\right). \tag{23.3}$$

Expressing $D(x/(d_1d_2))$ by (23.2) (which holds uniformly for $0 < x < \infty$) and inserting in (23.3), we get the remainder term directly, while the main term requires a little manipulation to get (23.1).

We shall need some facts about Kloosterman sums:

$$S(m,n;k) = \sum_{h\bar{h}\equiv 1\ (k)} e^{2\pi i\frac{mh+n\bar{h}}{k}}\,,\ [63] \tag{23.4}$$

[63] H.D. Kloosterman [1].

where m and n are integers and h runs over a complete set of residue classes relatively prime to k.

We have

$$|S(m,n;k)| \le (m,n,k)^{1/2} k^{1/2} \tau(k)\,.\text{[64]} \tag{23.5}$$

Lemma 17. *If either m or n is $\equiv 0 \pmod k$, we have*

$$S(m,n;k) - \frac{S(m,0;k)\,S(n,0;k)}{\varphi(k)} = 0\,, \tag{23.6}$$

while if m and n are both not $\equiv 0 \pmod k$ we have

$$\left| S(m,n;k) - \frac{S(m,0;k)\,S(n,0;k)}{\varphi(k)} \right| < 2\,(m,n,k)^{1/2}\, k^{1/2} \tau(k)\,. \tag{23.6'}$$

(23.6) here follows from $S(0,0;k) = \varphi(k)$, while (23.6′) follows easily from (23.5) and the fact that

$$S(m,0;k) = \sum_{\delta \mid (m,k)} \mu(k/\delta)\,\delta\,,$$

so

$$|S(m,0;k)| \le \varphi(m,k)\,.$$

We define for $|\alpha| \le \frac{1}{2}\pi$,

$$S_t(\alpha) = \sum_{0<\nu<t} e^{2\pi i \nu \alpha}\,, \tag{23.7}$$

and have then the inequality

$$|S_t(\alpha)| \le \frac{1}{2|\alpha|}\,. \tag{23.8}$$

Lemma 18. *For $(k,\ell) = 1$ and $k < x^2$ we have for any $\varepsilon > 0$, that*

$$\begin{aligned} D_{k,\ell}(x) &= \sum_{\substack{n \equiv \ell\ (k) \\ n \le x}} \tau(n) \\ &= \frac{\varphi(k)}{k^2}\, x \left(\log x + c + 2 \sum_{p \mid k} \frac{\log p}{p-1} \right) + O(k^{-1/4} x^{1/2+\varepsilon})\,. \end{aligned} \tag{23.9}$$

[64] The deepest part of this ($k = p$) is due to A. Weil [1], the rest then follows from the results of H. Salié [1] (for $k = p^r$, $r > 1$) and the multiplicative properties of Kloosterman sums.

We have the two identities

(23.10)

$$\int_1^x D_{k,\ell}(t)\frac{dt}{t} = \sum_{\substack{n\equiv \ell\,(k)\\ n\le x}} \tau(n)\,\log\frac{x}{n}$$
$$= \frac{1}{k^2}\sum_{-k/2<m,n\le k/2} S(m,n\ell;k)\int_1^x S_t\left(\frac{m}{k}\right)\,S_{x/t}\left(\frac{n}{k}\right)\,\frac{dt}{t},$$

and

(23.10′)

$$\int_1^x D_k(t)\frac{dt}{t} = \sum_{\substack{(n,k)=1\\ n\le x}} \tau(n)\,\log\frac{x}{n}$$
$$= \frac{1}{k^2}\sum_{-k/2<m,n\le k/2} S(m,0;k)S(n,0;k)\int_1^x S_t\left(\frac{m}{k}\right)\,S_{x/t}\left(\frac{n}{k}\right)\,\frac{dt}{t}.$$

These identities are easily proved by inserting the expressions for the Kloosterman sums and the $S_t(m/k)\,S_{x/t}(n/k)$ and carrying out the summations over m and n.

Observing that when $(k,\ell)=1$, we have $S(n,0;k)=S(\ell n,0;k)$, we get by dividing the second identity with $\varphi(k)$ and subtracting it from the first, that by Lemma 17,

(23.11)

$$\left|\int_1^x D_{k,\ell}(t)\frac{dt}{t} - \frac{1}{\varphi(k)}\int_1^x D_k(t)\frac{dt}{t}\right|$$
$$< \frac{2}{k^2}\sum_{0<|m|,|n|\le k/2}(m,n,k)^{1/2}k^{1/2}\tau(k)\frac{k^2}{4|mn|}\int_1^x\frac{dt}{t}$$
$$= 2k^{1/2}\tau(k)\log x\sum_{0<m,n\le k/2}\frac{(m,n,k)^{1/2}}{mn} < 6k^{1/2}\tau(k)\log^2(k+1)\log x$$

where we have used that

$$(m,n,k)^{1/2}\le\sum_{\substack{d\,|\,m\\ d\,|\,n\\ d\,|\,k}} d^{1/2},$$

so that

$$\sum_{0<m,n\le k/2}\frac{(m,n,k)^{1/2}}{mn}\le\sum_{d\,|\,k}d^{1/2}\left(\sum_{\substack{d\,|\,m\\ 0<m\le k/2}}\frac{1}{m}\right)^2<\sum_{d}d^{-3/2}\left(\sum_{0<m\le k/2}\frac{1}{m}\right)^2$$
$$<\zeta(3/2)\,\log^2(k+1)<3\log^2(k+1).$$

Using now the inequality valid for $0<\delta<1$,

$$\frac{1}{\delta}\int_{x\,e^{-\delta}}^{x} D_{k,\ell}(t)\frac{dt}{t}\le D_{k,\ell}(x)\le \frac{1}{\delta}\int_{x}^{x\,e^{\delta}} D_{k,\ell}(t)\frac{dt}{t}$$

as well as the formula for $D_k(t)$ given in Lemma 16, we get from (23.11) that

$$D_{k,\ell}(x)=\frac{\varphi(k)}{k^2}\,x\,\left(\log x+c+2\sum_{p\,|\,k}\frac{\log p}{p-1}\right)+O\left(\delta\,\frac{\varphi(k)}{k^2}\,x\,\log x\right)$$
$$+O\left(\frac{1}{\delta}k^{1/2}\tau(k)\log^2(k+1)\log x\right).$$

Choosing here the δ which makes the two remainder terms of the same order we get the result of the Lemma, with a sharper but more complicated remainder term.

Lemma 19. *We have for $(6k,\ell)=1$ and $6k$ squarefree,*
(23.12)

$$\sum_{\substack{n\equiv\ell\ (6k)\\ n\le x}} 2^{\nu(n)}=C_1\,x\prod_{p\,|\,k}\frac{p-2}{p(p-1)}\left(\log x+C'+2\sum_{p\,|\,k}\frac{\log p}{p-2}\right)+O(k^{-1/4}x^{1/2+\varepsilon}),$$

for any $\varepsilon>0$, here $\nu(n)$ is the number of prime factors of n, counted with multiplicity, and

$$C_1=\frac{1}{18}\prod_{p>3}\frac{(p-1)^2}{p(p-2)}. \tag{23.13}$$

Let
$$\sum\frac{b_d}{d^s}=\prod_{p>3}\frac{(1-p^{-s})^2}{1-2p^{-s}}=\prod_{p>3}\left(1+\frac{1}{p^s(p^s-2)}\right).$$
We see that $b_d\ge 0$, and that the first singularity on the real axis is a simple pole at $s=1/2$, thus we have

$$\sum_{d\le x}\frac{b_d}{\sqrt{d}}=O(\log x).$$

now for $(n,6)=1$, we have

$$2^{\nu(n)}=\sum_{d\,|\,n}b_d\tau(n/d),$$

which gives

$$\sum_{\substack{n\equiv\ell\ (6k)\\ n\le x}}2^{\nu(n)}=\sum_{(d,k)=1}b_d\sum_{\substack{dm\equiv\ell\ (6k)\\ m\le x/d}}\tau(m)=\sum_{(d,k)=1}b_dD_{6k,\ell\bar d}(x/d),$$

where $d\bar{d} \equiv 1\ (6k)$. Inserting here the expression for

$$D_{6k,\ell\bar{d}}(x/d)$$

from Lemma 18, and summing the series

$$\sum_{(d,k)=1} \frac{b_d}{d} \quad \text{and} \quad \sum_{(d,k)=1} \frac{b_d}{d} \log d\,,$$

we obtain the result of the Lemma after some simplification.

Lemma 20. *If f_1 and f_2 are multiplicative functions and ω_1 and ω_2 are additive functions on the positive integers, then for n squarefree and A a constant, we have that*

$$\sum_{uv=n} f_1(u)f_2(v)\,(A+\omega_1(u)+\omega_2(v)) = f_3(n)\,(A+\omega_3(n))\,, \tag{23.14}$$

where f_3 is multiplicative with

$$f_3(p) = f_1(p) + f_2(p)\,, \tag{23.15}$$

and ω_3 additive with

$$f_3(p)\omega_3(p) = f_1(p)\omega_1(p) + f_2(p)\omega_2(p)\,. \tag{23.15'}$$

This is simplest verified by considering for variable s the identity

$$e^{As} \sum_{uv=n} f_1(u)f_2(v)e^{s(\omega_1(u)+\omega_2(v))} = e^{As} \prod_{p\,|\,n} (f_1(p)\,e^{s\omega_1(p)} + f_2(p)\,e^{s\omega_2(p)})\,,$$

forming the derivative with respect to s and putting $s = 0$.

Lemma 21. *If f is multiplicative and ω additive, we have for squarefree d*

$$f(d)\omega(d) = \sum_{\rho\,|\,d} f'(\rho)\,\omega'(\rho)\,, \tag{23.16}$$

here $\omega'(\rho)$ is an additive function defined by

$$f'(p)\,\omega'(p) = f(p)\omega(p)\,. \tag{23.17}$$

This follows by considering $f_s(d) = f(d)\,e^{s\omega(d)}$ and writing

$$f_s(d) = \sum_{\rho\,|\,d} f_s'(\rho)$$

where $f_s'(p) = f_s(p) - 1$. Taking the derivative with respect to s and putting $s = 0$, we get the result of the Lemma.

We shall denote by $h_k(v,w)$ the function defined by (10.11) in the case of constant sifting density k, and have then

Lemma 22. *We have for $v \geq w$,*

$$h_k(v,w) = \frac{e^{-\gamma k}}{\Gamma(k+1)} \left(\frac{w}{v}\right)^k, \tag{23.18}$$

where γ is Euler's constant.

From the form of Lemma 7 it is clear that for constant sifting density k we have that for $v \geq w$

$$h_k(v,w) = c_k \left(\frac{w}{v}\right)^k,$$

to determine the constant c_k, we put $w = 1$ and consider for $v \geq 1$

$$c_k = v^k h_k(v,1) = \frac{v^k}{2\pi i} \int_{-i\infty}^{i\infty} e^{s-k\int_0^v \frac{1-e^{-st}}{t}\,dt} \frac{ds}{s} = \frac{v^k}{2\pi i} \int_{-i\infty}^{i\infty} e^{s-k\int_0^{sv} \frac{1-e^{-t}}{t}\,dt} \frac{ds}{s}.$$

Here we may write

$$\int_0^{vs} \frac{1-e^{-t}}{t}\,dt = \gamma + \log vs + \int_{vs}^{\infty} \frac{e^{-t}}{t}\,dt$$

and get

$$c_k = \frac{e^{-k\gamma}}{2\pi i} \int_{-i\infty}^{i\infty} e^{s-k\int_{vs}^{\infty}\frac{e^{-t}}{t}\,dt} \frac{ds}{s^{k+1}}.$$

Letting here $v \to \infty$, we get

$$c_k = \frac{e^{-k\gamma}}{2\pi i} \int_{-i\infty}^{i\infty} \frac{e^s}{s^{k+1}}\,ds = \frac{e^{-k\gamma}}{\Gamma(k+1)},$$

which proves the Lemma.

We now consider the two expressions

$$Q_1 = Q_1(\lambda) = \sum_{\substack{n \equiv -1\ (6) \\ x \leq n < 2x}} \{2^{\nu(n)} + 2^{\nu(n+2)}\} \times \left\{ \sum_{d \mid n(n+2)} \lambda_d \right\}^2, \tag{23.19}$$

and

$$Q_2 = Q_2(\lambda) = \sum_{\substack{n \equiv -1\ (6) \\ x \leq n < 2x}} \left\{ \sum_{d \mid n(n+2)} \lambda_d \right\}^2. \tag{23.20}$$

For Q_2 we have immediately

$$Q_2 = \frac{x}{6} \sum_{d_1,d_2 \le z} \frac{f(\kappa)}{f(d_1)f(d_2)} \lambda_{d_1}\lambda_{d_2} + O\left(\left(\sum_{d\le z} \tau(d)\,|\lambda_d|\right)^2\right), \tag{23.20$'$}$$

where $f(d) = d/\tau(d)$, and $\kappa = (d_1, d_2)$.

To obtain a similar expression for Q_1, we have first to evaluate

$$N_d = \sum_{\substack{d\,|\,n(n+2)\\ n\equiv -1\ (6)\\ x\le n<2x}} \{2^{\nu(n)} + 2^{\nu(n+2)}\} = N'_d + N''_d\,, \tag{23.21}$$

where

$$N'_d = \sum_{\substack{d\,|\,n(n+2)\\ n\equiv -1\ (6)\\ x\le n<2x}} 2^{\nu(n)}\,, \tag{23.21$'$}$$

and N''_d is a similar expression with $2^{\nu(n+2)}$ instead of $2^{\nu(n)}$.

We write

$$N'_d = \sum_{d_1 d_2 = d} N'_{d_1,d_2} \tag{23.22}$$

where

$$N'_{d_1,d_2} = \sum_{\substack{d_1\,|\,n\\ d_2\,|\,n+2\\ n\equiv -1\ (6)\\ x\le n<2x}} 2^{\nu(n)} = 2^{\nu(d_1)} \sum_{\substack{x/d_1\le m<2x/d_1\\ m\equiv \ell\ (6d_2)}} 2^{\nu(m)}\,,$$

where ℓ is the residue class mod $6d_2$ determined by $d_1\ell \equiv -1$ (6), $d_1\ell \equiv -2$ (d_2). Since $(\ell, 6d_2) = 1$, we have immediately from Lemma 19, that

$$N'_{d_1,d_2} = C_1\, x\, \frac{2^{\nu(d_1)}}{d_1 d_2} \prod_{p\,|\,d_2} \frac{p-2}{p-1} \left(\log\frac{x}{d_1} + C' + 2\sum_{p\,|\,d_2} \frac{\log p}{p-2}\right) + O\left(\frac{2^{\nu(d_1)}}{\sqrt{d_1}}\, d_2^{-1/4} x^{1/2+\varepsilon}\right). \tag{23.23}$$

We may now evaluate N'_d from (23.22) and (23.23) by using Lemma 20 with $f_1(p) = 2$, $\omega_1(p) = -\log p$, $f_2(p) = \frac{p-2}{p-1}$ and $\omega_2(p) = \frac{2\log p}{p-2}$, which gives

$$f_3(p) = \frac{2p-4}{p-1}$$

and

$$\omega_3(p) = -\frac{2p-4}{3p-4}\log p = -\frac{2}{3}\log p + \frac{4}{3}\,\frac{\log p}{3p-4}\,.$$

From this we get

$$N'_d = C_1 \frac{x}{d} \prod_{p\,|\,d} \frac{3p-4}{p-1} \left(\log x + C'' - \frac{2}{3}\log d + \frac{4}{3} \sum_{p\,|\,d} \frac{\log p}{3p-4} \right) + O(d^{-1/4}x^{1/2+\varepsilon})$$

for $\varepsilon > 0$. For N''_d we get the same expression so finally we have

$$(23.24)\ N_d = 2C_1 \frac{x}{d} \prod_{p\,|\,d} \frac{3p-4}{p-1} \left(\log x + C'' - \frac{2}{3}\log d + \frac{4}{3} \sum_{p\,|\,d} \frac{\log p}{3p-4} \right) + O(d^{-1/4}x^{1/2+\varepsilon})\,.$$

Writing f_4 for the multiplicative function with

$$f_4(p) = \frac{p(p-1)}{3p-4}\,,$$

and ω_4 for the additive function with

$$\omega_4(p) = \frac{4}{3}\,\frac{\log p}{3p-4}\,,$$

we now get
(23.25)

$$Q_1(\lambda) = 2C_1 x \sum_{d_1,d_2 \le z} \frac{f_4(\kappa)\,\lambda_{d_1}\lambda_{d_2}}{f_4(d_1)f_4(d_2)} \left(\log x + C'' - \frac{2}{3}\log\frac{d_1d_2}{\kappa} + \omega_4\left(\frac{d_1d_2}{\kappa}\right)\right) + O\left(x^{1/2+\varepsilon} \sum_{d_1,d_2\le z} \frac{\kappa^{1/4}}{(d_1,d_2)^{1/4}} |\lambda_{d_1}\lambda_{d_2}| \right).$$

We now try to make the ratio of Q_1/Q_2 as small as possible. Since this is a rather intractable minimum problem, we shall instead just choose our λ's so as to minimize

$$Q_1^* = \sum_{d_1,d_2\le z} \frac{f_4(\kappa)}{f_4(d_1)f_4(d_3)} \lambda_{d_1}\lambda_{d_2}$$

with $\lambda_1 = 1$. This is the standard problem. We find

$$(23.26)\qquad Q_1(\min) = \frac{1}{\displaystyle\sum_{\rho\le z} \frac{1}{f'_4(\rho)}} = \frac{1}{\sum_z}\,,$$

and

$$(23.27)\qquad \frac{\lambda_d}{f_4(d)} = \frac{\mu(d)}{\sum_z} \sum_{\substack{d\,|\,\rho \\ \rho\le z}} \frac{1}{f'_4(\rho)}\,.$$

Since f_4 corresponds to constant sifting density 3, we get from Lemma 22 and Lemma 7 that

$$\begin{aligned}
(23.28)\qquad \sum\nolimits_z &\sim \frac{e^{-3\gamma}}{6}\prod_{3<p\le z}\frac{f_4(p)}{f_4'(p)}\\
&= \frac{e^{-3\gamma}}{6}\prod_{3<p\le z}\frac{1}{(1-1/p)^3}\times\prod_{3<p\le z}\frac{f_4(p)}{f_4'(p)}\left(1-\frac{1}{p}\right)^3\\
&= \frac{e^{-3\gamma}}{6\cdot 27}\prod_{p\le z}\frac{1}{(1-1/p)^3}\times\prod_{3<p\le z}\frac{(p-1)^4}{p^2(p-2)^2}\\
&\sim \frac{\log^3 z}{2\cdot 9^2}\prod_{p>3}\frac{(p-1)^4}{p^2(p-2)^2}\\
&= 2C_1^2\log^3 z\,.
\end{aligned}$$

Since we have that with our choice of λ's,

$$\sum\frac{f_4(\kappa)\,\lambda_{d_1}\lambda_{d_2}}{f_4(d_1)f_4(d_2)}\,(\omega(d_1)+\omega(d_2))$$

equals zero for any $\omega(d)$ with $\omega(1)=0$, it only remains to evaluate

$$(23.29)\qquad \sum_{d_1,d_2\le z}\frac{f_4(\kappa)\lambda_{d_1}\lambda_{d_2}}{f_4(d_1)f_4(d_2)}\left(\frac{2}{3}\log\kappa-\omega_4(\kappa)\right),$$

by the use of Lemma 21, we can write

$$f_4(\kappa)\left(\frac{2}{3}\log\kappa-\omega_4(\kappa)\right)=\sum_{\substack{\rho\mid d_1\\ \rho\mid d_2}}f_4'(\rho)\omega'(\rho)$$

and so transform (23.29) to

$$\sum_{\rho\le z}f_4'(\rho)\,\omega'(\rho)\left\{\sum_{\rho\mid d}\frac{\lambda_d}{f_4(d)}\right\}^2=\frac{1}{\sum_z^2}\sum_{\rho\le z}\frac{\omega'(\rho)}{f_4'(\rho)}\,.$$

Here, since $\omega'(p)=\frac{2}{3}\log p+O((\log p)/p)$, we may replace $\omega'(\rho)$ by $\frac{2}{3}\log\rho$ without changing the asymptotic behavior. The resulting sum can then be handled by Lemma 7′ in much the same way as we handled $\sum_z$, we get that

$$\sum_{\rho\le z}\frac{\omega'(\rho)}{f_4'(\rho)}\sim C_1^2\log^4 z\,.$$

Combining our results, we get that the main term in (23.25) is

$$\sim\frac{1}{C_1}\,x\,\frac{\log x+\frac{1}{2}\log z}{(\log z)^3}\,.$$

Since our $|\lambda_d| \le 1$, the remainder term is

$$O(x^{1/2+\varepsilon}z^{3/2}),$$

thus if we choose $z = x^{1/3-\varepsilon}$, we have

$$Q_1(\lambda) \sim \frac{1}{C_1}\, x\, \frac{\log x + \frac{1}{2}\log z}{(\log z)^3}. \tag{23.30}$$

We next turn to $Q_2(\lambda)$, we have

$$Q_2 = \frac{x}{6}\sum_{\rho\le z} f'(\rho)\left\{\sum_{\rho\,|\,d}\frac{\lambda_d}{f(d)}\right\}^2 + O(z^{2+\varepsilon}). \tag{23.31}$$

We first wish to evaluate the expression in the curly bracket. From (23.27) we get

$$\frac{\lambda_d}{f(d)} = \frac{\mu(d)}{\Sigma_z}\frac{f_4(d)}{f(d)}\sum_{\substack{d\,|\,\sigma\\ \sigma\le z}}\frac{1}{f_4'(\sigma)},$$

thus

$$\sum_{\rho\,|\,d}\frac{\lambda_d}{f(d)} = \frac{1}{\Sigma_z}\sum_{\substack{\rho\,|\,d\,|\,\sigma\\ \sigma\le z}}\mu(d)\frac{f_4(d)}{f(d)}\frac{1}{f_4'(\sigma)}. \tag{23.32}$$

Here we first turn our attention to the summation over d with ρ and σ fixed. We have then

$$\begin{aligned}\sum_{\rho\,|\,d\,|\,\sigma}\frac{\mu(d)f_4(d)}{f(d)} &= \frac{\mu(\rho)f_4(\rho)}{f(\rho)}\sum_{\delta\,|\,\sigma/\rho}\frac{\mu(\delta)f_4(\delta)}{f(\delta)}\\ &= \frac{\mu(\rho)f_4(\rho)}{f(\rho)}\prod_{p\,|\,\sigma/\rho}\left(1-\frac{f_4(p)}{f(p)}\right)\\ &= \mu(\rho)\prod_{p\,|\,\rho}\frac{2(p-1)}{3p-4}\prod_{p\,|\,\sigma/\rho}\frac{p-2}{3p-4}\\ &= \mu(\rho)\prod_{p\,|\,\rho}\frac{2(p-1)}{p-2}\prod_{p\,|\,\sigma}\frac{p-2}{3p-4}.\end{aligned}$$

Inserting this in (23.32) we get

$$\begin{aligned}\sum_{\rho\,|\,d}\frac{\lambda_d}{f(d)} &= \frac{\mu(\rho)}{\Sigma_z}\prod_{p\,|\,\rho}\frac{2(p-1)}{p-2}\sum_{\rho\,|\,\sigma}\frac{1}{f_4'(\sigma)}\prod_{p\,|\,\sigma}\frac{p-2}{3p-4}\\ &= \frac{\mu(\rho)}{\Sigma_z}\prod_{p\,|\,\rho}\frac{2(p-1)}{p-2}\sum_{\substack{\rho\,|\,\sigma\\ \sigma\le z}}\prod_{p\,|\,\sigma}\frac{1}{p-2}\\ &= \frac{\mu(\rho)}{\Sigma_z}\prod_{p\,|\,\rho}\frac{2(p-1)}{(p-2)^2}\sum_{\substack{\sigma'\le z/\rho\\ (\sigma',\rho)=1}}\prod_{p\,|\,\sigma'}\frac{1}{p-2}.\end{aligned} \tag{23.33}$$

Here the last sum can be evaluated easily by standard elementary means, we get it is

$$= 6C_1 \prod_{p\,|\,\rho} \frac{p-2}{p-1} \log\frac{z}{\rho} + O\left(\prod_{p\,|\,\rho}\left(1+\frac{1}{\sqrt{p}}\right)\right).$$

Inserting this in (23.33) we get

$$(23.34)\quad \sum_{\rho\,|\,d} \frac{\lambda_d}{f(d)} = \mu(\rho)\frac{6C_1}{\sum_z} \prod_{p\,|\,\rho} \frac{2}{p-2} \log\frac{z}{\rho} + O\left(\frac{1}{\sum_z}\frac{\tau(\rho)}{\rho}\prod_{p\,|\,\rho}\left(1+\frac{2}{\sqrt{p}}\right)\right).$$

Going back to (23.31) with (23.34) and inserting the main term we get

$$(23.35)\qquad x\,\frac{6C_1^2}{\sum_z^2} \sum_{\rho\leq z} \frac{1}{f'(\rho)} \log^2\frac{z}{\rho}.$$

The sum occurring here is again easily evaluated by Lemma 7′ we find it is asymptotic to

$$\frac{1}{6}C_1 \log^4 z\,.$$

Using also the value of $\sum_z$ from (23.28) we get finally (since the remainder term in (23.34) is seen to give only a contribution of lower order) that

$$(23.36)\qquad Q_2(\lambda) \sim \frac{x}{4C_1\log^2 z}.$$

Combining this with (23.30) we see that

$$(23.37)\qquad \frac{Q_1(\lambda)}{Q_2(\lambda)} \sim 4\left(\frac{\log x}{\log z} + \frac{1}{2}\right),$$

since z can be taken as $x^{1/3-\varepsilon}$ with any positive ε we can bring this ratio as near to 14 as we like. This shows that we must have some n for which

$$(23.38)\qquad 2^{\nu(n)} + 2^{\nu(n+2)} < 16\,,$$

and since

$$16Q_2(\lambda) - Q_1(\lambda) \sim \frac{3\frac{1}{2} - \frac{\log x}{\log z}}{C_1 \log^2 z}\,x\,,$$

we see that the number of such n must be at least of order

$$\frac{x}{\log^2 x}.$$

But (23.38) means *that of n and $n+2$, one has at most 2, the other at most 3 prime factors.*

It is possible to reduce the ratio (23.37) somewhat, since we are not actually at the minimum of the ratio. By adjusting the λ_d's suitably Gerd Hofmeister was able to bring the ratio down from about 14 (which is what we get if we put $z = x^{1/3}$, so we can come arbitrarily close to 14) to about 13. This is not enough to improve the result qualitatively, for that one would need to bring it down below 12 (since $12 = 4 + 8$), but it is enough to exclude for instance the case $\nu(n) = 3$, $\nu(n+2) = 2$, this could be done by replacing the expression

$$2^{\nu(n)} + 2^{\nu(n+2)}$$

by

$$\frac{4}{3}\,2^{\nu(n)} + \frac{2}{3}\,2^{\nu(n+2)}\,,$$

in $Q_1(\lambda)$. This would not affect the ratio Q_1/Q_2, but one sees easily that if the ratio is brought below 40/3, this excludes the case $\nu(n) = 3$, $\nu(n+2) = 2$ without permitting any new possibility. The bound for our z essentially came about since the remainder term in (23.9) makes that formula effective only for $k < x^{2/3-\varepsilon}$. If one could prove that the main term dominates sufficiently (by a certain power of $\log x$ would be enough) for $k < x^{\alpha}$ one could choose z as $x^{\alpha/2-\varepsilon}$ and would end up with $8/\alpha + 2 + \varepsilon$ instead of $14 + \varepsilon$, thus $\alpha > 4/5$ would eliminate the possibility that of n and $n+2$ one has 2 and the other 3 prime factors. Of course better results could be obtained for a smaller α if we choose our λ's so as to get nearer the real minimum of the ratio Q_1/Q_2. Another possibility of improvement is to replace the expression

$$\sum_{d\,|\,n(n+2)} \lambda_d$$

in Q_1 and Q_2 with

$$\sum_{\delta_1\,|\,n,\ \delta_2\,|\,n+2} \lambda_{\delta_1,\delta_2}$$

where the $\lambda_{\delta_1,\delta_2}$ are not just dependent on the product $\delta_1\delta_2$, but can be chosen freely except that $\lambda_{1,1} = 1$. Since we can estimate N'_{d_1,d_2} and N''_{d_1,d_2} well enough as long as $d_1 d_2^{2/3}$ and $d_1^{2/3} d_2$ both are $\leq x^{2/3-\varepsilon}$, this means that we need to assume $\lambda_{\delta_1,\delta_2} = 0$ only for

$$\max(\delta_1\delta_2^{2/3}, \delta_1^{2/3}\delta_2) > x^{1/3-\varepsilon}\,,$$

instead of for $\delta_1\delta_2 > x^{1/3-\varepsilon}$. We thus have the choice of about $x^{2/5-\varepsilon}$ λ's instead of $x^{1/3-\varepsilon}$. This is much more difficult to handle, but should lead to better results. These would not however be as sharp as Jing–Run Chen's result in any case.

Bibliography

(Essentially lists only contributions believed to be somewhat relevant to the contents of these lectures)

Ankeny, N.C., *Applications of the sieve.* Proc. Sympos. Pure Math. vol. VIII, Amer. Math. Soc. Providence, RI 1965 pp. 113–118

Ankeny, N.C. and Onishi, H., *The general sieve.* Acta Arith. 10 (1964/65) pp. 31–62

Bombieri, E., *On the large sieve.* Mathematatika 12 (1965) pp. 201–225

Bombieri, E., *The asymptotic sieve.* Rend. Accad. Naz. XL(5) 1975/87 pp. 243–269

Brun, V., *Le crible d'Eratosthène et le theorémè de Goldbach.* Christiania Vidensk. Selsk. Skr. (1920) Nr 3, 36 pages

Buchstab, A. A., *New improvements in the method of the sieve of Eratosthenes.* Math. Sbornik (N.S.) 4(46) (1938) pp. 375–387

Chen, J. R., *On the representation of a large even integer as a sum of a prime and the product of at most two primes.* Sci. Sinica 16 (1973) pp. 157–176

Chen, J. R., *On the distribution of almost primes in an interval.* Sci. Sinica 18 (1975) pp. 611–627

Diamond, H.,G., *An elementary proof of the prime number theorem with a remainder term.* Inventiones Math. 1 (1970) pp. 199–258

Iwaniec, H., *Rossers sieve.* Acta Arith. 36 (1980) pp. 171–202

Iwaniec, H., *The half-dimensional sieve.* Acta Arith. 29 (1976) pp. 69–95

Iwaniec, H., *Almost primes represented by quadratic polynomials.* Inventiones Math. 47 (1978) pp. 171–182

Jurkat, W. B. and Richert, H. E., *An improvement in Selberg's sieve method. I.* Acta Arith. 11 (1965) pp. 217–240

Kloosterman, H. D., *On the representation of a number in the form* $ax^2 + by^2 + cz^2 + dt^2$. Acta Math. 49 (1926) pp. 407–464

Kuhn, P., *Zur Viggo Brunschen Siebmethode, I.* Norske Vid. Selsk. Forh. Trondhjem 14 (1941) No. 39, pp. 145-148

Landau, E., *Über die Einteilung der positiven ganzen Zahlen in vier Klassen nach der Mindestzahl der zu ihrer additiven Zusammensetzung erforderlichen Quadrate.* Arch. d. Math. u. Physik (3) 13 (1908) pp. 305–312

Montgomery, H., *The analytic principle of the large sieve.* Bull. Amer. Math. Soc. 84 (1978) pp. 547–567

Rademacher, H., *Beiträge zur Viggo Brunschen Methode in der Zahlentheorie.* Abh. math. Sem. Univ. Hamburg 3 (1924) pp. 12–30

Salié, H., *Über die Kloostermanschen Summen* $S(u, v; q)$. Math. Zeitschr. 34 (1931) pp. 91–109

Salié, H., *Zur Abschätzung der Fourier-koeffizienten ganzer Modulformen.* Math. Zeitscher 36 (1932) pp. 263–278

Selberg, A., *On an elementary method in the theory of primes.* Norske Vid. Selsk. Forh. Trondhjem 19 (1947) no. 18, pp. 64–67

Selberg, A., *On elementary methods in prime number theory.* C. R. 11 Skand. Math. Kong. Trondheim 1949 pp. 13–22

Selberg, A., *The general sieve method and its place in prime number theory.* Proc. Internat. Congr. Math. Cambridge Mass. 1950, vol. 1, pp. 286–292

Selberg, A., *Sieve Methods.* Proc. Sympos. Pure Math. vol. XX Amer. Math. Soc. Providence, RI, 1971 pp. 311–351

Selberg, A., *Remarks on Sieves*. Proc. 1972 Number Theory Conf. Univ. Colorado, Boulder, CO, pp. 205–216

Selberg, A., *Remarks on multiplicative functions*. Springer Lecture Notes, vol. 626, pp. 232–241

Selberg, A., *Sifting problems, sifting density and sieves. Number theory, trace formulas and discrete groups.* (Oslo 1987) Academic Press 1989 pp. 467–484

Titchmarsh, E. C., *A divisor problem*. Rend. Circ. Math. Palermo 54 (1930) pp. 414–429

Tsang, K. M., *Remarks on the sieving limit of the Buchstab-Rosser sieve. Number theory, trace formulas and discrete groups.* (Oslo 1987) Academic Press 1989 pp. 485–502

Weil, A., *On some exponential sums*. Proc. Nat. Acad. Sci. USA 34 (1948) pp. 204–207

Bibliography

1. Über einige arithmetische Identitäten. Avhandlinger utgitt av Det Norske Videnskaps-Akademi i Oslo. I. Mat.-Naturv. Klasse (1936), No. 8, 1–23
2. Über die Mock-Thetafunktionen siebenter Ordnung. Archiv for Mathematik og Naturvidenskab B. 41 (1938), Nr. 9, 1–15
3. Über die Fourierkoeffizienten elliptischer Modulformen negativer Dimension. C. R. Neuvième Congrès Math. Scandinaves, Helsingfors (1938), 320–322. Mercatorin Kirjapaino, Helsinki 1939
4. Bemerkungen über eine Dirichletsche Reihe, die mit der Theorie der Modulformen nahe verbunden ist. Archiv for Mathematik og Naturvidenskab B. 43 (1940), Nr. 4, 47–50
5. Beweis eines Darstellungssatzes aus der Theorie der ganzen Modulformen. Archiv for Mathematik og Naturvidenskab B. 44 (1941), Nr. 3, 33–44
6. Über ganzwertige ganze transzendente Funktionen. Archiv for Mathematik og Naturvidenskab B. 44 (1941), Nr. 4, 45–52
7. Über einen Satz von A. Gelfond. Archiv for Mathematik og Naturvidenskab B. 44 (1941), Nr. 15, 159–170
8. Über ganzwertige ganze transzendente Funktionen II. Archiv for Mathematik og Naturvidenskab B. 44 (1941), Nr. 16, 171–181
9. On the zeros of Riemann's zeta-function. Skrifter utgitt av Det Norske Videnskaps-Akademi i Oslo. I. Mat.-Naturv. Klasse (1942), No. 10, 1–59
10. On the zeros of Riemann's zeta-function on the critical line. Archiv for Mathematik og Naturvidenskab B. 45 (1942), No. 9, 101–114
11. On the zeros of the zeta-function of Riemann. Det Kongelige Norske Videnskabers Selskab Forhandlinger B. 15 (1942), No. 16, 59–62
12. On the normal density of primes in small intervals, and the difference between consecutive primes. Archiv for Mathematik og Naturvidenskab B. 47 (1943), No. 6, 87–105
13. On the remainder in the formula for N(T), the number of zeros of $\zeta(s)$ in the strip $0<t<T$. Avhandlinger utgitt av Det Norske Videnskaps-Akademi i Oslo. I. Mat.-Naturv. Klasse (1944), No. 1, 1–27
14. Bemerkninger om et multipelt integral. Norsk matematisk tidsskrift B. 26 (1944), 71–78
15. Contributions to the theory of the Riemann zeta-function. Archiv for Mathematik og Naturvidenskab B. 48 (1946), No. 5, 89–155
16. Contributions to the theory of Dirichlet's L-functions. Skrifter utgitt av Det Norske Videnskaps-Akademi i Oslo. I. Mat.-Naturv. Klasse (1946), No. 3, 1–62
17. The zeta-function and the Riemann hypothesis. C. R. Dixième Congrès Math. Scandinaves, Copenhague (1946), 187–200. Jul. Gjellerups Forlag, Copenhagen 1947

18. (with V. Brun, E. Jacobsthal, C. L. Siegel) En brevveksling om et polynom som er i slekt med Riemanns zetafunksjon. Norsk matematisk tidsskrift B. 28 (1946), 65 – 71
19. On an elementary method in the theory of primes. Det Kongelige Norske Videnskabers Selskab Forhandlinger B. 19 (1947), No. 18, 64 – 67
20. (with S. Chowla) On Epstein's zeta-function (I). Proceedings of the National Academy of Sciences of the USA, Vol. 35 (1949), 371 – 374
21. An elementary proof of Dirichlet's theorem about primes in an arithmetic progression. Annals of Mathematics, Vol. 50 (1949), No. 2, 297 – 304
22. An elementary proof of the prime-number theorem. Annals of Mathematics, Vol. 50 (1949), No. 2, 305 – 313
23. On elementary methods in primenumber-theory and their limitations. C. R. Onzième Congrès Math. Scandinaves, Trondheim (1949), 13 – 22, Johan Grundt Tanums Forlag, Oslo 1952
24. An elementary proof of the prime-number theorem for arithmetic progressions. Canadian Journal of Mathematics, Vol. 2 (1950), 66 – 78
25. The general sieve-method and its place in prime-number theory. Proceedings of the International Congress of Mathematicians, Cambridge, Mass. (1950), Vol. 1, 286 – 292, American Mathematical Society, Providence, R. I., 1952
26. Note on a paper by L. G. Sathe. J. Indian Math. Soc. B. 18 (1954), 83 – 87
27. Harmonic analysis and discontinuous groups in weakly symmetric Riemannian spaces with applications to Dirichlet series. J. Indian Math. Soc. B. 20 (1956), 47 – 87
28. Automorphic functions and integral operators. Seminars on Analytic Functions, Vol. II (1958), 152 – 161. The Institute for Advanced Study, Princeton, New Jersey
29. Some problems concerning discontinuous groups of isometries in higher dimensional symmetric spaces. Report of the Institute in the Theory of Numbers, University of Colorado, Boulder, Colorado (1959), 1 – 7
30. A new type of zeta functions connected with quadratic forms. Report of the Institute in the Theory of Numbers, University of Colorado, Boulder, Colorado (1959), 207 – 210
31. On discontinuous groups in higher-dimensional symmetric spaces. Contributions to function theory (International Colloquium on Function Theory), Bombay (1960), 147 – 164. Tata Institute of Fundamental Research, Bombay, 1960
32. Discontinuous groups and harmonic analysis. Proceedings of the International Congress of Mathematicians, Stockholm (1962), 177 – 189, Inst. Mittag-Leffler, Djursholm, 1963
33. On the estimation of Fourier coefficients of modular forms. Proc. Sympos. Pure Math. (Cal. Tech. Pasadena, Cal. 1963), Vol. VIII, 1 – 15. Amer. Math. Soc., Providence, R. I., 1965
34. (with S. Chowla) On Epstein's zeta-function. Journal für reine und angewandte Mathematik B. 227 (1967), 86 – 110
35. Recent developments in the theory of discontinuous groups of motions of symmetric spaces. Proceedings of the 15th Scandinavian Congress, (Oslo

1968), Lecture Notes in Mathematics, Vol. 118, 99–120, Springer, Berlin, 1970
36. Sieve methods. Proc. Sympos. Pure Math. (SUNY, Stony Brook, N.Y. 1969), Vol. XX, 311–351. Amer. Math. Soc., Providence, R. I., 1971
37. Remarks on sieves. Proc. of the 1972 Number Theory Conference held at the University of Colorado, Boulder, Colorado, August 14–18, 205–216
38. Remarks on multiplicative functions. Proc. Conf. Rockefeller Univ., New York (1976). Lecture Notes in Mathematics, Vol. 626, 232–241, Springer, Berlin, 1977
39. Harmonic analysis
40. Sifting problems, sifting density and sieves
41. Reflections around the Ramanujan centenary
42. Linear operators and automorphic forms
43. Remarks on the distribution of poles of Eisenstein series
44. Old and new conjectures and results about a class of Dirichlet series
45. Lectures on sieves

Afterword

Considering the paucity of my publications, it may not come as a surprise when I say that writing has never come easily for me. I do find it exceedingly difficult to produce something with which I am at least moderately content.

However, when my old friend Professor Chandrasekharan some years ago finally persuaded me to agree to a publication of my collected papers, it occurred to me that this might be a good occasion for writing up various things that over the years had remained unpublished. I therefore, somewhat rashly, promised to do so in time to have them included.

Many times since then have I sorely regretted this promise as I realised that I had vastly under-estimated the time and effort required. As it now stands there are several items that I originally had planned to include (among them the longer version, complete with proofs, of the lecture presented at the number theory conference in Amalfi in September 1989), which still remain unwritten. I hope to complete these for publication elsewhere in the coming years.

This second volume contains what I found time to finish, the major part being the "Lectures on Sieves" which I take some satisfaction in having completed. Even this would probably not have been finished yet were it not for the continued encouragement and prodding (nagging?) by Chandrasekharan and Hedi (my wife). For this I am now most grateful to them both. I am also indebted to Chandrasekharan for his generous assistance in proofreading the manuscripts.

The "Lectures on Sieves" was typed in TEX by Dr. Ilan Vardi while I was at Stanford University, he also did the introduction to the Göttingen lectures as well as the concluding remarks appended to them which are found in volume one. The remaining papers in this volume were done here at the Institute for Advanced Study by Mrs Dorothy Phares, who also did the final printouts of "Lectures on Sieves" here.

I wish to express my thanks to them both for their excellent work. I also wish to thank the Institute for Advanced Study and Stanford University for their cooperation in this, and the Springer Verlag for its patience.

Princeton, August 22, 1990 Atle Selberg